New Directions in the Study of Plants and People: Research Contributions from the Institute of Economic Botany

Ghillean T. Prance and
Michael J. Balick,
Editors

NYBG

The New York Botanical Garden
Bronx, New York, U.S.A.
1990

Volume 8, ADVANCES IN ECONOMIC BOTANY

This is Volume 8 of ADVANCES IN ECONOMIC BOTANY

Copyright © 1990 The New York Botanical Garden

Published by The New York Botanical Garden, Bronx, New York 10458, U.S.A.

Issued 31 January 1990

Library of Congress Cataloging-in-Publication Data

New directions in the study of plants and people : research contributions from
   the Institute of Economic Botany / Ghillean T. Prance and Michael J. Balick,
   editors.
       p.   cm.—(Advances in economic botany : v. 8.)
   Includes bibliographical references.
   ISBN 0-89327-347-3
   1. Botany, Economic—America.   2. Indians—Ethnobotany.
3. Botany—America.   4. Ethnobotany—America.   I. Prance, Ghillean T.,
1937–   . II. Balick, Michael J., 1952–   . III. Institute of Economic Botany
(New York Botanical Garden) IV. Series.
   SB108.A45N48   1989
   581.6'0973—dc20   89-13336
                                                                          CIP

   ISBN 0-89327-347-3
   Printed by Allen Press, Lawrence, Kansas, U.S.A.

Contributors

Pedro Acevedo-Rodriguez, New York Botanical Garden, Bronx, NY 10458-5126, U.S.A. Present address: Smithsonian Institution, NHB-166, Department of Botany, Washington, D.C. 20560, U.S.A.

William Balée, Departamento de Programas e Projetos, Museu Paraense Emílio Goeldi, Cx. Postal 399, 66.000 Belém, Pará, Brazil.

Michael J. Balick, Institute of Economic Botany, New York Botanical Garden, Bronx, New York 10458-5126, U.S.A.

Hélio H. C. Bastien, 77 Zacateros, San Miguel de Allende, GTO., Mexico.

Hans T. Beck, Institute of Economic Botany, New York Botanical Garden, Bronx, New York 10458-5126, U.S.A.

Brian M. Boom, New York Botanical Garden, New York 10458-5126, U.S.A.

Douglas C. Daly, New York Botanical Garden, Bronx, New York 10458-5126, U.S.A.

John E. Earhart, World Wildlife Fund and the Conservation Foundation, 1250 24th Street, N.W., Washington, D.C. 20037, U.S.A.

Stanley N. Gershoff, Dean, School of Nutrition, Tufts University, Medford, Massachusetts 02155, U.S.A.

Elysa J. Hammond, School of Forestry and Environmental Studies, Yale University, New Haven, Connecticut 06511, U.S.A.

Wil de Jong, Institute of Economic Botany, New York Botanical Garden, Bronx, New York 10458-5126, U.S.A.

Steven R. King, Institute of Economic Botany, New York Botanical Garden, Bronx, New York 10458-5126, U.S.A. Present address: Latin American Science Program, The Nature Conservancy, 1815 N. Lynn St., Arlington, Virginia 22209, U.S.A.

Mats Lundberg, Department of Cultural Anthropology, University of Uppsala, Sweden. Present address: International Council for Research in Agroforestry, Nairobi, Kenya.

Peter H. May, Honorary Research Associate, Institute of Economic Botany, New York Botanical Garden, and Program Officer for Rural Poverty and Resources, The Ford Foundation, Praia do Flamengo, 100, 12°-andar, 22210 - Rio de Janeiro-RJ, Brazil.

John D. Mitchell, Herbarium and Institute of Economic Botany, New York Botanical Garden, Bronx, New York 10458-5126, U.S.A.

Scott A. Mori, Associate Scientist, Institute of Economic Botany, New York Botanical Garden, Bronx, New York 10458-5126, U.S.A.

Christine Padoch, Institute of Economic Botany, New York Botanical Garden, Bronx, New York 10458-5126, U.S.A.

Charles M. Peters, Institute of Economic Botany, New York Botanical Garden, Bronx, New York 10458-5126, U.S.A.

Ghillean T. Prance, Founding Director, Institute of Economic Botany, New York Botanical Garden, Bronx, New York 10458-5126, U.S.A. Present address: Director, The Royal Botanic Gardens, Kew, Richmond, Surrey TW9 3AB, United Kingdom.

Anne E. Reilly, Institute of Economic Botany, New York Botanical Garden, Bronx, New York 10458-5126, U.S.A.

Jan Salick, Assistant Scientist, Institute of Economic Botany, New York Botanical Garden, Bronx, New York 10458-5126, U.S.A. Present address: Botany Dept., Ohio University, Athens, Ohio 45701, U.S.A.

Judith G. Schmidt, Institute of Economic Botany, New York Botanical Garden, Bronx, New York 10458-5126, U.S.A.

Jeremy Strudwick, Department of Biology, Lehman College-CUNY, C.U.N.Y. Graduate School, Bronx, New York 10468, U.S.A.

David E. Williams, Institute of Economic Botany, New York Botanical Garden, Bronx, New York 10458-5126, U.S.A.

Contents

Foreword

This volume is a collection of papers by some of the staff members and associates of The New York Botanical Garden's Institute of Economic Botany (IEB). The Institute, founded in 1981, works to direct a greater portion of the Garden's research and teaching efforts to economic botany and stimulate new thinking and new activities in this field. The IEB began with an emphasis on the search for food and fuel plants, later expanding into the area of medicinal plants, as well as the management of rain forest ecosystems. The research staff currently includes ten scientists, two graduate students and numerous resident and non-resident research associates from around the world.

The variety of topics discussed in this volume reflect the diversity of research activities that are being addressed by the Institute's staff and associates. It has focused largely on research that will contribute alternative solutions to the present modes of economic development that have led to the acute deforestation in the humid tropics. Our observations on the failure of many large-scale development projects in the tropics led us to believe that solutions are to be found much more in the wisdom of indigenous people who have lived for many generations in the forest, and in the encouragement of the small-scale farmers who have adopted many of these indigenous techniques as part of their lifestyle. We have therefore chosen to study the enthnobotany of various indigenous groups, as well as the agroforestry techniques and other management systems of these peoples. We have also selected a number of lesser-known crop plants from the tropics that we believe to hold much greater potential than realized to date, as well as those that may be appropriate for sustainable systems of land use.

One of the approaches that we have taken is the study of certain rain forest plants which, unlike the majority, occur in dense and frequently even unispecific stands. Examples of this approach are our work on the babassu palm (*Orbignya phalerata*) and the camu camu (*Myrciaria dubia*). Both species occur in monodominant stands, the former in the drier transition forests between the Amazon rain forest and the cerrado of central Brazil, and the latter in lake and river margins in acid waters. In these cases the local ecology suggests that aggregations of a single species will work, but for the majority of rain-forest areas the local ecology tells us that diversity is the rule. Therefore, agroforestry and extraction systems that are modeled on the diversity normal to these areas are much more likely to succeed than monocultures of *Gmelina, Eucalyptus* or pasture grasses.

Our concern for the fate of the tropical rain forest has dominated the research topics covered by the Institute of Economic Botany. The map in Figure 1 shows the geographic range of our research in the Neotropics during the first seven years of our Institute. However, we have not confined our efforts to that topic and have worked with the woodland Indians of the northeastern states, commenced quantitative ecological inventories on secondary forests in the Virgin Islands, as well as initiated several projects with school children of the New York metropolitan area. This important work enables us to demonstrate the interconnectedness of all ecosystems. For example, we have been surprised (but gratified) to note the increasing wealth of tropical products to be found in the fruit markets of New York City.

Another important aspect of our research (although not reported on in this volume) is the collection of germplasm of useful species for use in breeding programs and for the conservation of genetic diversity before it becomes extinct. In all our work we have given a high priority to the preservation of germplasm, and Figure 2 shows both sites of germplasm collection and deposit. The wild relatives of many crop species are rare and threatened with extinction and merit our urgent attention.

When we organized the Institute, there was little outside interest in medicinal plants, therefore our initial emphasis was on food and fuel. Initial efforts to interest chemical companies in the continued search for new drugs from plants were unsuccessful. However, there has been a resurgence of interest in research on plants as sources of new drugs. We are now working with private industry as well as the U.S. National Can-

IEB Field Sites and Scientists

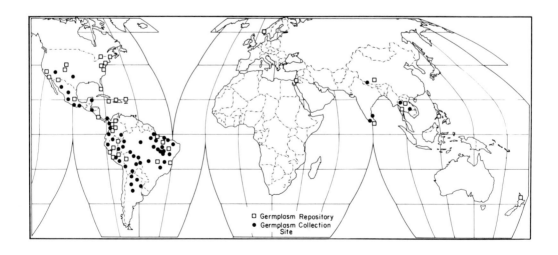

□ Germplasm Repository
● Germplasm Collection Site

cer Institute in the collection of plant material for screening as treatments for a broad range of human ailments.

The New York Botanical Garden has for many years furthered the publication of research in economic botany through its journal *Economic Botany*. The official journal of the Sociey of Economic Botany, it is for the publication of shorter research papers. With the founding of our Institute of Economic Botany we established *Advances in Economic Botany* for the publication of longer manuscripts, symposia and other collections of papers such as this volume. With the present work, the series has published eight volumes, all connected with the work of the IEB. They are proving to be significant contributions to knowledge in the field of economic and ethnobotany. Over the next few years we hope to expand this series by reporting research undertaken by those outside of our Institute. We would therefore like to remind our readers that *Advances in Economic Botany* is open to authors of monographs from any other institution in the world.

The Editors

The Occurrence of Piscicides and Stupefactants in the Plant Kingdom

Pedro Acevedo-Rodríguez

Table of Contents

Abstract

ACEVEDO-RODRÍGUEZ, P. (New York Botanical Garden, Bronx, New York 10458-5126, U.S.A.). The occurrence of piscicides and stupefactants in the plant kingdom. Advances in Economic Botany 8: 1–23. 1990. General aspects of the use of plants to poison fish are discussed. These include different methods of poison preparation and the different compounds present in ichthyotoxic plants. A list of 935 species of plants employed for fish poisoning is presented in alphabetical order of families, for a total of 103 families. The geographical distribution of their use and the active principles, if known, are given. The Legume family contains the greatest number of species utilized as fish poisons; it is followed by the Sapindaceae and the Euphorbiaceae.

Key words: Ichthyotoxins; isoflavonoids; saponins; cardiac glycosides; alkaloids; tannins; cyanogenic compounds; ichthyoctereol; Fabaceae; Sapindaceae; Euphorbiaceae.

Resumen

El present artículo discute varios aspectos sobre las plantas utilizadas como barbascos. Entre estos se discute la preparación de los venenos, así como los compuestos químicos presentes conocidos. Un total de 935 especies utilizadas como barbascos son enumeradas en una tabla por orden alfabético de familia, para un total de 103 familias. De estas, tres familias contienen 46% de las especies registradas como barbascos. La familia de las Leguminosas contiene el mayor numero, a esta le sigue las Sapindaceas y las Euphorbiaceas.

Introduction

Fish-poison plants are those used to stupefy or kill fish. The practice of poisoning the water in pools rich in fish secures a large quantity of food with relatively little effort. Ichthyotoxic plants, of course, must not render the fish poisonous to humans since that would be inappropriate and dangerous. The appropriate employment of these plants depends on the kind of plants used, the condition of the water, and the amount of plant material employed.

Ichthyotoxic plants are utilized on almost all continents, but more commonly so in tropical than temperate regions. Tropical America shows the greatest number of species used for this purpose, followed by Australasia and tropical Africa (including Madagascar). This fact agrees with that found by Prance (1977) for the species richness for the three geographical areas. It seems that this circumstance is a consequence of the floristic richness of the area rather than being indicative of the prevalence of this fishing practice.

Historical Background

The practice of using fish-stupefying plants is likely to have originated independently in the New and Old worlds. Schultes (1970) has suggested that in the Americas this practice has had an Amazonian origin, later spreading to Central America, the West Indies, and finally to North America. This is supported by the fact that this practice is not recorded for the central part of North America, suggesting diffusion from other areas. Hamlyn-Harris and Smith (1916) proposed independent origins of fish drugging for Australia and Melanesia, while Heizer (1953) believed that it was a diffusion phenomenon (see Heizer, 1953 for a discussion on the distribution and development of fish poisoning).

Chevalier (1937) has suggested that the use of fish-poison plants goes back to paleolithic times. Such a claim is not supported by archeological evidence because there are no vestiges in material cultures. The origin of fish drugging might have been accidental and perhaps depended on the empirical recognition of the toxic principles by humans. Many different situations might have led to the origin of this practice. Branches may have been used as fishing tools, and the poisonous effects of some of them may have been noted. It might also have been a consequence of using plants with sudsing properties for bathing. This is very likely because plants used as soaps are rich in saponins, one of the most common fish poisons.

The first written evidence for the use of ichthyotoxic plants seems to be from the classical Greeks. Aristotle mentioned in his "Historia Animalum" (quoted by Howes, 1930) the use of a plant to poison fish [perhaps a *Verbascum* sp. in the Scrophulariaceae]. Pliny also mentioned another ichthyotoxic plant used in Italy and Spain (Kerharo et al., 1959). The use of fish poisons in Europe during medieval times is mentioned by various authors. Kerharo et al. (1959) mention that King Federico II of Spain prohibited the use of fish-poison plants by 1212. Kamen-Kaye (1977) pointed out that in 1453 King Juan II of Spain had also prohibited such practices.

In the Americas, the first written evidence of the use of fish-poison plants comes from the chronicles of the Spanish conquerors in the seventeenth century. The practice, as documented, seems to have been a prevalent and ritualistic phenomenon among the Amerindians.

In more recent times, the use of ichthyotoxic plants has remained a common practice, as evidenced by its prohibition in different countries. In 1828, a law was passed in Venezuela prohibiting the use of barbascos or any other kind of fish-poison plant because the poisoned waters could endanger not only fish populations but also domestic animals. In the Mariana Islands there was a prohibition on the use of fish-poison plants during the Spanish dominion, but when the United States took over at the beginning of the century, it was lifted. At present, the use of ichthyotoxic plants is prohibited in Malaysia. The Malaysian Government can detect violation of this law by testing for the presence of rotenone in all fish caught (Sinnapa & Thuan, 1971).

The use of fish poisons is still prevalent among aboriginal peoples of tropical areas as evidenced by the work of many ethnobotanists. Prance (1972) has shown that this practice is common among four Amazonian Indian tribes. Boom (1987) has found it to be common among the Chácobo Indians of Bolivia. Moretti and Genard (1982) testify that it is a well-known practice in French Guiana.

The Sources of Ichthyotoxic Plants

Fish-poison plants are found in a wide range of habitats, including rain forests and open vegetation. A great number of them are spontaneous weeds or cultivated plants, such as many species of *Lonchocarpus, Phyllanthus, Euphorbia, Tephrosia,* and *Clibadium* (Archer, 1934; Killip & Smith, 1935; Moretti & Genard, 1982; Prance, 1972).

Use and Preparation of Fish Poisons

Fish-poison plants are usually employed in small rivers or small pools with slow-moving water. In larger rivers, dams are created in order to restrict the movement of water and fish with the result that the poison remains at high concentration, acting upon fish. The use of plants to stupefy fish is a delicate practice since its inappropriate use may kill all fish present in a particular area. Different fishing techniques have been developed by aboriginal tribes in order to prevent this problem. The Makú Indians of Amazonia practice fish poisoning in cycles of several years, allowing the fish populations to recover (Prance, pers. comm.). Dam constructions are designed to allow small fish to escape from the poisoned area (Howes, 1930).

The preparation of fish poisons varies from place to place. Some of these methods, while commonly employed in different areas of the world, presumably have originated independently nonetheless. Others are restricted to particular regions. The techniques used for the preparation of the poison vary according to the kind of plant used and the condition of the water (i.e., fresh or salty, running or still, deep or shallow).

The simplest and perhaps the most common method of poison preparation consists of crushing up the poisonous plant with a rock or a piece of stem and throwing the macerated tissue into the water. A variant of this method consists of placing the macerated material in small baskets in the water, allowing the dispersion of poison to occur evenly.

In many tropical areas, fish are fed with poisonous plant material, causing them to become stupefied. This practice has been reported for Australia where *Barringtonia asiatica* (L.) Kurz. (Lecythidaceae) seeds are crushed and fed to fish (Hamlyn-Harris & Smith, 1916). In tropical America a variant of this technique consists of mixing the crushed poisonous plant with food, often balls of *Manihot* flour, and using them as fish bait. Some ichthyotoxic plants such as *Tephrosia* spp. (Fabaceae) and *Clibadium sylvestre* (Aubl.) Baill. (Asteraceae) do not make the fish poisonous for human consumption (Howes, 1930). However, it has been reported that *Anamirta cocculus* (L.) Wight & Arn. (Menispermaceae), when prepared in this way, confers some toxicity to the fish (Howes, 1930). Other plants may be added to the mixture to give some spicy flavor to the fish meat.

In South America and the Malay Peninsula, the active principles are sometimes extracted by crushing the plant material and placing it in a receptacle with a little water. The liquid is spread into the river (Howes, 1930; Kamen-Kaye, 1977). This technique has also been observed by Hamlyn-Harris and Smith (1916) in the Mariana Islands.

Schultes (quoted by Kamen-Kaye, 1977) and Kawanishi et al., (1986) reported a practice in Colombia which consists of pounding the fruit of *Caryocar* spp. (Caryocaraceae) in a mudhole. The resulting mixture of fruit pulp and mud is then cast into the water. Similar practices have been described for some African poisonous plants (Howes, 1930).

A process in which the plants are cooked or fermented occurs in the New World and the Old World. Schultes reported on a preparation used in Colombia that consisted first of fermenting the leaves of an aroid (*Philodendron craspedodromum* R. E. Schultes) and later throwing them in the water. The roots of *Manihot esculenta*

Crantz. (Euphorbiaceae) are cooked or soaked in water when used as a fish poison. These processes, which may alter the toxicity of the plant, are parallel to that found in Australia and Africa, where the toxic plants *Jagera pseudorhus* (A. Rich.) Radlk. (Sapindaceae) and *Adenia cissampeloides* Harms (Passifloraceae), respectively, are roasted before being used as fish poison. The heating may liberate HCN (prussic acid or hydrogen cyanide), especially since the last three genera are listed as cyanophoretic (Hamlyn-Harris, 1916). Another technique in which heat is involved is practiced in Australia. The plant material is rubbed onto preheated stones; the stones are then thrown into the water where the poison diffuses (Hamlyn-Harris, 1916).

An unusual technique in which whole branches of *Euphorbia cotinifolia* L. (Euphorbiaceae) are thrown into the water and then beaten with a stick to liberate the toxic principles has been reported by Prance (1972) for the Makús Indians of the Amazon. In this way, direct contact with the caustic sap is avoided. A similar method has been reported by Teixeira et al. (1984) where the branches of *Serjania lethalis* St. Hil. (Sapindaceae) are tied together, thrown into the water and then beaten until they become completely smashed.

Howes (1930) has reported on the practice of tapping the poisonous tree *Hura crepitans* L. (Euphorbiaceae) to obtain its caustic sap which is thrown directly into the water. Such a practice is known only for Central and South America.

In Asia and Australia, whole branches of *Eucalyptus* spp. are thrown directly into water holes. The poisoning from these plants occurs very slowly because the action is due to tannins and essential oils (Howes, 1930).

The Active Principles

The principles responsible for poisoning fish may be classified into a few chemical groups, although the chemical constituents are not known in all cases. The main active compounds responsible for stupefying and killing fish are: (1) rotenone; (2) saponins; (3) cardiac glycosides; (4) alkaloids; (5) tannins; (6) cyanogenic compounds; and (7) ichthyocthereol.

Rotenone and its derivatives, tephrosine and lonchocarpin, are extremely active isoflavonoids. They kill fish by interfering with respi-

ration at the mitochondrial level (Fukami et al., 1967; Lindahl & Oberg, 1960). They are very active in cold-blooded animals, but less active in warm-blooded ones. Their toxicity is much higher when applied directly to the bloodstream than when taken orally. Due to its low toxicity when ingested, fished stupefied by rotenone can be eaten by humans without any adverse reaction. Rotenone occurs almost exclusively in legumes, especially in the genera *Lonchocarpus, Tephrosia, Derris,* and *Mundulea.*

Saponins are glycosides of triterpene and steroid compounds. They are surface-active agents with soap-like properties with the ability to lyse blood cells. Fish in contact with saponins die from asphyxiation because the saponins modify the surface tension of the water and thereby block respiration at the gills. As some saponins are toxic to humans, they are used carefully because fish caught by this means may become toxic (Moretti & Genard, 1982). The oral intake of saponins by humans provokes, in many cases, congestion of the digestive tract accompanied by hemorrhages and acute nephritis (Kerharo et al., 1959).

Cardiac glycosides are triterpenoid compounds, most of which are toxic. They act as cardiac poisons with associated effects on the central nervous system and the nerve mechanism of the heart muscle (Burkill, 1985). I have not been able to find any claim confirming toxiciy to humans from fish poisoned by plants containing cardiac glycosides. Perhaps the amount present in the poisoned fish is too low to produce any reaction in humans. Cardiac glycosides are common in the Apocynaceae, Asclepiadaceae and Moraceae.

Alkaloids are heterocyclic nitrogenous compounds which produce a strong physiological reaction primarily affecting the nervous system. They are abundant and variable throughout the plant kingdom, producing completely different syndromes after ingestion. The actions may be quite specific for different organs. Alkaloids may act on the central cortex, the central medulla, the nerve termini or the sympathetic nervous system. Death may result from a variety of disorders, including respiratory failure as the result of complete paralysis of muscle. Usually, fish poisoned by alkaloids do not become toxic for human consumption, perhaps because the quantity present in the fish is too low and because cooking may destroy some of these toxic alkaloids. However, poisoning has been reported from the ingestion of fish stupefied with plants containing alkaloids. Howes (1930) mentioned that fish poisoned with *Anamirta cocculus* (L.) Wight & Arn. (Menispermaceae), which contains curare alkaloids, may become poisonous. According to Prance (pers. comm.) fish poisoned with *Ryania* spp. (Flacourtiaceae) (which contains the poisonous alkaloid ryanodine) may become poisonous for human consumption depending on the concentration of the poison poured into the waters. These examples may be due to the high toxicity of the alkaloids involved. Alkaloids have a wide distribution among flowering plants and are especially common in the families Solanaceae, Loganiaceae, and Menispermaceae.

Tannins are phenolic compounds which have the ability to cross-link with proteins and render them resistant to proteolytic enzymes. They may act upon fish by precipitation of the protein in the gill tissues, forming a layer which interferes with respiration so the fish die from asphyxiation. Poisoning by tannins usually takes place in a very slow way, and their efficacy depends on their employment in relative abundance. Tannins present in captured fish are usually too low to cause any physiological reaction in people consuming them. Tannins are widely distributed throughout the plant kingdom.

Cyanogenic compounds are glycosides which release hydrocyanic acid by a hydrolytic process. Hydrocyanic acid is highly toxic, inhibiting the action of cytochrome oxidase (Kingsbury, 1964). This enzyme is a terminal respiratory catalyst linking atmospheric oxygen with metabolic respiration. Thus, HCN poisoning provokes asphyxiation at the cellular level. Fish so poisoned are apparently not toxic to humans, perhaps due to the small quantity present or to the high reactivity of cyanide, which may combine with fish hemoglobin to become harmless. Cyanogenic compounds are usually found in the Rosaceae, Flacourtiaceae and Euphorbiaceae.

Ichthyocthereol is a polyacetylenic alcohol. This very toxic substance acts on the oxidative phosphorylation of the mitochondria, thus interfering with respiration (Clark, 1969). Ichthyocthereol has been found to be very toxic to fish but not to humans. Fish caught by this means are safe for human consumption. Ichthyocthereol is present in several genera in the Asteraceae

(Gorinski et al., 1973). Plants containing this compound are mixed with food and fed to the fish; in this way, the poison is administrated via the intestinal tract.

Other Applications of Ichthyotoxic Plants

Most of the plants with ichthyotoxic activity have been used for many other purposes also. These include medicines, arrow poisons, insecticides, soap substitutes, molluskicides, antitumor agents, and hormone precursors. In a screening of fish-poison plants that included 264 species in 164 genera and 64 families, Spjut and Perdue (1976) found that 38% of the species and 65% of the genera showed antitumor activity.

Rotenone-containing plants are widely used as insecticides and molluskicides. For this purpose, the genera *Lonchocarpus* and *Derris* have been widely planted throughout the tropics, becoming important crops in many countries. Rotenone has also been used to control the growth of undesirable species of fish in water reservoirs and fish hatcheries (Bhuyan, 1967; Burmakin, 1958; Fukami et al., 1967). In the Amazon, it has been used as a selective poison to control populations of piranha (including their eggs), leaving valuable fish unaffected (Goulding, 1979). In Africa, it has contributed to the eradication of schistosomiasis by killing the host snail (Howes, 1930; Kerharo et al., 1959). Rotenone is one of the most potent natural insecticides; it has very low toxicity to warm-blooded animals and has a short residual life. As a stomach poison or a contact poison it is effective in killing crop pests and parasites (Higbee, 1947). Many species of the genus *Derris* are used in the preparation of arrow poisons (Roark, 1932).

Saponins have been used as soap substitutes. Their detergent action makes them useful for delicate fabrics (e.g., those from *Agave* spp.). The steroid configuration in others allows them to be used as precursors of sex hormones (e.g., from *Dioscorea* spp.; Kingsbury, 1964). Some saponins have a toxic effect on insects; therefore, they have been used as insecticides (Secoy & Smith, 1983).

The usefulness of alkaloids in medicine is widely known. Their applications are so diverse as to constitute subjects for whole books. Toxic alkaloids, like those found in fish-poison plants, have been used as insecticides and arrow poisons. Ryanodine, a very toxic alkaloid present in the genus *Ryania,* is an effective insecticide. It has low toxicity to warm-blooded animals in comparison with other common insecticides (Monachino, 1949). Similarly, nicotine, another toxic alkaloid, has been used as an insecticide, but it is not as effective as other substances such as rotenone (Howes, 1930).

Cardiac glycosides have been used to strengthen the action of weakened heart muscle. The use of tannins to convert animal skins into leather is an old worldwide practice. Tannins have also been used as astringents and antiseptics.

The Systematic Occurrence of Ichthyotoxic Plants

In this section, a list of 935 ichthyotoxic plants is presented in a table. These 935 species of ichthyotoxic plants are distributed in 393 genera and 103 families sensu Cronquist, 1981 (except for the legumes which are treated as a single family). Only three families with ichthyotoxic plants (Gnetaceae, Taxaceae, and Zamiaceae) belong to the Gymnosperms, each one having only one representative species. No Pteriodophytes, Bryophytes, Lichens or Fungi are reported as ichthyotoxic in any of the references cited.

Within the Angiosperms, the families which contain the highest number of representatives are listed below in decreasing order of importance.

A. FABACEAE

229 species (25% of total) and 74 genera (25% of total). These are distributed in the following way:

Caesalpinioideae
    18 species and 9 genera
Mimosoideae
    41 species and 15 genera
Papilionoideae
    170 species and 50 genera

B. SAPINDACEAE

110 species (11.5% of total) and 14 genera (4.4% of total).

C. EUPHORBIACEAE

92 species (10% of total) and 32 genera (10% of total).

These three families constitute 46.5% of the total species used as fish-poisons. The remaining 100 families constitute the remaining 53.5%.

Species reported to be ichthyotoxic (but without any apparent physiological activity) which are used in conjunction with known ichthyotoxic plants were deleted from the list. Examples of this are various species of *Piper, Pothomorphe peltata* (L.) Miq., *Cissus debilis* Planch. and *Carica papaya* L.

D. NOTES ON TABLE I.

The first name to appear (in bold) is the family name, sometimes followed by a reference number. On the next line follow the generic and specific names. If a species has a cited synonym this appears in parenthesis following the species name. This is followed by the reference number, geographical area, part of the plant employed, and the chemical compound if known. Abbreviations for information provided in the table appear after the table as an appendix.

Acknowledgments

Many people have contributed to the realization of this paper. Lois Brako made the first grammatical corrections; Hans Beck provided additional references for the palms; Ghillean T. Prance and Michael Balick made comments on the first draft; Rupert Barneby helped with the nomenclature of the Fabaceae; John Pruski checked the nomenclature of the Asteraceae; Scott A. Mori reviewed the nomenclature for the Lecythidaceae and the final version of the manuscript. This is publication number 134 in the series of the Institute of Economic Botany of the New York Botanical Garden.

Literature Cited

(1) **Adams, D., K. Magnus & C. Seaforth.** 1963. Poisonous plants of Jamaica. Department of Extra-Mural Studies, Univ. of the West Indies. Jamaica.
(2) **Airy Shaw, H. K.** 1975. The Euphorbiaceae of Borneo. Kew Bull. Add. ser. IV, London.

(3) **Allen, P. H.** 1943. Poisonous and injurious plants of Panama. Amer. J. Trop. Dis. Prev. Med. **1**: 1–76.

(4) ———. 1977. The rain forest of Golfo Dulce. Stanford University Press, Stanford, California.

(5) **Archer, W. A.** 1934. Fish poison plants of British Guiana. Agric. J. British Guiana **5**: 204–206.

(6) **Ayala Flores, F.** 1984. Notes on some medicinal and poisonous plants of Amazonian Peru. *In* G. T. Prance and J. Kallunki (eds.), Ethnobotany in the Neotropics. Adv. Econ. Bot. **1**: 1–8.

(7) **Bally, P. R. O.** 1937. Native medicinal and poisonous plants of East Africa. Kew Bull. 10–26.

(8) **Bates, H. W.** 1892. The naturalist on the River Amazon. J. Murray, London.

(9) **Bhuyan, B. R.** 1967. Eradication of an unwanted fish from ponds by using indigenous plant fish poisons. Sci. & Cult. **33**: 82–83.

(10) **Bloom, H.** 1962. Poisonous plants of Venezuela. Harvard University Press, Cambridge, Massachusetts.

(11) **Boom, B. M.** 1987. Ethnobotany of the Chácobo Indians. Beni, Bolivia. Adv. Econ. Bot. 4.

(12) **Bouquet, A. & M. Debray.** 1974. Plantes Médicinales de la Côte D'Ivoire. ORSTOM, Paris.

(13) **Brady, G. & H. R. Clauser.** 1977. Material handbook. MacGraw-Hill, New York.

(14) **Brown, W. H.** 1954. Useful plants of the Philippines. Dept. Agric. & Nat. Resour. Techn. Bull. 10. Vol 2. Bureau of Printing, Manila.

(15) **Burkill, H. M.** 1985. The useful plants of West Tropical Africa. Vol. 1. Royal Botanic Garden, Kew.

(16) **Burkill, I. H.** 1935. A dictionary of economic products of the Malay Peninsula. Vol. 1. Crown Agents for the Colonies, London.

(17) **Burmakin, E. V.** 1958. Ichthyotoxic substances and their utilization on the control of undesirable fish. Rechnogo Rybn. Khoz. **47**: 3–26.

(18) **Bye, R. A. & D. Burgess.** 1975. Ethnobotany of the western Tarahumara of Chihuahua, Mexico. Bot. Mus. Leafl. **24(5)**: 85–112.

(19) **Chevalier, A.** 1937. Plantes ichtyotoxiques des genres *Tephrosia* et *Mundulea*. Rev. Int. Bot. Appl. Agric. Trop. **17**: 9–27.

(20) **Chopra, R. N., S. L. Nayar & I. C. Chopra.** 1956. Glossary of Indian medicinal plants. Council of Scientific and Industrial Research, New Delhi.

(21) **Clark, J. B.** 1969. Effect of a polyacetylenic fish poison on the oxidative phosphorylation of rat liver mitochondria. Biochem. Pharmacol. **18**: 73.

(22) **Cox, P. A.** 1979. Use of indigenous plants as fish poison in Samoa. Econ. Bot. **33**: 397–399.

(23) **Cronquist, A.** 1981. An integrated system of classification of flowering plants. Columbia University Press, New York.

(24) **Dahlgren, B. E.** 1944. Economic products of palms. Trop. Woods. **78**: 10–34.

(25) **Dalziel, J. M.** 1937. Useful plants of West Tropical Africa. The Crown Agents for the Colonies, London, Great Britain.

(26) **Dastur, J. F.** 1952. Useful plants of India and Pakistan. B. D. Taraporevala and Sons Co. Ltd, Bombay, India.

(27) **De Wilde, W. J. J. O.** 1971. A monograph of the genus *Adenia* Fors. (Passifloraceae). H. Veenman & Zonen N.V.-Wageningen.

(28) **Drury, H.** 1858. The useful plants of India. Asylum Press, Mount Road, India.

(29) **Ducke, A.** 1939. As leguminosas da Amazônia Brasilera. Serv. Publ. Agric. R. J., Brasil.

(30) ———. 1942. *Lonchocarpus* subgenus *Phacelanthus* Pittier, in Brasilian Amazonia. Trop. Woods **69**: 2–7.

(31) **Duke, J. A.** 1970. Ethnobotanical observations on the Chocó Indians. Econ. Bot. **24**: 344–366.

(32) ———. 1975. Ethnobotanical observations on the Cuna Indians. Econ. Bot. **29**: 278–293.

(33) **Eshleman, A.** 1977. Poison plants. Houghton Mifflin Company, Boston.

(34) **Fanshawe, D. B.** 1953. Fish poisons of British Guiana. Kew Bull. 239–240.

(35) **Fawcett, W.** 1891. Economic plants of Jamaica. Government Printing Establishment, Kingston, Jamaica.

(36) **Forero, E.** 1983. Flora Neotropica Monograph No. 36. Connaraceae. Organization for Flora Neotropica.

(37) **Fukami, J., I. Yamamoto & J. E. Casida.** 1967. Metabolism of rotenone in vitro by tissue homogenates from mammals and insects. Science **155**: 713.

(38) **Gautier, E.** 1953. A pharmacognostic study of *Piscidia erythrina*. Econ. Bot. **7**: 270–284.

(39) **Gembala, G.** 1955. Contribuçao para a caracterizão da essencia de *Ocotea pretiosa* Mez. Thesis. Rio de Janeiro, Brasil. (Not consulted directly.)

(40) **Gorinski, C., W. Templeton & S. A. H. Zaidi.** 1973. Isolation of ichthyochthereol and its acetate from *Clibadium sylvestre*. Lloydia **36**: 352–353.

(41) **Goulding, M.** 1979. Ecología da Pesca do Rio Madeira. INPA. Manaus, Brasil.

(42) **Gutiérrez, G.** 1947. Estudio sobre los principales barbascos colombianos. Revista Fac. Nac. Agron. Bogotá **5**: 77–93.

(43) **Hamlyn-Harris, R. & F. Smith.** 1916. On fish poisoning and poisons employed among the aborigines of Queensland. Mem. Queensland Mus. **5**: 1–22.

(44) **Hegnauer, R.** 1963. Chemotaxonomie der Pflanzen. Birkhauser Verlag, Basel. Vols. 1–6.

(45) **Heizer, R. F.** 1953. Aboriginal fish poisons. U.S. Amer. Ethn. Bureau Bull. 151 No. **38**:

231–283. U.S. Printing Off. Washington 5: 277–281.

(46) **Herrarte, M. P.** 1933. Plantes employées comme "barbasco" au Guatemala. Rev. Int. Bot. Appl. Agric. Trop. **13**: 351–353.

(47) **Higbee, E. C.** 1947. *Lonchocarpus*—a fish poison insecticide. Econ. Bot. **1**: 427–436.

(48) **Hodge, W. H.** 1961. Nature's biggest bouquet. Principes **5**: 125–134.

(49) ——— **& C. Sneath.** 1956. The Mexican Candelilla plant and its wax. Econ. Bot. **10**: 134–154.

(50) ——— **& D. Taylor.** 1957. The ethnobotany of the Island Caribs of Dominica. Webbia **12**: 513–644.

(51) **Hooper, D.** 1937. Useful plants and drugs of Iran and Iraq. Field Mus. Nat. Hist. Bot. Ser. **3**: 75–241.

(52) **Howes, F. N.** 1930. Fish poison plants. Kew Bull. **4**: 129–153.

(53) **Jablonski, E.** 1974. Catalogus Euphorbiarum. Phytologia **28**: 121–187.

(54) **Irwin, H. S. & R. Barneby.** 1982. The American Cassiinae. Memoirs of NYBG. Vol. 35, parts 1 & 2. New York Botanical Garden.

(55) **Jayasuriya, M.** 1984. A systematic revision of the genus *Dioscorea* in the Indian Subcontinent. Ph.D. Thesis. City University of New York.

(56) **Johnston, I.** 1945. The vegetation of San José Island, Perlas Group. Republic of Panama. S.J.P.R. 88. Headquarters San José Project Division. New Orleans, Louisiana.

(57) **Kamen-Kaye, D.** 1977. Ichthyotoxic plants and the term barbasco. Bot. Mus. Leafl. **25**: 71–89.

(58) **Kawanishi, K., R. F. Raffauf & R. E. Schultes.** 1986. The Caryocaraceae as a source of fish poison in the Northwest Amazon. Bot. Mus. Leafl. Vol. **30(4)**: 247–254.

(59) **Kerharo, J. & A. Bouquet.** 1950. Plantes Médicinales et Toxique de la Côte D'Ivoire. Haute-Volta, Vigot. Freres, Paris.

(60) ———, **F. Guichard & A. Bouquet.** 1959–1961. Les végétaux ichtyotoxiques (poison de peche). Bull. et Mém. Fac. Med. Pharm. Dakar. **8**: 314–329; **9**: 355–386; **10**: 223–242.

(61) **Killip, E. P. & A. C. Smith.** 1930. The identity of the South American fish poisons "cube" and "timbo." J. Wash. Acad. Sci. **20**: 75–81.

(62) ——— **& ———.** 1935. Some American plants used as fish poisons. Bureau of Plant Industry, U.S.D.A.

(63) **Kingsbury, J. M.** 1964. Poisonous plants of the U.S. and Canada. Prentice-Hall, Inc., Englewood Cliffs, New Jersey.

(64) **Krukoff, B. A. & A. C. Smith.** 1937. Rotenone-yielding plants of South America. Amer. J. Bot. **24**: 573–587.

(65) ——— **& J. Monachino.** 1942. The American species of *Strychnos*. Brittonia **4**: 248–322.

(66) **Kunar, S. K.** 1970. Nicotine as a fish poison. Progr. Fish Cult. **32**: 103–104.

(67) **Lamba, S. S.** 1970. Indian piscicidal plants. Econ. Bot. **24**: 134–136.

(68) **Lawle, J. L.** 1984. Echnobotany of the Orchidaceae. *In* J. Arditti (ed.), Orchid biology, reviews and perspectives III. Cornell University Press, Ithaca, New York.

(69) **Leenhoots, P. W.** 1952. Revision of the Burseraceae of the Malaysian area in a wider sense. Blumea VII: 154–160.

(70) **Lewis, W. H. & M. P. F. Elvin-Lewis.** 1977. Medical botany. Wiley Interscience Publications, Canada.

(71) **Lindahl, P. E. & K. E. Oberg.** 1960. Mechanism of physiological action of rotenone. Nature **187**: 784.

(72) **Little, E. & F. H. Wadsworth.** 1964. Common trees of Puerto Rico and the Virgin Islands. U.S.D.A., Washington, D.C.

(73) **Macbride, F.** 1956. Flora of Peru. Field Mus. Nat. Hist. Publ., Bot. Ser., Vol. 13 (3a) No. 2: 291–744.

(74) **Martyn, E. B. & R. R. Follett Smith.** 1936. The fish poison plants of British Guiana, with special reference to the genera *Tephrosia* and *Lonchocarpus*. Agric. J. British Guiana 7: 154–159.

(75) **Monachino, J.** 1949. A revision of *Ryania* (Flacourtiaceae). Lloydia **12**: 1–29.

(76) **Moretti, C. & P. Genard.** 1982. Les nivrés ou plantes ichtyotoxiques de la Guyane Française. J. Ethnopharm. **6**: 139–160.

(77) **Mori, S. A.** 1979. *Gustavia*. Pages 128–197 *in* G. T. Prance & S. A. Mori, Lecythidaceae—I. Fl. Neotrop. Monogr. 21.

(78) **Mors, W. B., M. C. Nascimiento, J. R. Valle & J. A. Aragão.** 1973. Ichthyotoxic activity of plants of the genus *Derris* and compounds isolated therefrom. Cienc. e Cult. **25**: 647–648.

(79) **O'Connel, J., P. Latz & P. Barnett.** 1983. Traditional and modern plant use among the Alyawara of Central Australia. Econ. Bot. **37**: 80–109.

(80) **Oliver-Bever, B.** 1983. Medicinal plants in tropical West Africa. III Anti-infection therapy with higher plants. J. Ethnopharm. **9**: 1–83.

(81) **Payens, J. P. D. W.** 1967. A monograph of the genus *Barringtonia* (Lecythidaceae). Blumea **15**: 157–263.

(82) **Pennington, C.** 1958. Tarahumav fish stupefaction plants. Econ. Bot. **12**: 95–102.

(83) **Perry, L. M. & J. Metzger.** 1980. Medicinal plants of East and Southeast Asia: Attributed properties and uses. MIT Press, Cambridge, Massachusetts.

(84) **Pio Correa, M.** 1931–1978. Diccionario das Plantas Uteis do Brasil. Vols. II (pp. 304–309) and VI (pp. 229–251). Rio de Janeiro. Inst. Bras. de Desenvolvimiento Florestal.

(85) **Pipoly, J. J.** 1983. Contributions toward a monograph of *Cybianthus* (Myrsinaceae): III. A revision of subgenus *Laxiflorus*. Brittonia **35**: 61–80.

(86) **Pires, J. M.** 1978. Plantas ichthyotoxicas: Aspecto da botânica systemâtica. Ciencia e Cult. Supl. (Some specific names lacking in the article were later provided by J. M. Pires, pers. comm.)

(87) **Prance, G. T.** 1972. An ethnobotanical comparison of four tribes of Amazonia Indians. Acta Amaz. **11**: 7–28.

(88) ———. 1977. Floristic inventory of the tropics: Where do we stand? Ann. Missouri Bot. Gard. **64**: 659–684.

(89) ———. 1978. The poisons and narcotics of the Dení, Paumarí, Jamamadí and Jarawara Indians of the Purus river region. Rev. Bras. Bot. **1**: 71–82.

(90) ——— & **M. F. Silva.** 1973. Flora Neotropica Monograph No. 12. Caryocaraceae. Organization Flora Neotropica. Hafner Publishing Co., New York.

(91a) **Radlkofer, L. T.** 1886. Ueber fischvergiftende pflanze. Sitzung. Math.-Phys. C. Bayer Akad. Wiss. **16**: 345–416.

(91) ———. 1931–1934. *In* A. Engler (ed.) Das Pflanzenreich. IV 165 Sapindaceae. Vols. I and II. Verlag von W. Engelmann, Leipzig.

(92) **Rao, V.** 1949. Fish poison from wood of *Adina cordifolia.* J. Sc. Indust. Res. India **8 B**: 95–96.

(93) **Rickard, P. P. & P. A. Cox.** 1986. Use of *Derris* as a fish poison on Guadalcanal, Solomon Islands. Econ. Bot. **40**: 479–484.

(94) **Rizzini, G. T. & W. Mors.** 1976. Botânica Económica Brasileira São Paulo, EPU, Ed. da Universidade do São Paulo. Pages 96–99.

(95) **Roark, R. C.** 1932. A digest of the literature of *Derris* species as insecticides. USDA Misc. Publ. **120**: 1–86.

(96) **Russell, T. A.** 1963. The *Raphia* palm of West Africa. Kew Bull. **19**: 173–196.

(97) **Schultes, R. E.** 1970. De plantas toxicariis e mundo novo tropicale commentationes VII. Bot. Mus. Leafl. **22**: 133–164 & 345–351.

(98) ———. 1977. De plantes toxicariis e mundo novo tropicale commentationes XVI Bot. Mus Leafl. **25**: 109–127.

(99) ———. 1978. De plantas Toxicaris e mundo novo tropicale commentationes XXV. Bot. Mus. Leafl. **26**: 267–275.

(100) ——— & **J. Cuatrecasas.** 1972. De plantas toxicariis e mundo novo tropicale commentationes IX. Bot. Mus. Leafl. **24**: 129–136.

(101) ——— & **R. F. Raffauf.** 1986. De plantis toxicariis e mundo novo tropicale commentationes XXXVII. Bot. Mus. Leafl. **30**: 255–285.

(102) **Secoy, D. M. & A. E. Smith.** 1983. Use of plants in control of agricultural and domestic pests. Econ. Bot. **37**: 28–57.

(103) **Sheng-Ji, P.** 1985. Preliminary study of ethnobotany in Xishuans Banna, People's Republic of China. J. Ethnopharm. **13**: 121–137.

(104) **Sillans, R.** 1952. Sur quelques plantes ichtytoxique de l'Afrique Centrale. Rev. Int. Bot. Appl. Agr. Trop. **32(351–352)**: 54–65.

(105) **Sinnapa, S. & C. E. Thuan.** 1971. Identification of rotenone in fish poisoning by *Derris* root resin. Malayan Agric. J. **48**: 24–25.

(106) **Spjut, R. W. & R. E. Perdue.** 1976. Plant folklore: A tool for predicting sources of antitumor activity. Cancer Treatment Reports **60**: 979–985.

(107) **Standley, P. A. & J. A. Steyermark.** 1949. Flora of Guatemala. Fieldiana Bot. **24(VI)**: 254–255.

(108) **Steyermark, J. A.** 1984. Flora de Venezuela. Piperaceae. Ediciones Fundación Ambiental, Venezuela.

(109) **Teixeira, J. R. M., A. J. Lapa, C. Souccar & J. R. Valle.** 1984. Timbos: Ichthyotoxic plants used by Brasilian Indians. J. Ethnopharm. **10**: 311–318.

(110) **Uphof, J. C. Th.** 1968. Dictionary of economic plants. Verlag von J. Cramer.

(111) **Valle, J. R. & N. P. Silva** 1973. Ichthyotoxicity of cannabinoids. Cienc. e Cult. **25**: 647.

(112) **VonReis, S. & F. J. Lipp.** 1982. New plant sources for drugs & foods from the New York Botanical Garden Herbarium. Harvard University Press, Cambridge, Massachusetts.

(113) **VonReis, S.** 1973. Drugs & foods from little-known plants. Notes in Harvard University Herbaria. Harvard University Press, Cambridge, Massachusetts.

(114) **Walker, A.** 1952. Une nouvelle Legumineuse du Gabon servant á narcotiser le poisson. Rev. Int. Bot. Appl. Agric. Trop. **351**: 327.

(115) **Wilbaux, R.** 1934. Composition et propriétés toxique des graines et des feuilles de *Tephrosia vogelii* Hook. f. Rev. Int. Bot. Appl. Agr. Trop. **XIV**: 1019–1034.

(116) **Williams, E. P.** 1962. Algunos datos sobre el Barbasco. Bol. Soc. Venez. Cienc. Nat. **6**: 21–34.

(117) **Williams, L. O.** 1981. The useful plants of Central America. Ceiba **24**: 1–381.

(118) **Williams, R. O.** 1941. The useful and ornamental plants of Trinidad and Tobago. A. L. Rhodes, M. B. E. Government Printer, Trinidad and Tobago.

(119) **Williamson, J.** 1972. Useful plants of Malawi. The Government Printer, Zomba, Malawi.

(120) **Woodson, R. E.** 1930–1936. Studies in the Apocynaceae. IV The American genera of Echitoideae. Ann. Missouri Bot. Gard. **20**: 605–790.

(121) **Yamada, T., H. Auki & M. Namiki.** 1970. Studies on the saponin from seeds of *Camellia sasanqua* Thunb. Nippon Noger Kagakv Raisi **44**: 58–586. (Not consulted directly.)

Table I
Plants used for the preparation of fish poisons

ACANTHACEAE

Adhatoda bucholzii (Lind.) S. Moore; (15); WAF; PL; ALK.

Duvernoya dewevrei De Wild. & Th. Dur.; (60); WAF; NR; NK.

Dyschoriste walkerii R. Benoist.; (15,25,60); WAF; NR; NK.

Eremomastax speciosa (Hochst) Cufod.; (*E. polysperma* Benth., *Paulo-Wilhelmia polysperma* Benth.); (15,25,52,59); WAF; B, PL, S; ALK, SAP.

Justicia extensa T. Anders; (25,15); WAF; PL; ALK, SAP.

J. laxa T. Anders; (15,25,60); WAF; PL; ALK.

Rhinacanthus virens (Nees) Milne-Redhead; (*R. communis* Nees); (15,25,44); WAF; PL; COU.

AGAVACEAE

Agave americana L.; (10, 26); CAM, IND; L; SAP.

A. bovicornuta Gentry; (18,82); MEX; L, R; SAP.

A. lechuguilla Torr.; (82); MEX; L, R; SAP.

A. mexicana Lam.; (45); MEX, CUB; NR; SAP.

A. schottii Engelm.; (82); MEX; L, R; SAP.

Dracaena thalioides C. Morren.; (44,60); GAB; NR; SAP.

Furcraea foetida Haw.; (76); GUI; R; SAP.

AIZOACEAE

Aïzoön canariensis L.; (25,44,52); WAF; PL; SAP.

ANACARDIACEAE

Anacardium excelsum (Kunth) Skeels; (4); CRI; B; ANA.

Pleiogynium timonense (DC.) Leenh.; (*P. solandri* Engl.); (60,69); AUS; NR; CAD.

Toxicodendron mexicanum Vahl; (nomen nudum); (60), MEX; NR; NK.

ANNONACEAE

Annona reticulata L.; (45); BRA; NR; ALK.

A. spinescens Mart.; (44,45,62); SAM; FR; ALK.

A. squamosa L.; (44,46,60); CAM; L; ALK.

Fissistigma glaucescens Merr.; (113); CHI; R; NK.

Xylopia brieyi De Wild.; (44,104); CAF; B, L; ALK.

X. vallotii Hutch. & Dalz.; (44,104); CAF; B, L; ALK.

APIACEAE

Conium maculatum L.; (82); MEX; R; ALK.

Hydrocotyle javanica Thunb. & Webb; (60,113); AUS; B; CYA.

Leptotaenia dissecta A. Gray; (57); NAM; NR; NK.

Oenanthe crocata L.; (52,60); EUR; NR; PHY.

APOCYNACEAE

Adenium multiflorum Kl.; (52,60,110); WAF; NR; CAR.

A. obesum (Forssk.) Roem. & Schult.; (*A. hongel* A. DC.); (52,70); R; CAR.

A. speciosum Fenzl.; (52); CAF; R; CAR*.

Ambelania lopezii Woodson & R. E. Schult.; (97); SAM; L; NK.

Aspidosperma sessiliflora Allen; (62); SAM; J; ALK*.

Cerbera odollam Gaertn.; (*C. manghas* L.); (45,60,110); PHI; S; CAR.

Kibatalia blancoi (Rolfe.) Merr.; (110); PHI; B, L; NK.

Mandevilla illustris (Vell.) Woodson; (*Dipladenia illustris* A. DC.); (5,120); SAM; B, L; NK.

Melodinus monogynus Roxb.; (44,45,60); IND; NR; TRI*.

Nerium odorum Solander; (45); IND; B, R; NK.

Odontadenia puncticulosa (Rich.) Pulle; (*O. cururu* K. Schum., *Echites urucu* Mart.); (76,86); SAM; NK.

Picralima nitida Th. & Hel. Dur.; (60); SAM, TAF; FR, S; ALK.

Rauwolfia serpentina Benth. ex Kurz; (60); IND; NR; ALK, TRI*.

Strophanthus gracilis K. Schum. & Pax; (15,25); CAF; S, W; CAR.

S. hispdus DC.; (60); WAF; S; CAR.

Tabernaemontana mauritiana Poir.; (45); CAI; B; ALK.

Tabernaemontana sp.; (113); BRA; NR; ALK*.

Thevetia ahouai (L.) A. DC.; (45,62); INO; FR, L, S, W; CAR.

T. peruviana (Pers.) Schum.; (*T. neriifolia* Juss.); (10,25,45,62); WAF, SAM; PL, W; CAR.

Urechites lutea (L.) Britt.; (*U. suberecta* Muell. Arg.); (110,120); TAM; NR; NK.

AQUIFOLIACEAE

Ilex verticillata Gray; (45); SEU; PL; NK.

ARACEAE

Alocasia macrorhiza Schott; (43,45,60); WAF; PL; CYA.

Amorphophallus dracontoides N. E. Brown; (60); WAF; R; SAP.

Arisaema triphyllum Schott; (57); NAM; NR; NK.

Colocasia esculenta Schott; (44,60); WAF; L; SAP*.

Culcasia scandens P. Beauv.; (25); CAF; J; ALK.

Homalomena cordata Schott; (60); MOL; NR; NK.

H. rubra Harsk.; (60); CEL; NR; NK.

Philodendron craspedodromum R. E. Schult.; (44,57); SAM; PL; CYA*.

Table I
Continued

Xanthosoma mafaffa Schott; (cited by 60 as *X. sagittaefolium* Schott); (5,25); WAF; L; NK.

ARALIACEAE

Pentapanax angelicifolium Griseb.; (112); BRA; L; NK.
Schefflera blancoi Merr.; (44,112); PHI; L; SAP*.
S. oblongifolia Merr.; (44,112); PHI; L; SAP*.

ARECACEAE

Cryosophila warscewiczii (Wendl.) Bartt.; (31); PAN; BD; SAP.
Corypha lecomtei Becc.; (60,83); IND; FR; SAP.
C. umbraculifera L.; (20,26,45,48,60,118); IND; FR; SAP.
Raphia gigantea A. Chev.; (12,60); TAF; FR, S; NK.
R. sassandrensis A. Chev.; (60); TAF; FR, S; NK.
R. vinifera P. Beauv.; (24,60,96); TAF; FR, S; NK.
Syagrus weddelliana Becc.; (86); AMA; S; NK.

ARISTOLOCHIACEAE

Apama tomentosa Engl.; (44,114); ASI; NR; ALK*.
Aristolochia clematis L.; (44,52,60); NRE; ALK, TAN.
A. rotunda L.; (60); EUR; RH; ALK*.
A. pallida Willd.; (60); EUR; RH; ALK.

ASCLEPIADACEAE

Asclepias curassavica L.; (43,45,46,60); GUA, AUS; PL; CAR.
Cynanchum sarcostematoides K. Schum.; (45); ZAN; NR; NK.
Marsdenia efulensis N. E. Brown; (104); CAF; PL; NK.
Omphalogonus nigritanus N. E. Brown; (60); WAF; NR; NK.
Pergularia extensa (Jacq.) N. E. Brown; (44,104); CAF; PL; TRI.
Sarcostemma viminale (L.) R. Brown; (25,44,70); TAF; J; CAR.

ASTERACEAE

Artemisia vulgaris L.; (45); IND; L, B; NR; NK.
Baccharis glutinosa Pers.; (82); MEX; S, L; SAP.
Cacalia decomposita A. Gray; (45,82); MEX; PL; NK.
Clibadium appressipilum S. F. Blake; (113); PAN; L; ICH.
C. abadieri Girsch.; (60); SAM; L; ICH.
C. barbasco DC.; (45); BRA; NR; NK.
C. erosum DC.; (60); SAM; L; ICH.
C. heterotrichum S. F. Blake; (45,62); SAM; L; ICH.
C. polygonum S. F. Blake; (42,62); TAM; L; ICH.
C. strigillosum S. F. Blake; (45,62); SAM; FR, L; ICH.

C. surinamensis L.; (*C. asperum* DC.); (10,42,45,60, 62,76,113); SAM; L; ICH.
C. sylvestre (Aubl.) Baill.; (10,42,45,50,60,62,76); SAM; FR, L; ICH.
C. vargasii DC.; (45); GUI; PL; ICH.
Dyssodia anomala (Camby & Rose) Robins.; (44,82); MEX; L, R; SAP*.
Eupatorium odoratum L.; (44,67); IND; PL; SAP*.
Ichthyothere peruviana Baker; (113); PER; PL; ICH.
I. terminalis (Spreng.) S. F. Blake (*I. cunabi* Mart.); (42,45,62); SAM; PL, L; ICH.
Neurolaena lobata (L.) R. Brown; (49,50,70); CAR; NR; ALK.
Salmea scandens (L.) DC.; (42,62,70); CAM; NR; NK.
Senecio hartwegii Benth.; (44,82); MEX; L, R; ALK*.
Sphaeranthus indicus L.; (60); IND; S; NK.
Spilanthes acmella Murr.; (112); TRO; PL; NK.
S. paniculata DC.; (nomen nudum); (45); BUR; NR; NK.
Stevia salicifolia Cav.; (82); MEX; R, S; NK.
Vernonia cinera (L.) Less.; (60); WAF; NR; ALK, CYA, TRI.
V. macrocyanus O. Hoffm.; (*V. primulina* Dalziel); (15,25); PL; ALK, SAP*.
V. podocoma Schultz; (7); EAF; J; ALK.
V. thomsoniana Oliv. & Hiern; (15,25,60); WAF; NR; ALK.
Viguiera decurrens A. Gray; (82); MEX; R; NK.
Zexmenia podocephala A. Gray; (82); MEX; L, R; NK.

BERBERIDACEAE

Berberis aristata DC.; (44,60); IND; B; ALK, CYA.

BIGNONIACEAE

Bignonia capreolata L.; (*B. crucigera* L.); (45); IND; R; NK.
Dolichandrome falcata Seem.; (45,60); IND; NR; NK.
Jacaranda copaia (Aubl.) D. Don; (44,62); GUI; NR; ALK*, SAP.
J. procera Spreng.; (45); GUI; NR; NK.
Tecoma lecoxylon Mart. ex DC.; (45); JAM; NR; ALK*.
T. radicans Juss. (91a); TAM; ALK*.
T. stans Juss.; (44,60); WIN; B, L; ALK*.

BOMBACACEAE

Patinoa ichthyotoxica R. E. Schult. & Cuatr.; (57,100); SAM; FR; NK.

BORAGINACEAE

Anchusa azurea Mill.; (45); GRE; NR; NK.
Cordia dichotoma Forst. f.; (113); CHI; FR; NK.
Ehretia navesii Vidal; (113); PHI; B; NK.

Table I

Continued

BRASSICACEAE

Lepidium bidentatum Montin; (45); SIS; NR; NK.
L. draba L.; (60); IND, AUS; NR; CYA.
L. oleraceum Forst.; (45); NZE; NR; NK.
L. sativum L.; (25); WAF; PL; CYA.
Nasturtium indicum DC.; (60); SEA; NR; NK.

BUDDLEJACEAE

Buddleja asiatica Lour.; (60); ASI; B; SAP.
B. brasiliensis Jacq.; (45,62); SAM; NR; SAP*.
B. curvifolia Hook. & Arn.; (45,60,110); ASI, JAP; BR, L; SAP.
B. verticillata H.B.K.; (45); MEX; PL; NK.

BURSERACEAE

Canarium australasium F. Muell.; (43,44,45,60); AUS; W; TRI*.

CACTACEAE

Lophocereus scotti (Engl.) B. & R.; (45); MEX; PL; NK.
Machaerocereus gummosus (Engl.) B. & R.; (45); MEX; PL; NK.

CAMPANULACEAE

Lobelia mucronata Cav.; (*Tupa barterii* DC.); (60); GUI; PL; ALK;
L. persicaefolia DC.; (60); (*Tupa persicaefolia* (Lam) DC.); GUI; PL; ALK*.
L. tupa L.; (44,62); CHI, PER; NR; ALK*.

CANELLACEAE

Canella winterana (L.) Gaertn.; (44,72); WIN; L; ESS, CYA.

CANNABINACEAE

Cannabis sativa L.; (45,111); LAB, IND; BR, L; CAN.

CAPPARACEAE

Cleome gynandra L.; (cited by 25 as *Gynandropsis pentaphylla* DC.); (15,25,79); WAF; PL; CYA.
C. rosea Vahl; (62); SAM; NR; HET*.
C. spinosa Jacq.; (10,62); SAM; NR; HET*.

CARYOCARACEAE

Anthodiscus obovatus Wittmack; (58); NRE; NR; SAP ?.

A. peruanus Baill.; (58); NRE; NR; SAP ?.
Caryocar amygdaliferum Mutis; (44,58,62); SAM; NR; SAP*.
C. amygdaliforme G. Don; (*C. tessmannii* Pilger.); (113); COL; FR; SAP*.
C. glabrum (Aubl.) Pers.; (45,58,62,112); SAM; FR; SAP*.
C. microcarpum Ducke; (44,58,112); GUY; FR; SAP*.

CARYOPHYLLACEAE

Agrostemma githago L.; (45); EUR; S; NK.
Gypsophila paniculata L.; (44,51); EAS; NR; SAP.

CHENOPODIACEAE

Chenopodium polyspermum L.; (91a); EUR; NK.

CLUSIACEAE

Calophyllum inophyllum L.; (60); SEA; NR; NK.
C. montanum Planch. & Trian.; (45); NCA; NR; NK.
C. rubiginosum H. & W.; (112); BOR; J; NK.
Garcinia cherriyi F. M. Bailey; (43,45); AUS; B; SAP.
Mammea africana G. Don; (60); TAF; B; MAM.
M. americana L.; (1); TAM; NR; MAM.
Pentadesma butyracea Sab.; (25,60); WAF; B; NK.
Vismia cayennensis Pers.; (112); SUR; B; NK.

COMBRETACEAE

Combretum nigricans Lepr.; (cited by 25 as *C. elliotii* E. & D.); (15,25,49,60); WAF; BR; ALK, TAN.
C. ternifolium Engl. & Diels; (44,119); TAF; NR; ALK, TAN.
Terminalia bellerica Roxb.; (45); IND; B; TAN.
T. sericocarpa F. Muell.; (43,45,60); INO, AUS; FR; TAN.

CONNARACEAE

Connarus coriaceus Schell.; (*C. opacus* Schell.); (97); SAM; NR; ALK*.
C. sprucei Baker; (97); COL; R, S; ALK*.
C. tuber (P. & E.) Planch.; (97); SAM; NR; ALK*.

CONVOLVULACEAE

Merremia tuberosa (L.) Rendle; (*Ipomoea tuberosa* L.); (45); MEX; R; NK.

CORIARIACEAE

Coriaria ruscifolia L.; (62); CHI; NR; COR.
C. thymifolia H.B.K.; (10); VEN; NR; COR.

Table I
Continued

CORNACEAE

Cornus amomum Mill.; (45); NAM; B; TRI*.

CUCURBITACEAE

Echinocystis fabacea Naud.; (45); USA; R; SAP*.
E. horrida Cong.; (45); USA; R, S; SAP*.
E. oregana Cong.; (45); USA; NR; SAP*.
Lagenaria breviflora (Benth.) Roberty; (*Adenopus breviflorus* Benth.); (15,25); WAF; S; ALK.
L. siceraria (Molina) Standl.; (45); HAW; FR; CYA.
Luffa acutangula Roxb.; (15,25,60); WAF, ASI; FR; SAP*.
L. cylindrica Roem.; (15,25,60); AUS; FR; SAP.

DATISCACEAE

Datisca glomerata Baill.; (45); USA; NR; NK.

DICHAPETALACEAE 90

Dichapetalum toxicarium (G. Don) Baill.; (*Chailletia toxicaria* G. Don); (15,25,44,45,60,110); WAF; S; ALK, SAP.
Tapura amazonica Poepp. & Endl.; (62); SAM; NR; NK.
T. guianensis Aubl.; (34,45,62) TAM; BR; NK.

DILLENIACEAE

Tetracera alnifolia Willd.; (25,60); WAF; NR; HET, TAN.
T. assa DC.; (44,45,60); WAF; NR; HET, TAN*.
T. indica Merr.; (44,60); ASI; NR; HET, TAN*.
T. macrophylla A. Chev.; (44,60); WAF; NR; HET, TAN*.

DIONCOPHYLLACEAE

Habropetalum dawei (Hutch. & Dalz.) A. Shaw; (*Dioncophyllum dawei* Hutch. & Dalz.); (15,25); WAF; L; PHE.

DIOSCOREACEAE

Dioscorea barletii Morton; (10); MEX; R; ALK, SAP*.
D. bulbifera L.; (45); SRI; R; ALK, SAP*.
D. dumetorum (Kunth) Pax; (60); GUI; RH; ALK, SAP.
D. frazeri Prain & Burkill; (60); IND; RH; ALK, SAP*.
D. hirsuta Blum.; (60); MAD; RH; ALK, SAP*.
D. hispida Dennst.; (60); PHI; RH; ALK.
D. macrostachya Benth.; (60); TAM; RH; ALK, SAP*.
D. piscatorum Prain & Burkill; (10,16,55); MAL; RH; ALK, SAP.

D. poilaei Prain & Burkill; (60); ASI; NR; ALK, SAP*.
D. remotiflora Kunth; (113); MEX; R; ALK, SAP*.
D. rupicola Kunth; (60); SAF; NR; ALK, SAP*.
D. sansibarensis Pax; (*D. macroura* Harms); (15,25); WAF; BU, RH; ALK.
D. tokoro Makino; (45); JAP; R; NR; ALK, SAP*.

EBENACEAE

Diospyros atropurpurea Gurke; (25); WAF; NR; PLU*.
D. cauliflora De Wild.; (104); CAF; B, L; PLU*.
D. ebenaster Retz.; (16,50); DOM; B, FR; PLU*.
D. ebenum Koenig; (60); IND; NR; PLU*.
D. hebecarpa A. Cunn.; (43,52,60) AUS; FR; PLU*.
D. lucida Wall.; (16); SUM; FR; PLU*.
D. montana Roxb.; (26,52,60); IND, PHI; NR; PLU*.
D. paniculata Dalte.; (60); IND; NR; PLU*.
D. piscatoria Gurke; (60); TAF; NR; PLU*.
D. rhombifolia Hemsl.; (113); CHI; FR; PLU*.
D. rufa King; (16); MAL; W; PLU*.
D. samoensis A. Gray; (60); POL; NR; PLU*.
D. toposoides King; (16,45,52); MAL; FR; PLU*.
D. toxicaria A. Gray; (60); IND, PHI; NR; PLU*.
D. wallichii King & Gamble; (16); MAL; FR; PLU*.
D. xanthoclamys Gurke; (60); TAF; NR; SAP, TAN, PLU*.

ERICACEAE

Rhododendron barbatum Wall. ex. G. Don; (20,44,60); IND; NR; DIT.
R. caucasicum Pall.; (45); ARM; NR; DIT.
R. davuricum L.; (45); SIB; NR; DIT.
R. falconeri Hook. f.; (20,45,60,67); IND; FL; DIT.

EUPHORBIACEAE 52

Anthostema aubryanum Baill.; (60); GAB; B; NK.
A. senegalensis A. Juss.; (60); WAF; BR, L; NK.
Antidesma laciniatum Muell. Arg.; (104); CAF; L, R; NK.
Cleistanthus collinus Benth.; (45,52,60,110); IND; B, FR; SAP, TAN.
Croton sylvaticus Hochst.; (44,60); SAF; NR; TOX*.
C. tachibangensis Pellegr.; (60); PHI; B; TOX*.
C. tiglium L.; (2,14,45,52,60); ASI, PHI; FR, L; ALK, TOX.
Drypetes arborescens Hutch.; (104); CAF; B, L; NK.
D. dinklagei Hutch.; (104); CAF; B, L; NK.
D. gossweileri S. Moore; (60); GAB; B, FR; NK.
D. klainei Pierre; (60); GAB; B, FR; NK.
D. urophylla Pax; (104); CAF; B, L; NK.
Elaeuphorbia drupifera Stapf.; (25,52,60); WAF; FR, J, L; NK.
Eremocarpus setigerus (Hook.) Benth.; (*Croton setigerus* Hook.); (45,52,60); NAM; PL; NK.
Euphorbia aleppica L.; (44,52,60); EUR; NR; TRI*.
E. amygdaloides L.; (60); EUR; S; TRI*.

Table I

Continued

E. antiquorum L.; (16,60); ASI; J; TRI*.
E. baga A. Chev.; (60); TAF; J, PL; TRI*.
E. candelabrum Tremaux.; (60); TAF; NR; TRI*.
E. caracasana Boiss.; (42,45,62); TAM; BR; TRI*.
E. chamaesyce L.; (52,60); MED; NR; TRI*.
E. coralloides L.; (45); ENG; NR; TRI*.
E. cotinifolia L.; (E. cotinoides Miq.); (5,10,34,42, 45,52,60,62,76,94); TAM; BR; TRI*.
E. dendroides L.; (52,60); GRE; NR; TRI*.
E. esula L.; (45,52,60) EUR; NR; TRI*.
E. hyberna L.; (45); ENG; L; TRI*.
E. kerrii Craib; (60); ASI; R; TRI*.
E. laro Drake; (45,60); MAD; NR; TRI*.
E. lathyris L.; (45); EUR; NR; TRI*.
E. mellifera Ait.; (45,52,60); IND; NR; TRI*.
E. neglecta N. E. Brown; (60); EAF; J; TRI*.
E. neriifolia L.; (16,45); MOL; L; TRI*.
E. pilulifera L.; (45); PHI; PL; TRI*.
E. piscatoria Ait.; (45,52,60); NAF; W; TRI*.
E. platyphylla L.; (45,60); EUR, NAF; NR; TRI*.
E. plumerioides Teijsm.; (113); NGU; L; TRI*.
E. poissonii Pax; (60); TAF; PL; TRI*.
E. pulcherrima Willd. ex Klotz.; (45); CHI; NR; TRI*.
E. punicea Sw.; (45); JAM; L; FR; TRI*.
E. resinifera Berg; (60); NAF; NR; TRI*.
E. royleana Boiss.; (45); IND; NR; TRI*.
E. schlechtendalii Boiss.; (112); MEX; J; TRI*.
E. sibthorpii Boiss.; (45); GRE; B, L; TRI*.
E. tirucalli L.; (16,60,80,119); SAF; J; TRI*.
E. trigona Haw.; (16); MAL; J; TRI.
E. unispina N. E. Brown; (60); WAF; PL; TRI*.
Excoecaria agallocha L.; (60); ASI; NR; NK.
Fluggea leucopyrus Wight.; (20); IND; NR; NK.
Fontainea pancheri Heckel; (113); NHE; PL; NK.
Hippomane mancinella L.; (42,46,60,62); WIN; NR; ALK.
Homalanthus nutans Benth.; (113); SAL; L; NK.
H. populneus Pax; (113); PHI; L; NK.
Hura crepitans L.; (Hura brasiliensis Willd.); (42,45,52, 60,62,76); TAM, PER; J; TOX.
H. polyandra Baill.; (3,52,60); MEX; NR; TOX.
Jatropha curcas L.; (10,25,42,60,62); TRO; NR; TOX.
Joannesia princeps Vell.; (45,62); BRA; B, S; ALK*.
Macaranga spinosa Muell. Arg.; (60); TAF; NR; NK.
M. vedeliana Muell. Arg.; (60); NCA; NR; NK.
Mallotus apelta Muell. Arg.; (60,110); ASI, CHI; S; NK.
Manihot esculenta Crantz; (42,45,57,62); TAM; CYA.
M. utilissima Pohl.; (91a); BRA; CYA.
Neoalcornea sp.; (6); PER; NR; NK.
Petalostigma quadriloculare F. Muell.; (43,60); AUS; FR; NK.
Phyllanthus acuminatus Vahl; (34,42,62,113); TAM; L; CYA, TRI.
P. brasiliensis (Aubl.) Muell. Arg.; (34,62,94,113); AMA; FR, L; CYA*, TRI.
P. cladotricus Muell. Arg.; (45,52); BRA; NR; CYA*, TRI.
P. distichus Muell. Arg.; (45); CER; R; CYA*, TRI.

P. epiphyllanthus L.; (52,60); WIN; NR; CYA*, TRI.
P. ichthyometius Rusby; (42,62); SAM; L; CYA*, TRI.
P. niruri L.; (P. amarus Schum. & Thonn.); (60,113); TAS; NR; CYA*, TRI.
P. piscatorium H.B.K.; (42,52,112,113); COL, VEN; L; CYA*, TRI.
P. pseudoconami Muell. Arg.; (112,113); PER, COL; NR; CYA*, TRI.
P. simplex Retz.; (45); SMO; NR; CYA*, TRI.
P. subglomerulatus Poir.; (P. conami Sw.); (42,45,52, 62,76); TAM; BR; CYA*, TRI.
P. urinaria L.; (60,62); GUI, MAD; NR; CYA*, TRI.
Phytocrene blancoi Merr.; (113); PHI; FR; NK.
Piranhea trifoliata Baill.; (45,62); AMA; NR; NK.
Plagiostyles africana Prain.; (60); GAB; L; NK.
Pycnocoma macrophylla Benth.; (25,52,60); WAF; B; NK.
Ricinus communis L.; (60); IND; NR; ALK, TOX.
Sapium appendiculatum (Muell. Arg.) Pax; (82,113); MEX; NR; NK.
S. biglandulosum Muell.; (46,60); GUA; NR; NK.
S. biloculare (Wats.) Pax; (Sebastiania bilocularis Wats.); (45,82); MEX; J; NK.
S. indicum Willd.; (Exoecaria indica Muell. Arg.); (45,60,113); IND; S, FR; NK.
Sebastiania pavoniana Muell. Arg.; (45); MEX; J; NK.
S. pringlei Wats.; (82,113); MEX; B; NK.
S. ramirezii Maury; (113); MEX; NR; NK.
Securinega leucopyrus Muell. Arg.; (91a); ALK*.
S. virosa Baill.; (44,60); TAF; B; ALK*.
Spirostachys africanus Sond.; 112; WAF; NR; NK.
Synadenium sp.; (7); EAF; J; NR; NK.

FABACEAE

CAESALPINIOIDEAE 54

Bauhinia glabra Jacq.; (112); VEN; NR; TAN*.
B. guianensis Aubl.; (5,45,62,112); GUI; NR; TAN*.
Bussea occidentalis Hutch.; (48); TAF; B; NK.
Caesalpinea nuga Ait.; (45,60,67); IND; FR; NR; TAN*.
C. pulcherrima Swartz; (60); GUA; B; CYA.
C. tsoongii Merr.; (113); CHI; PL; CYA*.
Cassia sieberiana DC.; (25,60); WAF; B, L; CYA*.
Erythrophleum guineense G. Don; (60,104); TAF; B, L; ALK, SAP.
Macrolobium acaciaefolium (Baill.) Benth.; (62); SAM; NR; NK.
Pachyelasma tessmannii (Harms) Harms; (60); WAF; B, L; NK.
Senna alata (L.) Roxb.; (Cassia alata L., Hepertica alata (L.) Raf.); (25,42,60,62); COL; NR; NK.
S. angustisiliqua (Lam.) Irwin & Barneby; (Cassia semperflorens DC.); (45); TAM; NR; NK.
S. arereh Del.; (Cassia arereh Del.); (25); TAF; R, S; NK.

Table I
Continued

S. didymobotrya (Fres.) Irwin & Barneby; (*Cassia didymobotria* Fres.); (7,60); EAF; L; NK.

S. hirsuta (L.) Irwin & Barneby; (*Cassia venenifera* Rodsch., *Cassia hirsuta* L., *Ditremexa hirsuta* (L.) Britt. & Rose); (45); GUI; R; NK.

S. pallida (Vahl) Irwin & Barneby; (*Cassia biflora* L., *Pieranisia biflora* (L.) Pitt.); (42,62); TAM; NK.

Swartzia fistuloides Harms; (7,60); NRE; B; NK.

S. madagascarensis Desv.; (5,119); AFR; FR, R, S; SAP.

MIMOSOIDEAE

Acacia falcata Willd.; (52,60); AUS; NR; TAN.

A. melanoxylon R. Brown; (60); AUS; NR; TAN.

A. pennata (L.) Willd.; (26,45,67); PTR; B, FR, L; TAN*.

A. penninervis Sieb.; (45,52); AUS; L; TAN.

A. pruinescens Kurz; (45); BUR; B; NK

A. salicina Lind.; (43,52); AUS; NR; TAN.

Adenanthera abrosperma F. Muell.; (45); AUS; B; NK.

Albizia acle Merr.; (45); PHI; NR; SAP*.

A. chinensis Merr.; (60); IND; NR; SAP.

A. coriaria Welw.; (104); CAF; B, L; SAP*.

A. procera Benth.; (26,43,60,67); AUS, IND; B; SAP.

A. saponaria Bl.; (45); PHI; B; SAP.

A. stipulata Boiv.; (45); JAV; B; SAP.

Desmanthus virgatus (L.) Willd.; (42,62); PER; NR; NK.

Elephantorrhiza goelzei (Harms) Harms; (118); TAF; NR; NK.

Entada africana Guill. & Perr.; (60); AFR; L; SAP*.

E. phaseloides Merr.; (45); PHI; BR; SAP*.

E. pusaetha DC.; (60); PTR; L, B; ALK, ROT, SAP.

E. scandens (L.) Benth.; (32,45); PAN; NR; SAP*.

E. sudanica Schweinf.; (60); AFR; L; SAP*.

Enterolobium cycloscarpum Gris.; (45); GUA; NR; NK.

E. timbouva Mart.; (45,62,84); SAM; BR; NK.

Leucaena leucocephala (Lam.) De Wit.; (*L. glauca* Benth.); (60); AUS; S; NK.

L. odoratissima Hassk; (nomen nudum ?); (45); LIB; NR; NK.

L. trichodes Benth.; (111); PER; PL; NK.

Mimosa dysocarpa Benth.; (82); MEX; R; NK.

M. himalayana Gamble; (45); IND; B; NK.

Parkia biglobosa Benth.; (60); AFR; FR; ALK, CYA, TAN.

P. filicoidea Welw.; (60); AFR; FR; ALK, TAN.

Pentaclethra macroloba Kunth; (111); SUR; B; SAP*.

P. macrophylla Benth.; (25,52,104); AFR; B, S; SAP.

Piptadenia africana Hook. f.; (104); AFR; NR; NK.

P. psilostachya (DC.) Benth.; (84); FRG; NR; NK.

P. recurvata Ducke; (84); BRA; NR; NK.

Pithecellobium bigeminum Mart.; (45,60,67); IND; B, L, S; ALK, SAP.

P. ellipticum Hassk.; (113); INO; R; NK.

P. jupunba (Willd.) Urb.; (50); CAR; NR; NK.

Prosopis africana Taub.; (60); WAF; S; NK.

Tetrapleura chevalieri E. G. Baker; (60); WAF; B; SAP*.

T. tetraptera Taub.; (60); TAF; NR; SAP*.

T. thonningii Benth.; (45); TAF; NR; NK.

PAPILIONOIDEAE 29,30,73

Abrus pulchellus Wall.; (45); MOL; NR; NK.

Alexa imperatricis (Schomb.) Baker; (5,45,75,112); GUI; B; NK.

Andira inermis (Wright) H.B.K.; (10,42,62); SAM; B; ALK.

A. rosea Mart.; (62); BRA; W; ALK.

A. surinamensis Splitg. ex Pulle; (62); SAM; NR; ALK.

Antheroporum pierrei Gagnep.; (60); SEA; S; ROT.

Apurimacia incarum Harms; (45,62); PER; NR; NK.

A. michelii Harms; (112); PER; R; NK.

Baphia nitida Lodd.; (52); NRE; NR; NK.

B. polygalacea Baker; (25); WAF; NR; NK.

Barbieria pinnata (Pers.) Baill.; (42,62); BRA; B; NK.

Bowdichia virgilioides H.B.K.; (34,45,62); SAM; NR; NK.

Cadia ellisiana Baker; (60); MAD; NR; NK.

Centrosema plumierii (Turp.) Benth.; (42,45,62); TAM; B; NK.

Cicer arietinum L.; (45); EUR; PL; NK.

Clathrotropis brachypetala (Tul.) Kleinh.; (45,74,112); GUY; NR; NK.

C. macrocarpa Ducke; (84,112); SAM; B; NK.

Clitoria amazonum Mart. ex Benth.; (45,62); AMA; BR; NK.

C. arborescens Ait.; (42,45,60,62); TAM; L, R; NK.

C. guianensis (Aubl.) Benth.; (42,62); TAM; NR; NK.

Crotalaria arenaria Benth.; (60); NRE; NR; NK.

C. retusa L.; (7); EAF; L; NK.

Dahlstedtia pinnata (Benth.) Malme; (*Camptosema pinnatum* Benth.); (34,62); BRA; NR; NK.

Dalbergia stipulacea Roxb.; (60); IND; B, R; ROT.

Derris amazonica Killip; (*Lonchocarpus negrensis* Benth.); (64,76,84); GUI; NR; ROT.

D. amoena Benth.; (16); MAL; NR; ROT*.

D. araripensis Ducke; (78); NRE; NR; ROT*.

D. cebuensis Merr.; (113); PHI; PL; ROT.

D. chinensis Benth.; (60); JAP; R; ROT.

D. elegans Benth.; (16); MAL; NR; ROT.

D. elliptica (Wall.) Benth.; (13,16,20,45,52,67); MAL; RB, S; ROT, TEP.

D. ferruginea Benth.; (67); IND; R; ROT.

D. grandifolia Kunth; (*Piscidia communis* I. M. Johnst.); (45,46,67); IND, USA; R, S; ROT.

D. guianensis Benth.; (45,62); GUI; NR; ROT.

D. involuta Sprague; (60); AUS; NR; ROT, SAP.

D. koolgibberah M. F. Bailey; (45,52); NRE; NR; ROT.

D. malascensis Prain; (52,67,93,112); MAL, IND; R, RB; ROT.

D. mindorensis Perkins; (112); PHI; S; ROT*.

D. negrensis Benth.; (45,62,84); AMA; L, S; ROT.

D. paniculata Benth.; (*D. benthamii* Thwa.); (45); SRI; R; ROT.

D. philippinensis Merr.; (45); PHI; R; ROT.

D. polyantha Perk.; (45,67); IND; NR; ROT.

Table I

Continued

D. pterocarpa (DC.) Killip; (45); NRE; NR; ROT.
D. scandens (Roxb.) Benth.; (16,45,67); IND, MAL; J, R; ROT.
D. tonkinensis Gagnepain; (60); CHI; NR; ROT.
D. trifoliata Lour.; (16,45,67); PHI; R, S; ROT.
D. uliginosa Benth.; (13,45,52); MAL; R, RB, S; ROT.
Desmodium sp.; (45); LOY; NR; NK.
Dolichos lupuniflorus N. E. Brown; (60); AFR; R; SAP.
D. pseudopachyrhizus Harms; (25); MAA; R; SAP.
Eriosema griseum Bak.; (25); WAF; R; NK.
Erythrina berteroana Urb.; (72); SAM; S; NK.
E. glauca Willd.; (10); CAM; S; NK.
E. poeppigiana (Walp.) Cook.; (72); SAM; S; NK.
Fordia filipes Dunn; (111); BOR; R; NK.
Gliricidia sepium Steud.; (45); PHI; NR; NK.
Indigofera lepedezoides H.B.K.; (45); SAM; NR; NK.
I. pascuorum Benth.; (115); SAM; NR; NK.
I. suffruticosa Mill.; (10,45); SAM; PL; NK.
Leptoderris congolensis Dunn.; (60); GAB; NR; ROT.
Lonchocarpus benthamianus Pittier; (50,113); CAR; R; ROT*.
L. chrysophyllus Kleinh.; (34,62,64,76,112,113); SAM; R, S; ROT*.
L. cyanescens Benth.; (60); AFR; NR; ROT*.
L. densiflorus Benth.; (45,52,62,74); SAM; NR; ROT*.
L. denudatus Benth.; (62,112); BRA; R; ROT*.
L. dominguensis DC.; (112); WIN; R; ROT*.
L. ernestii Harms.; (*Derris ernestii* (Harms.) Ducke); (86); AMA; NR; ROT*.
L. floribundus Benth.; (*L. nitidulus* Benth.); (42,45,62, 76,84,112); SAM; R; ROT*.
L. glabrescens Benth.; (*Derris glabrescens* (Benth.) Macbr.); (86); AMA; NR; ROT*.
L. guilleminiana Tul.; (*Derris guilleminiana* (Tul.) Macbr.); (86); AMA; NR; ROT*.
L. hedyosma Miq.; (*Derris hedyosma* (Miq.) Macbr.); (86); AMA; NR; ROT*.
L. izalabamus S. F. Blake.; (46); GUT; NR; ROT*.
L. latifolius (Willd.) H.B.K.; (cited by 84 as *L. discolor* Hub.); (42,50,62,78,84,112); GUI, VEN; NR; ROT, SAP.
L. madagascariensis Viguier; (60); AFR; NR; ROT*.
L. martynii A. C. Smith; (34,64,112,113); GUY; R; ROT*.
L. minimiflorus D. Smith; (46); GUA; NR; ROT*.
L. mutabilis ?; (nomen nudum); (111); PER; S; ROT*.
L. nicou (Aubl.) DC.; (*Derris nicou* (Aubl.) Macbr.); (6,62,74,78,94,116); SAM; NR; ROT*.
L. oxycarpus DC.; (60); WIN; NR; ROT*.
L. peckolti Wawra; (62); BRA; NR; ROT*.
L. pterocarpus DC.; (*Derris pterocarpus* (DC.) Killip); (34,62,76); BRA; NR; ROT*.
L. rariflorus Mart. ex Benth.; (5,45,62,112); AMA, GUI; ROT*.
L. rubiginosus Benth.; (60); GUY; NR; ROT*.
L. rufescens Benth.; (45); GUI; NR; ROT*.
L. sericeus H.B.K; (*L. hemidyosmus* Miq., *Derris sericea* (H.B.K.) Ducke); (10,46,78); TAM; NR; LON.

L. spiciflora Mart.; (*Derris spiciflora* (Mart.) Macbr.); (86); AMA; NR; ROT*.
L. spruceanus Benth.; (*Derris spruceana* (Benth.) Ducke); (76,86); FRG; L, R; NR; ROT*.
L. stipularis C. T. White; (112); AUS; NR; ROT*.
L. sylvestris A. C. Smith; (*Derris sylvestris* (A. C. Smith) Macbr.); (64,86); AMA; NR; ROT*.
L. urucu Killip & Smith; (*Derris urucu* (Killip & Smith) Macbr.); (62,64,78,94,112); AMA; NR; ROT*.
L. utilis A. C. Smith; (13,42,67); TAM; NR; ROT*.
L. velutinus Benth.; (112); PAN; NR; ROT*.
L. violaceus (Jacq.) H.B.K; (45,52,62); TAM; NR; ROT*.
Lupinus crotalarioides Mart.; (62); BRA; NR; ALK.
L. mutabilis Sweet; (10,43,62); PER; NR; ALK.
Machaerium macrophyllum Mart.; (84); BRA; NR; NK.
Millettia atropurpurea Benth.; (60); JAV; S; SAP*.
M. barteri Dunn.; (44,60); TAF; NR; ROT, SAP*.
M. cafra Meissn.; (45); SAF; NR; ROT*.
M. chapelieri Baill.; (60); MAD; NR; ROT, SAP*.
M. dasyphylla (Miq.) Boerl.; (45); JAV; NR; ROT*.
M. ferruginea Hochst.; (45,60); ABY; NR; ROT, SAP*.
M. ichthyoctona Drake; (60); ASI; S; W; ROT, SAP*.
M. lasiopetala Merr.; (112,113); CHI; R; ROT, SAP*.
M. lenneoides Vakte; (60); MAD; NR; ROT, SAP*.
M. pachycarpa Benth.; (45,67); IND; NR; ROT, SAP*.
M. piscidia Wight; (60); IND; R; NR; ROT, SAP*.
M. sericea Wight & Arn.; (45); SUM; NR; ROT, SAP*.
M. taiwaniana Hayata; (45); TAI; R; ROT, SAP*.
Muellera frutescens (Aubl.) Standl.; (45,62,76); GUI; NR; ROT, SAP*.
M. moniliformis L.; (*Derris moniliformis* (L.) Ducke); (62,86); AMA; B; ROT, SAP*.
Mundulea monantha (Baker) P. Boiteau; (44,60); MAD; NR; ROT, SAP*.
M. pauciflora Baker; (19,45); EAF; J; ROT, SAP*.
M. sericea (Willd.) A. Chev.; (19,52,60,67,119); TAF, ASI; B, L, R; ALK, ROT.
M. striata Dub. & Dop.; (45,60); MAD; S; ROT*.
M. suberosa Benth.; (20,45,52); IND; NR; ROT*.
M. telfairii Baker; (19); MAD; ROT*.
Neorautanenia coriacea C. A. Sim.; (60); SAF; NR; NK.
Nissolia fruticosa Jacq.; (45); TAM; NR; NK.
Ormosia sp.; (86); AMA; B; NK.
Orobus piscidia Spruce; (45); NCA; NR; NK.
Ostryocarpus riparius Hook. f.; (25,60); LIB; B; NK.
Ougeinia dalbergioides Benth.; (45,102); IND; B; NK.
Pachyelasma tessmannii Harms; (45,60); GAB; B, FR; NK.
Pachyrhizus angulatus Rich.; (45); BRA; RH; ROT, SAP*.
P. erosus (L.) Urb.; (10,62,70); GUI, VEN; L, RH; ROT, SAP*.
P. tuberosus Spreng.; (62); TRO; NR; ROT, SAP*.
Phaseolus lathyroides L.; (45,62); WIN; S; NK.
Phylloxylon ensifolius Baill.; (60); MAD; W; NK.
P. pierrieri Drake; (60); MAD; NR; NK.

Table I
Continued

Piscidia guaricensis (Pitt.) Pittier; (*Lonchocarpus guarinensis* Pittier); (10,62); VEN; B; ROT.
P. piscipula (L.) Sarg.; (*P. carthaginensis* Jacq., *P. erythrina* L., *Ichthyometia piscipula* (L.) Hitchc.); (10,38,42,45,57,62,84,112); CAR; B, L; ROT.
Pongamia glabra Vent.; (45); IND; R, S; ROT*.
P. pinnata (L.) Merr.; (cited by 43 as *P. glabrata* Vent.); (26,43,60,67,70); AUS; R; ALK, ROT.
Pueraria peduncularis R. Grah.; (112); CHI; R; NK.
Rhynchosia minima (L.) DC.; (1,56); JAM; NR; NK.
Sesbania pubescens DC.; (52); NRE; NR; NK.
Sophora occidentalis L.; (25); WAF; NR; ALK.
S. tomentosa Baker; (7); WAF; NR; NK.
Tephrosia aequilata Bak.; (119); NRE; NR; ROT*.
T. anselii Hook. f.; (60); CON; NR; ROT*.
T. astragaloides R. Br.; (45); AUS; L; ROT*.
T. barbigera Welw.; (60); ASI, NR; ROT*.
T. candida DC.; (19,45,52,67); IND; B, L; ROT*.
T. cinerea (L.) Pers.; (*T. litoralis* Pers.); (10,19,45,62); TAM; L, S; ROT.
T. coronillaefolia DC.; (91a); ROT.
T. densiflora Hook. f.; (19,52); SAM; ROT*.
T. elegans Schum. & Thonn.; (60); NRE; NR; ROT*.
T. emarginata H.B.K.; (45); BRA; NR; ROT*.
T. huilensis Welw.; (60); CON; NR; ROT*.
T. hydeana (Rydb.) Standl.; (112); GUA; R; ROT*.
T. ichthyneca Bertol.; (45); MOZ; ROT*.
T. leiocarpa A. Gray; (82); MEX; R; ROT.
T. macropoda Harv.; (45,52); AFR; R; ROT*.
T. monantha Baker; (45); EAF; NR; ROT*.
T. multifolia Rose; (113); CAM; NR; ROT*.
T. nitens Benth. ex Seem.; (*Cracca nitens* Benth.); (45,52,60); AMA; ROT*.
T. periculosa Baker; (45,52); WAF; NR; ROT*.
T. piscatoria Pers.; (22,60); PAC; R; ROT*.
T. purpurea (L.) Pers.; (19,45,52,60); SAM, IND; R, S; TEP.
T. rosea F. Muell.; (45,52,60); AUS; BR, R; ROT*.
T. sessiliflora (Poir.) Hassl.; (42,62); SAM; ROT*.
T. sinapou (Buchholz) A. Chev.; (*T. toxicaria* (Sw.) Pers., *Cracca toxicaria* Sw.); (5,19,45,52,60,62,76); TAM; NR; ROT, TEP*.
T. talpa (Wats.) Rose; (*Cracca talpa* Wats.); (19,45,82); MEX; R; ROT*.
T. tomentosa Pers.; (45); ARA; L; ROT*.
T. toxifera ?; (nomen nudum); (45); CAM; PL; ROT*.
T. virginiana (L.) Pers.; (45); NAM; NR; ROT.
T. vogelii Hook. f.; (7,19,60,80,119); WAF; S; ROT, TEP.
Vataireopsis speciosa Ducke; (112); SUR; R; NK.
Willardia mexicana Rose; (113); MEX; R; NK.

FLACOURTIACEAE

Casearia graveolens Dalz.; (20,45,60); IN; FR; NK.
C. tomentosa Roxb.; (20,45); IND; S; NK.

Gynocardia odorata R. Brown; (52,60,67); IND; FR, S; CYA.
Hydnocarpus anthelmintica Pier; (60); NRE; NR; CYA*.
H. castanea Hook. & Thoms.; (45); BUR; FR; CYA*.
H. heterophylla Bl.; (45); JAV; NR; CYA*.
H. kurzii Warb.; (60); IND; NR; CYA*.
H. laurifolia Sleumer; (45,60); IND; NR; CYA*.
H. venenata Geartn.; (44,60); ASI; FR; CYA.
H. wightiana Blume; (67); ASI; FR; CYA*.
Pangium edule Reinw.; (60); JAV; B; CYA.
Ryania angustifolia (Turcz.) Mon.; (10,44); SAM; NR; ALK.
R. speciosa Vahl; (75,89); NRE; NR; ALK*.
Scottelia kamerunensis Gilg; (104); CAF; B; NK.

FRANKENIACEAE

Frankenia ericifolia C. Sm.; (52,60); TAF; PL; NK.

GNETACEAE

Gnetum scandens Roxb.; (16,45,67); IND; L; SAP*.

HIPPOCASTANACEAE

Aesculus arguta Burkill; (110); NRE; S; SAP*.
A. californicus Nutt.; (45,57,60); NAM; BR, L; COU, SAP*.
A. chinensis Bunge; (110); CHI; B; SAP*.
A. glabra Willd.; (45); USA; FR; SAP*.
A. pavia L.; (57); NAM; NR; SAP*.

HUMIRIACEAE

Saccoglottis gabonensis Urb.; (44,60); GAB; B; NK.

ICACINACEAE

Humirianthera duckei Huber; (86); AMA; NR; NK.
H. rupestris Ducke; (86); AMA; NR; NK.
Phytocrene blancoi Merr.; (44,113); PHI; FR; ALK*.

IRIDACEAE

Sisyrinchium arizonicum Rothr.; (44,82); MEX; NR; SAP.

JUGLANDACEAE

Carya illinoensis (Wang) K. Koch; (44,82); MEX; B, L; TAN*.
Engelhardtia polystachya Radlk.; (110); IND; B; NK.

Table I
Continued

Juglans nigra L.; (45); IND; FR; NK.
J. regia L.; (26,45); IND; NR; JUG.
J. rupestris Engelm.; (44,82); MEX; B, L; TAN*.
Pterocarya tonkinensis Dode; (60,103); SEA; NR; NK.

LAMIACEAE

Eremostachys superba Royle ex Benth.; (60); ASI; NR; NK.
E. vicaryi Benth.; (67); IND; PL; NK.
Marrubium supinum L.; (60); W; NR; DIT.
Trichostema lanceolatum Benth.; (45); USA; NR; NK.

LAURACEAE

Ocotea pretiosa Mez; (39); BRA; O; ESS.
Umbellularia californica Nutt.; (45); USA; L; NK.

LECYTHIDACEAE 81

Barringtonia acutangula Gaertn; (44,45,52,60); IND, POL; FR; SAP*.
B. asiatica (L.) Kurz.; (*B. speciosa* L.); (42,52,60,67); ASI, AUS; B, L, S; SAP*.
B. calyptrata R. Brown ex Bailey; (52,67); AUS; B; SAP.
B. macrocarpa Haskarl.; (*B. insignis* Miq.); (110); IND; RB; SAP.
B. macrostachya Kurz; (*B. balabacensis* Merr., *B. cylindrostachya* Kurz.); (45,100); MAL; B, FR; SAP.
B. racemosa Bl.; (45,67,110,113); TRO; FR, S; SAP.
Careya arborea Roxb.; (44,60,113); AUS; FR, L; TAN, SAP*.
C. australis F. Muell.; (43,60); AUS; S; SAP.
Chydenanthus exelsus (Bl.) Miers.; (*Barringtonia vriesei* Teij. & Binn.); (44,60); PTR; NR; SAP*.
Gustavia augusta L.; (45,62); SAM; FR; SAP*.
G. hexapetala (Aubl.) Smith; (*G. brasiliana* DC.); (45,62,77); SAM; FR; SAP*.
Planchonia careya Kunth; (45); AUS; R, B; NK.

LILIACEAE

Chlorogalum pomeridianum Pomel.; (45,57); NAM; BU; ALK, SAP.
Crinum asiaticum L.; (45); IND; R, L; ALK*.
C. yuccaefolium Salib.; (44,60); WAF; NR; ALK*.
C. zeilanicum (L.) L.; (25); WAF; S; ALK*.
Haemanthus foetida Haw.; (76); FGU; R; SAP.
Ornithogalum sp.; (112); GUY; NR; ALK.
Scadoxus multiflorus Martyn) Raf.; (25); WAF; BU; ALK.
Schoenocaulon officinale A. Gray; (45); MEX; NR; NK.
Smilacina sessilifolia Baker.; (45); NAM; NR; SAP.
Veratrum album L.; (45); SAP; NR; NK.
Yucca decipiens Trel.; (82); MEX; NR; SAP.

LOGANIACEAE 65

Antonia ovata Pohl; (5,45,76); GUI; NR; NK.
Gelsemium elegans Benth.; (45,60); SEA; S; ALK*.
G. sempervirens (L.) Ait.; (57); USA; NR; ALK*.
Scyphostrychnos talboti S. Moore; (104); CAF; R; NK.
Spigelia anthelmia L.; (80,113); SUR; NR; ALK.
Strychnos aculeata Solered.; (25,60,104); WAF; FR; ALK*.
S. colubrina L.; (60); IND; R, S, W; ALK*.
S. erichsonii Schomb.; (112); SUR; NR; ALK*.
S. icaja Baill.; (104); AFR; B, L; ALK*.
S. nux-blanda A. W. Hill; (113); BUR; FR; ALK*.
S. nux-vomica L.; (26,60); IND; FR; ALK*.
S. rondeletioides Spruce; (112); BRA; B, L; ALK*.

MALPIGHIACEAE

Byrsonima crassifolia (L.) H.B.K.; (10,34,42,44,45); VEN; BR; SAP*.
Mascagnia sp.; (86); AMA; S; NR; NK.

MALVACEAE

Thespesia populnea Soland.; (43,60); NRE; NR; HET.

MELIACEAE

Carapa guianensis Aubl.; (86); AMA; NR; NK.
Lovoa klaineana Pierre; (104); CAF; B, L; NK.
Melia azedarach L.; (3,10,45,60); TRO; B; SAP.
Pseudocedrela kotschyi Harms.; (60); WAF; F; ALK, SAP, TAN.
Turraeanthus africanus Pell.; (60); WAF; B; NK.
Walsura piscidia Roxb.; (60,67,113); IND; B, FR; NK.

MENDONCIACEAE

Mendoncia aspera (R. & P.) Nees; (99); COL; R; NK.
M. pedunculata Leonard.; (101); COL; R; NK.

MENISPERMACEAE

Abuta imene (Mart.) Eichl.; (44,62); SAM; NR; ALK.
Anamirta cocculus (L.) Wight & Arn.; (*A. paniculata* Colebr.); (28,45,51,52); MAL, PER; B, FR, S; ALK.
Arcangelisia sp.; (113); PHI; NR; NK.
Chasmanthera welwitschii Troopin; (60); GAB; L; NK.
Cissampelos pareira L.; (cited by 60 as *C. owariensis* DC.); (60,62); WIN; NR; ALK.
Cocculus carolinus (L.) DC.; (57); NAM; NR; ALK*.
C. fernandianus Gaud.; (45); HAW; BR; ALK*.
C. imene Mart.; (45); BRA; NR; ALK*.
C. indicus Wehmer; (45); JAV; NR; ALK*.
Pachygone ovata Miers.; (44,60); PTR; NR; ALK*.

Table I
Continued

Stephania hernandifolia Walp.; (43,45); PTR; L; ALK, SAP.
Tinomiscum philippinense Diels.; (110); PHI; NR; NK.
Tinospora cordifolia Miers; (45); IND; BR; NK.

MORACEAE

Antiaris toxicaria Lesch.; (44,60); ASI; B; CAR.

MYRICACEAE

Myrica nagi Thunb.; (67); IND; B; NK.

MYRSINACEAE

Aegiceras majus Gaertn.; (45); AUS; NR; SAP.
A. minus Gaertn.; (44,45); AUS; NR; SAP*.
Ardisia foetida Rom. & Schult.; (*Conomorpha magnolifolia* Mez); (76); SAM; NR; SAP*.
Cybianthus fulvopulverulentus (Mez) Agost.; (85); NR; SAM; SAP.
Geissanthus andinus Mez; (85); AND; NR; SAP*.
Maesa denticulata Mez; (45); PHI; NR; SAP*.
M. indica Wall.; (67); IND; B, L; SAP*.
M. novo-caledonica Mez; (113); NCA; NR; SAP*.
Stylogyne longifolia (Mart. & Miq.) Mez; (85); SAM; NR; SAP*.

MYRTACEAE

Eucalyptus microtheca F. Muell.; (52,60); AUS; BR; ESS, TAN.

OCHNACEAE

Wallacea insignis Spruce ex Benth.; (94); AMA; NR; NK.

OLEACEAE

Ligustrum porteri C. & R.; (82); MEX; R; NK.

ORCHIDACEAE

Bletia verecunda R. Brown; (44,68); BAH; BU; NK.

PAPAVERACEAE

Bocconia pearcei Hutch.; (44,113); BOL; NR; ALK*.

PASSIFLORACEAE

Adenia cissampeloides (Hook. f.) Harms; (*Ophiocaulon cissampeloides* Hook. f.); (25,27,45,52,57,60); TAF; R; CYA.

A. gracilis Harms; (60,104); CAF; L, S; CYA.
A. lobata Engler; (25,45,60); WAF; BR, L; CYA.
Barteria barteri Hook. f.; (60); GAB; PL; NK.

PHYTOLACCACEAE

Didymotheca cupressiformis Walt.; (60); AUS; NR; NK.
Petiveria allicea L.; (46,60); TAM; L, S; NK.
P. tetrandra Gomez; (62,10); BRA; NR; NK.
Phytolacca americana L.; (57,70); NAM; NR; ALK, SAP.
P. rivinoides Kunth & Bouch; (42); TAM; NR; SAP.

PIPERACEAE

Piper bartinglianum (Miq.) C. DC.; (108); SAM; NR; ALK*.
P. darienense C. DC.; (45,108); SAM; NR; ALK*.
P. hispidum H.B.K.; (112); CUB; L; ALK*.
P. methysticum Forst.; (45); HAW; PL; ALK*.
P. piscatorum Trel. & Yuncker; (108); SAM; NR; ALK*.
P. riolimonense Trel.; (45); VEN; NR; ALK*.
P. tuberculatum Jacq.; (112); VEN; NR; ALK*.

PITTOSPORACEAE

Pittosporum arborescens Rich.; (44,113); FIJ; FR; SAP*.
P. brackenriegei A. Gray; (44,113); FIJ; FR; SAP*.
P. javanicum Bl.; (45); JAV; FR; SAP*.
P. pickeringii A. Gray; (44,113); FIJ; FR; SAP*.
P. rhytidocarpum A. Gray; (44,113); FIJ; FR; SAP*.
P. tuberculatum Zeyh.; (44,113); WIN; NR; SAP*.

PLUMBAGINACEAE

Statice pectinata Ait.; (44,45,52,60); NRE; R; NK.

POLEMONIACEAE

Ipomopsis macombii (Torr.) Grant; (*Gilia macombi* Torr.); (82); MEX; NR; SAP*.
I. thurberi (Torr.) Grant; (82); MEX; NR; SAP*.

POLYGONACEAE

Polygonum acre H.B.K.; (44,45,60); SAM; PL; CYA.
P. acuminatum H.B.K.; (45); TAM; NR; CYA*.
P. erythrodes Miq.; (45); SUM; J; CYA*.
P. glabrum Willd.; (45); SAM; PL; CYA.
P. hispidum H.B.K.; (62); SAM; PL; CYA.
P. hydropiper L.; (43,45,60); SEA; NR; TAN.
P. lapathifolium L.; (45); USA; PL; CYA*.
P. orientale L.; (60); ASI; NR; CYA*.
P. pennsylvanicum L.; (82); MEX; PL; CYA*.

Table I
Continued

P. punctatum Ell.; (82,112); MEX; PL; CYA*.
P. strigosum R. Brown; (60); NAM; NR; CYA*.
Ruprechtia laurifolia C. A. Mey.; (45,62); SAM; PL; NK.

PRIMULACEAE

Anagallis arvensis L.; (45,60,67); PTR; NR; SAP.
Cyclamen graecum Link.; (45); EUR; NR; SAP.
C. elegans L.; (60); EUR; BU; SAP.
C. europaeum L.; (45,60); EUR; BU; SAP.
C. latifolium Sibth.; (67); IND; R; SAP.
C. neopolitanum Ten.; (52); EUR; J; SAP.
C. persicum Mill.; (45,67); IND; R; SAP.
C. vernum Lob.; (60); EUR; BU; SAP.

RANUNCULACEAE

Aconitum sp.; (45); GER; NR; NK.
Helleborus sp.; (44,60); EUR; NR; CAR, SAP.

RHAMNACEAE

Columbrina asiatica (L.) Brongn.; (110); POL; FR; NK.
Gouania polygama (Jacq.) Urb.; (56); PAN; PL; SAP.
Ziziphus jujuba Lam.; (26,44,55); AFR, AUS; FR; SAP*.

ROSACEAE

Prunus capuli Cav.; (44,82); MEX; B, L; CYA.
P. persica (L.) Batsch; (82); MEX; L; CYA.
Purshia tridentata (Pursh.) DC.; (82); MEX; L; NK.
Pygeum gardnerii Hook. f.; (60); IND; S; CYA.

RUBIACEAE

Adina cordifolia Benth. & Hook.; (92); IND; NR; NK.
Cinchona sp.; (45); GUI; B; NR; NK.
Gardenia lutea Fresen; (25); EAF; FR; SAP*.
G. curranii Merr.; (113); PHI; FR; SAP*.
Mitragyna ciliata Aubr. & Pell.; (44,60); WAF; B; ALK*.
Morelia senegalensis A. Rich.; (60); WAF; FR; NK.
Pausinystalia yohimba (K. Schum.) Pierre; (60); WAF; NR; NK.
Randia acuminata Benth.; (25,44,60); TAF; FR, L; SAP*.
R. armata (Sw.) DC.; (*Bassanacantha armata* Hook. f.); (45); MAR; FR; SAP*.
R. dumetorum Lam.; (26,45,52,60,96); IND; FR, R; SAP*.
R. nilotica Stapf.; (52,60); WAF; FR; SAP*.
R. spinosa (Jacq.) Karst.; (62); TAM; FR; SAP*.
R. uliginosa DC.; (60); IND; FR; SAP*.
R. walkeri Pell.; (60) WAF; FR; SAP*.

Sarcocephalus cordatus Miq.; (43,44,60); AUS; B; ALK*.
Vanqueriopsis discolor Robyns; (60); WAF; W; NK.

RUTACEAE

Acronychia pedunculata (L.) Miq.; (60); AUS, IND; NR; NK.
Casimiroa edulis Llave & Lex.; (44,45,82); MEX; B, PL; ALK*.
C. sapota Oerst.; (45,82); MEX; B, PL; ALK*.
Cusparia trifoliata Benth.; (42,45,65); SAM; B, S; ALK*.
Fagara angolense Engl.; (104); CAF; B; ALK*.
F. heitzii Aubrév. & Pellegr.; (60); WAF; NR; ALK*.
F. leimairei De Wild.; (104); CAF; B; ALK*.
F. macrophylla (Oliv.) Engl.; (104); CAF; B; ALK*.
F. zanthoxyloides Lam.; (60); WAF; B; ALK*.
Zanthoxylum alatum Roxb.; (45,60,67); PHI; B, FR; ALK, ESS.
Z. nitidum DC.; (111); PHI; PL; ALK, ESS*.
Z. piperitum DC.; (45); JAP; L; ESS.

SALICACEAE

Salix sp.; (45); MEX; L; NK.

SAPINDACEAE

Blighia laurentii De Wild.; (44,60); CAF; NR; SAP*.
B. welwitschii (Hieron.) Radlk.; (110); WAF; FR, J; SAP*.
Cardiospermum grandiflorum Sw.; (45); SAM; PL; CYA.
C. halicacabum L.; (111); GUY; NR; CYA.
Dodonaea viscosa (L.) Jacq.; (60); AUS; TAM, SAP.
Ganophyllum falcatum Bl.; (45); PHI; B; SAP.
Harpullia arborea (Blanco) Radlk.; (45); PHI; NR; SAP*.
H. cupanioides Roxb.; (60); MAD; SAP*.
H. thanatophora Bl.; (45); NGU; B; SAP*.
Jagera pseudorhus (A. Rich.) Radlk.; (*Cupania pseudorhus* A. Rich.); (43,45,52,110); AUS; B; SAP.
Magonia pubescens St. Hil.; (*M. glabrata* St. Hil.); (5,62,84); BRA; B, L, R, S; NK.
Paullinia acuminata Uitt.; (76); GUI; NR; SAP*.
P. alata (R. & P.) G. Don; (45,62); TAM; R; SAP*.
P. aspera Radlk.; (84); BRA; NR; SAP*.
P. carpopodea Camb.; (62,84); BRA; NR; SAP*.
P. costata Schlecht. & Cham.; (45); MEX; NR; SAP*.
P. cupana H.B.K.; (42,45,62); SAM; NR; ALK*.
P. cururu L.; (10,42,45,60); TAM; NR; SAP*.
P. dasyphylla Radlk.; (84); BRA; NR; SAP*.
P. echinata Hub.; (84); TAM; NR; SAP*.
P. elegans Camb.; (42,62,84); SAM; NR; SAP*.
P. fuscescens H.B.K.; (46,62,76); TAM; NR; SAP*.
P. imberbis Radlk.; (84); BRA; NR; SAP*.
P. jamaicensis Macf.; (45); JAM; NR; SAP*.

Table I
Continued

P. *macrophylla* H.B.K.; (42,62); COL; NR; SAP*.
P. *meliaefolia* Juss.; (45,62,84); BRA; NR; SAP*.
P. *neglecta* Radlk.; (84); SAM; NR; SAP*.
P. *pachycarpa* Benth.; (112); BRA; S; SAP*.
P. *pinnata* L.; (10,35,42,45,46,62,96,99,117); TAF, TAM; F, L, S; ALK*.
P. *rhizantha* Poepp.; (84); BRA; SAP*.
P. *rhomboidea* Radlk.; (84); BRA; NR; SAP*.
R. *rubiginosa* Camb.; (45,62); BRA; NR; SAP*.
P. *scarlatina* Radlk.; (107); GUA; NR; SAP*.
P. *seminuda* Radlk.; (62); BRA; NR; SAP*.
P. *spicata* Benth.; (76,94); SAM; NR; SAP*.
P. *subnuda* Radlk.; (84); AMA; NR; SAP*.
P. *thalictrifolia* Juss.; (45,62); NR; SAP*.
P. *tomentosa* Jacq.; (107); GUA; NR; SAP*.
P. *trigonia* Vell.; (45,62,84); BRA; NR; SAP*.
P. *xestophylla* Radlk.; (84); BRA; NR; SAP*.
Phialodiscus unijugatus (Baker) Radlk.; (25,52,104); TAF; B, PL; NK.
Sapindus marginatus Willd.; (102); NAM; NR; SAP.
S. *mukorosii* Gaertn.; (67); IND; NR; SAP.
S. *saponaria* L.; (10,35,42,45,46,52,62,67,102); SAM; FR, R, S; SAP.
S. *trifoliatus* L.; (26,45,67); IND; FR, S; SAP.
Serjania acoma Radlk.; (84); BRA; NR; SAP*.
S. *aculeata* Radlk.; (84); BRA; NR; SAP*.
S. *acuminata* Radlk.; (45,94); BRA; NR; SAP*.
S. *atrolineata* Wright; (46); CAM; S; SAP*.
S. *caracasana* (Jacq.) Willd.; (42,62); BRA; NR; SAP*.
S. *chaetocarpa* Radlk.; (84); BRA; NR; SAP*.
S. *clematidifolia* Camb.; (62); BRA; NR; SAP*.
S. *communis* Camb.; (45,62,84); SAM; NR; SAP*.
S. *confertiflora* Radlk.; (84); BRA; NR; SAP*.
S. *curassavica* (L.) Radlk.; (60); NRE; NR; SAP*.
S. *cuspidata* Camb.; (45,62); BRA; NR; SAP*.
S. *dentata* (Vell.) Radlk.; (45,62,84); BRA; NR; SAP*.
S. *diversifolia* (Jacq.) Radlk.; (84); NRE; NR; SAP*.
S. *erecta* Radlk.; (*Paullinia grandiflora* Camb.); (45,62,84,91,94); BRA; NR; SAP*.
S. *faveolata* Radlk.; (84); BRA; NR; SAP*.
S. *fuscifolia* Radlk.; (62,84,94); BRA; NR; SAP*.
S. *fusca* Radlk.; (84); BRA; NR; SAP*.
S. *glabrata* H.B.K.; (45,62,73); SAM; NR; SAP*.
S. *glutinosa* Radlk.; (62,84); BRA; NR; SAP*.
S. *goniocarpa* Radlk.; (113); MEX; NR; SAP*.
S. *grandiflora* Camb.; (62,84); BRA; NR; SAP*.
S. *grandifolia* Sagot; (76); GUI; NR; SAP*.
S. *hebecarpa* Benth.; (84); BRA; NR; SAP*.
S. *herterii* Ferruci; (*Paullinia australis* St. Hil.); (45,62,84); BRA; NR; SAP*.
S. *hirsuta* Camb.; (84); BRA; NR; SAP*.
S. *ichthyoctona* Radlk.; (45,62,84); BRA; NR; SAP*.
S. *inebrians* Radlk.; (45,46); NRE; NR; SAP*.
S. *inscripta* Radlk.; (73); PER; NR; SAP*.
S. *laruotteana* Camb.; (62,84); BRA; NR; SAP*.
S. *leptocarpa* Radlk.; (112); BRA; NR; SAP*.
S. *lethalis* St. Hil.; (cited by 11 as S. *trirostris* Radlk.); (11,45,62,84,91,94); SAM; PL; SAP*.
S. *marginata* Casar.; (84); SAM; NR; SAP*.

S. *meridionalis* St. Hil.; (84); BRA; NR; SAP*.
S. *mexicana* (L.) Willd.; (60); CAM; PL; SAP*.
S. *nodosa* (Jacq.) Radlk.; (60); SAM; NR; SAP*.
S. *noxia* Camb.; (45,84); BRA; NR; SAP*.
S. *ovalifolia* Radlk.; (62,84); BRA; NR; SAP*.
S. *orbicularis* Radlk.; (84); BRA; NR; SAP*.
S. *paludosa* Camb.; (84); BRA; NR; SAP*.
S. *paradoxa* Radlk.; (84); BRA; NR; SAP*.
S. *paucidentata* DC.; (5,10,34,45,52,62,76,84); SAM; NR; SAP*.
S. *perulacea* Radlk.; (84); SAM; NR; SAP*.
S. *piscatoria* Radlk.; (45,62,84); BRA; NR; SAP*.
S. *piscineans* Gutierrez; (nomen nudum); (42); COL; NR; SAP*.
S. *platycarpa* Benth.; (84); BRA; NR; SAP*.
S. *polyphylla* (L.) Radlk.; (42,45,62); WIN; NR; SAP*.
S. *purpurascens* Radlk.; (45); SAM; NR; SAP*.
S. *pyramidata* Radlk.; (cited by 11 as S. *tenuifolia* Radlk.); (5,11,45); SAM; NR; SAP*.
S. *regnelli* Schlecht.; (84); BRA; NR; SAP*.
S. *reticulata* Camb.; (84); BRA; BR; SAP*.
S. *rubicaulis* Benth.; (45,62); PER; NR; SAP*.
S. *rubicunda* Radlk.; (112); BRA; S; SAP*.
S. *rufa* Radlk.; (45,62); SAM; BR; SAP*.
S. *serrata* Radlk.; (45,62); SAM; NR; SAP*.
S. *tricostata* Radlk.; (84); SAM; NR; SAP*.
S. *tristis* Radlk.; (84); BRA; NR; SAP*.
S. *viridissima* Radlk.; (84); BRA; NR; SAP*
Talisia esculenta (St. Hil.) Radlk.; (45,62); BRA; NR; SAP*.
T. *furfuracea* Sandwith; (34); GUY; NR; SAP*.
T. *hexaphylla* Vahl.; (NY-specimen); GUY; NR; SAP*.
T. *squarrosa* Radlk.; (34,45,112); GUI; J; SAP*.
T. *stricta* (K. & T.) Triana & Planch.; (84); BRA; NR; SAP*.
Thinouia obliqua (R. & P.) Radlk.; (NY-specimen, M. *Baker 5993*); ECU; NR; NK.
T. *paraguyensis* (M. & B.) Radlk.; (45); SAM; NR; NK.
Tripterodendron filicifolium (Lind.) Radlk.; (45,62); BRA; NR; NK.

SAPOTACEAE

Donella klainei (Pierre) Engl.; (104); CAF; B, L; NK.
Lecomtedoxa nogo (A. Chev.) Aubr.; (60); WAF; S; NK.
Madhuca butyracea (Roxb.) Roxb.; (*Bassia butyracea* Roxb.); (45); IND; B; SAP*.
M. *latifolia* Roxb.; (*Bassia latifolia* Roxb.); (44,67); IND; S; SAP.
Manilkara sp.; (44,104); CAF; B, L; SAP.

SCROPHULARIACEAE

Digitalis thapsi L.; (45); EUR; NR; NK.
Scrophularia sp.; (45); GER; NR; NK.
Verbascum blattaria L.; (60); EUR; S; SAP*.

Table I

Continued

V. crassifolium Hoffmanns.; (45,52); EUR; S; SAP*.
V. lychnitis L.; (60); EUR; S; SAP*.
V. nigrum L.; (45,60); EUR; S; SAP*.
V. phlomoides L.; (45,52); EUR; S; SAP*.
V. sinnatum L.; (45,52); EUR; PL, S; SAP*.
V. speciosum Schrad.; (113); EUR; S; SAP*.
V. ternacha Hochst.; (91a); ABY; R; SAP*.
V. thapsiforme Schard.; (45); RUS; NR; SAP*.
V. tapsoides L.; (91a); POR; S; SAP*.
V. thapsus L.; (45,52); EUR, IND; S; SAP*.

SIMARUBACEAE

Balanites aegyptiaca Del.; (25,52,60,70); CAF; B, FR, R; SAP.
B. roxburghii Planch.; (45,60); IND; FR; SAP.

SMILACACEAE

Smilax medica Schlecht. & Cham.; (60); MEX; R; SAP.

SOLANACEAE

Acnistus arborescens (L.) Schlecht.; (62); WIN; NR; NK.
Cestrum laevigatum Schlecht.; (45,62); BRA; NR; NK.
Datura metel L.; (45); PHI; PL; ALK.
D. stramonium L.; (46); NRE; NR; ALK.
Duboisia myoporensis R. Brown; (110); AUS; PL; NK.
Hyoscyamus niger L.; (60); SPA; NR; ALK.
Nicotiana rustica L.; (60); AFR; PL; ALK.
N. tabacum L.; (45,60); IND; PL; ALK.
Physalis heterophylla Nees.; (84); GLY.
Schwenckia americana L.; (52,60); TAF; PL; PHE.
Solanum incanum L.; (60); NAF; PL; ALK.
S. marginatum L.f.; (45); ABY; BR; ALK.
S. rugosum Dun.; (112); BRA; FR; ALK.

STYRACACEAE

Styrax argenteus Presl; (44,113); MEX; B; SAP*.
S. officinalis L.; (45); PAL; S; SAP*.

SYMPLOCACEAE

Symplocos racemosa Roxb.; (45); SUM; B; NK.

TAXACEAE

Taxus baccata L.; (44,45,52,60); EUR; BR, L; ALK.

THEACEAE

Camellia japonica L.; (121); NRE; NR; SAP.
C. sasanqua Pierre; (121); ASI; S; SAP.

C. sinensis (L.) Kunth; (121); ASI; S; SAP.
Laplacea sp.; (113); NGU; B; NR; NK.
Schima wallachi (DC.) Choisy; (45,113); IND; B; NK.
Ternstroemia cherryi Merr.; (60); AUS; L; SAP.
T. toquian F. Villar; (112); PHI; FR; SAP*.

THEOPHRASTACEAE

Jacquinia arborea Vahl; (45); WIN, CAM; FR, L; SAP*.
J. aristata Jacq.; (45); VEN; NR; SAP*.
J. armillaris Jacq.; (44,60,119); JAM; NR; SAP*.
J. aurantiaca Ail.; (42,45,60); SAM; NR; SAP*.
J. axillaris Oerst.; (46,60); SAM; NR; SAP*:
J. barbasco Mez; (45); WIN; FR, L; SAP*.
J. brasiliensis Mez; (62); BRA; NR; SAP*.
J. caracasana H.B.K.; (42); COL; NR; SAP*.
J. donell-smithii Mez; (46); GUA; NR; SAP*.
J. gracilis Mez; (42,62); COL; NR; SAP*.
J. mucronata Blake; (45,62); VEN; NR; SAP*.
J. panamensis Lundell; (3); PAN; PL; SAP*.
J. pungens A. Gray; (45); MEX; B; SAP*.
J. racemosa DC.; (60); SAM; NR; SAP*.
J. revoluta Jacq.; (42,45,62); CAR; NR; SAP*.
J. seleriana Urb. & Loesn.; (45); MEX; J; SAP*.
J. sprucei Mez; (62); ECU; NR; SAP*.

THYMELAEACEAE

Daphne cneorum L.; (45,52); EUR; FR, L; HET.
D. gnidium L.; (44,45,52); EUR; NR; COU.
D. mezereum L.; (45); ASI; NR; COU.
D. oleoides Schreb.; (52); NRE; NR; COU*.
Edgeworthia gardnerii Meissn.; (60); AUS; NR; NK.
Gonystylus sp.; (112); BOR; FR; NK.
Lasiosiphon eriocephalus Dcne.; (25,45,60); WAF, IND; B, L; NK.
L. hoepfnerianus Vatke; (60); IND; NR; NK.
L. kraussianus Meissn.; (25,60); WAF; NR; NK.
Linostoma decandrum Wall.; (45,67); IND; B, BR, FR, L; NK.
Schoenobiblus peruvianus Standl.; (6); PER; R; NK.
Wikstroemia chamaedaphne Meissn.; (8,45,113); CHI; NR; NK.
W. foetida A. Gray; (45); SOI; NR; NK.
W. indica C. A. Mey.; (67,112); IND; B, R; NK.
W. uva-ursi A. Gray; (113); HAW; NR; NK.
W. viridiflora Meissn.; (112); NHE; B; NK.

TILIACEAE

Grewia asiatica L.; (45); IND; NR; NK.
G. mallococca L.f.; (91a); NK.
Microcos sp.; (113); NGU; B; NK.

VERBENACEAE

Callicarpa longifolia Lam.; (45,60); AUS; L; NK.
Clerodendrum inerme Gaertn.; (44,112); CIS; NR; NK.

Table I
Continued

Faradaya splendida F. Muell.; (44,45,60); AUS; B; NK.
Vitex leucoxylon L.; (8,44,112); SRI; FR; NK.

Cissus cordifolia L.; (84); BRA; NR; NK.
C. quadrangularis L.; (45); CAF; NR; NK.

VIOLACEAE

Rinorea dentata Kunth; (45); CAR; B; NK.

WINTERACEAE

Illicium religiosum Sieb.; (60); JAP; L; NK.

VITACEAE

Ampelocissus pentaphylla G. & P.; (25,52); WAF; L; NK.

ZAMIACEAE

Zamia furfuracea L.f.; (45); GUA; NR; NK.

Table II
Appendix

Geographical Regions:

ABY = Abyssinia; AFR = Africa; AMA = Amazon; ARA = Arabia; ARM = Armenia; ASI = Asia; AUS = Australia; BAH = Bahamas; BOR = Borneo; BRA = Brazil; BUR = Burma; CAF = Central Africa; CAI = Canary Islands; CAM = Central America; CAR = Caribbean; CEL = Celebes; CER = Ceram; CHI = China; CHL = Chile; CIS = Caroline Islands; CME = Cameroon; COL = Colombia; CON = Congo; CRI = Costa Rica; CUB = Cuba; DOM = Dominica; EAF = East Africa; EAS = Eurasia; ECU = Ecuador; ENG = England; ETI = Ethiopia; EUR = Europe; FIJ = Fiji Islands; FRG = French Guiana; GER = Germany; GRE = Greece; GUA = Guatemala; GUI = Guianas; GUY = Guyana; HAW = Hawaii; IND = India; INO = Indonesia; JAM = Jamaica; JAP = Japan; JAV = Java; LAB = Laboratory; LIB = Liberia; LOY = Loyalty Islands; MAA = Malawi; MAD = Madagascar; MAL = Malay Peninsula; MAR = Martinique; MED = Mediterranean; MEX = Mexico; MOL = Moluccas; MOZ = Mozambique; NAM = North America; NCA = New Caledonia; NGU = New Guinea; NHE = New Hebrides; NRE = Not reported; NZE = New Zealand; PAC = Pacific; PAL = Palestine; PAN = Panama; PER = Peru; PHI = Philippines; POL = Polynesia; POR = Portugal; PTR = Paleotropics; RUS = Russia; SAF = South Africa; SAM = South America; SEA = Southeast Asia; SEU = Southeastern U.S.; SIB = Siberia; SIS = Society Islands; SMO = Samoa; SRI = Sri Lanka; SPA = Spain; SUM = Sumatra; SUR = Surinam; TAI = Taiwan; TAM = Tropical America; TAF = Tropical Africa; TRO = Tropics; USA = United States; VEN = Venezuela; WAF = West Africa; WIN = West Indies; ZAN = Zanzibar.

Parts of Plants:

B = bark; BD = buds; BR = branches; BU = bulbs, bulbils; FL = flowers; FR = fruits; J = sap, latex; L = leaves; NR = not reported; PL = whole plant; R = root; RB = root bark; RH = rhizome; S = seeds; W = wood.

Poisonous Principles:

ALK = alkaloids; ANA = anacardic acid; CAD = cardol; CAN = cannabidiol; CAR = cardiac glycoside; COR = coriamyrtine; COU = coumarin; CYA = cyanogenic compounds; DIT = diterpene; ESS = essential oil; HET = heterosides; ICH = ichthyocthereol; JUG = juglone; MAM = mammein; NK = not known; PHE = phenol; PHY = phytotoxin; PLU = plumbagin; ROT = rotenone; SAP = saponins; TAN = tannins; TEP = tephrosine; TOX = toxialbumin; TRI = triterpene; * = reported for the genus.

Resin Classification by the Ka'apor Indians

William Balée and Douglas C. Daly

Table of Contents

Abstract

BALÉE, W. (Depto. de Programas e Projetos, Museu Paraense Emílio Goeldi, Cx. Postal 399, 66.000 Belém, PA, Brazil) AND D. C. DALY (New York Botanical Garden, Bronx, New York 10458-5126, U.S.A.). Resin classification by the Ka'apor Indians. Advances in Economic Botany 8: 24–34. 1990. The Ka'apor Indians of Amazonian Brazil, who speak a Tupi-Guarani language, possess a special purpose classification of resins, saps, and latexes, which is highly reminiscent of general folk classification. Ka'apor resin classification is hierarchical, with six identifiable ranks, thus comparing favorably to a general classification of plants based on habit. Yet the two systems do not overlap. Evidence from syntax, the lexicon, myth, and Ka'apor classificatory procedures suggest that the two classification systems are cognitively distinct in Ka'apor culture. The chief difference between the systems of resin and plant classification concerns the semantic range of the phenomena being classified, namely, combustible plant substances versus plants themselves.

Key words: Ka'apor; resins; saps; latexes; Brazil; combustion; illumination; *Protium*.

Introduction

Numerous studies analyze the folk botanical domain in terms of gross morphological discontinuities between plants (e.g., Berlin, 1973; Berlin et al., 1973, 1974; Brown, 1977, 1984; Conklin, 1954; Hays, 1979, 1983). These discontinuities are seen as being "objective" and "natural," since gross morphological differences between organisms are easily and universally discernible. Folk taxonomies of this sort are considered to be "general," since they involve similarities and differ-

Advances in Economic Botany 8: 24–34, 1990
© 1990 The New York Botanical Garden

24

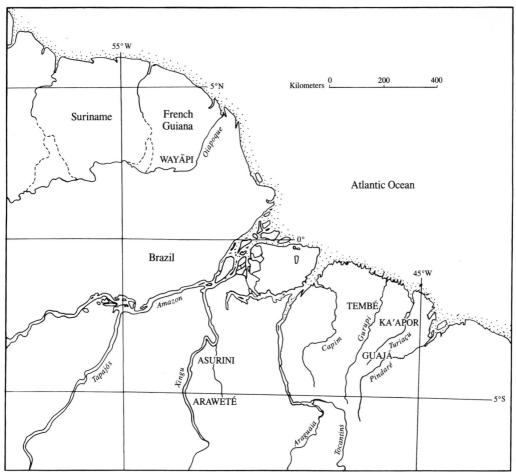

MAP 1.   Situation map of the Ka'apor showing locations of some other Tupi-Guarani groups of eastern Amazonia.

ences between organisms as juxtaposed among themselves, regardless of their relationship to the land, to human beings, or to other life forms (Atran, 1983: 60; 1985: 308). Folk taxonomies that stress, on the other hand, the relationships of organisms to human beings, or to the land, or to some factor other than morphology, are perceived as being more arbitrary than general purpose ones, since these special purpose taxonomies tend to focus narrowly on one "powerful factor" in order to analyze that factor (Gilmour, 1961: 34; cf. Atran, 1985: 304–305).

The objective of this chapter is to circumscribe the special factor of "usefulness," especially as this applies to resins, saps, and latexes of trees, to the Ka'apor Indians of extreme eastern Am-azonian Brazil (cf. Gianno, 1986). Although resins, saps, and latexes certainly represent components of particular plants, we propose that Ka'apor classification of these reflects also utilitarian priorities in Ka'apor culture (cf. Hunn, 1982).

The Ka'apor

The Ka'apor Indians have been also referred to as the *Urubus* ("vultures") and *Urubu-Kaapor* (Huxley, 1957; Ribeiro, 1976). They call themselves *Ka'apor,* which is analyzable as "footprints of the forest." The Ka'apor speak a language of the Tupi-Guarani language family, which is one of the most geographically dispersed indigenous language families in South America. The Ka'apor occupy what is principally *terra firme* dense forest (Pires & Prance, 1985) in northern Maranhão state, Brazil (see map). In 1928, when the Ka'apor population was 2000 or more, the Brazilian government "pacified" them (Ribeiro, 1970: 181–182). The Ka'apor view this event somewhat differently; their ancestors "pacified" (*mu-katu*) Brazilian society, after a period of more than 100 years of warfare (see Balée, 1988). Afterwards, the Ka'apor suffered from introduced respiratory infections and their population steadily declined (Balée, 1984: 53–64). At present, Ka'apor population is only about 500. The people are scattered in approximately 15 small settlements across a forested expanse of 5305 square km.

Ka'apor Ethnobotany

The economic infrastructure of Ka'apor society draws on horticulture, hunting, fishing, and gathering. Plants figure prominently in all these activities, even hunting and fishing, since the principal tools (bows, arrows, and fish poisons) are derived from plants. The Ka'apor cultivate more than 50 species of plants with some 83 named varieties (Balée & Gély, 1989; cf. Ribeiro, 1976). These include typical cultigens of lowland South America, such as manioc, sweet potato, peanut, yam, cotton, and cashew. Ka'apor culture is not merely a "forest" culture, since the Ka'apor manipulate and alter the forest in clearing and burning swiddens, as do many other Tupi-Guarani speaking peoples (e.g., F. Grenand & Haxaire, 1977). Horticulture among Tupi-Guarani speaking peoples is ancient. One can reconstruct from modern Tupi-Guarani languages the names for many cultivars in Proto-Tupi-Guarani, the mother language (Lemle, 1971). Ka'apor horticulture is, no doubt, traceable to this remote period.

In addition to the traditional and intensive use of cultivated plants (for food, fiber, medicine, and the like), the Ka'apor also utilize forest plants in many ways. Some kind of use and/or activity signature (Hunn, 1982) can be elicited from informants, for example, for the vast majority of forest trees (Balée, 1986; Prance et al., 1987). Such uses include food for people, game animal food, glue, pottery slip, dye, lashing material,

lumber for construction, medicine, magic, and firewood. Many of the general objectives in the use of such plants cannot be attained solely with the cultivated plants of the Ka'apor (such as lumber for construction, pottery slip, glue, and firewood). In other words, many non-domesticated plants are indispensible in Ka'apor economy.

Although in the Ka'apor language, "plant" in the botanical sense (i.e., "vascular plants") is unnamed, there is nevertheless a covert category which covers this semantic range in Ka'apor ethnobotany (Berlin et al., 1973, 1974). This category exists by virtue of numerous words that pertain exclusively to plants and plant products (see Berlin et al., 1973: 214; 1974: 30; Berlin, 1976: 383–384). These terms include "stem" (*'i*), "leaf" (*ho*), "root" (*hapo*), "thorn" (*yu*), and "sap" (*hɨk*).

The Ka'apor subdivide the domain of vascular plants into three major morphological discontinuities, i.e., life forms. These are labeled by words for "tree" (*mira*), "herb" (*ka'a*), and "vine" (*sɨpo*). In Ka'apor, these terms are polysemous. "Tree" (*mira*) is polysemous with "wood"; "herb" (*ka'a*) is polysemous with "forest" and with the verb "defecate"; and "vine" (*sɨpo*) is polysemous with lashing material obtained from vines. The botanical life form terms in Ka'apor are cognate with semantically equivalent terms in other Tupi-Guarani languages and hence are reconstructable in Proto-Tupi-Guarani. In other words, the major life form labels in Tupi-Guarani languages have been retained, not diffused (see Table I). This suggests the long-term recognition of several gross morphological discontinuities among plants in Tupi-Guarani culture.

Below the folk taxonomic rank of life form are folk generics—names for individual taxa of trees, vines, and herbs as well as plant taxa which are unaffiliated with any of the life forms (such as manioc and other cultigens, which are not considered by informants to be constituents of any life form class). Of the 404 Ka'apor folk generic plant names thus far collected, more than 80% are believed to pertain to one of the three major life form classes. Although there is a dichotomy between cultivated and unmanaged plants in Ka'apor ethnobotany (Balée, in press), most of the folk generic names are to be found in the domain of unmanaged plants.

In addition to having a general system of classification of plants which is based mostly on plant

Table I

Botanical life form labels in various Tupi-Guarani languages with reconstructed forms in Proto-Tupi-Guarani (PTG)

Language[a]	"Tree"	"Herb"	"Vine"
Ka'apor	mɨra	ka'a	sɨpo
Araweté	iwirã	ka'ã	ihipa
Asurini	iwɨra	ka'a	iipa
Guajá	wɨra	ka'a	wɨpo
Tembé	wɨra	ka'a	wɨpo
Wayãpi	wɨla	ka'a	ɨpo
PTG[b]	*ɨβɨra	*ka'a	*ɨwɨpo

[a] Data from the modern languages were gathered by Balée (unpubl. data), excpet those pertaining to Wayãpi, which are recorded in Grenand (1980).
[b] The reconstruction of the PTG term for "herb" (ka'a) is from Lemle (1971: 122, n. 125). The other reconstructions are from Aryon D. Rodrigues (pers. comm., 1988).

Table II

Terms denoting the semantic field of resins, saps, and latexes in various Tupi-Guarani languages[a] with reconstructed form in Proto-Tupi-Guarani (PTG)

Language	Term for resin, sap, latex
Ka'apor	hɨk
Araweté	hi
Asurini	hɨka
Guajá	hɨkə
Tembé	hɨkə
PTG	*hɨkə

[a] Data from the modern languages were gathered by Balée (unpubl. data).

habit, the Ka'apor also have numerous special purpose classification systems in which plants and plant parts are highly salient features. Perhaps because of a failure to distinguish between general and special purpose systems of classification, Friedberg (1974: 31) assumed that the Bunaq of Timor had numerous botanical systems of reference, which rendered impossible the construction of a hierarchical, classificatory tree based on stem habit. For example, she (1974: 332) indicated that the Bunaq lumped some latex-producing herbs with latex-producing trees in their classification of latex-producing trees. In other words, possession of latex was seen to be more important than the herbaceous habit. The point, however, should be that one system (based, for example, on habit) need not supersede another (based, for example, on some special purpose, as with useful latex-producing trees). The two kinds of classification may intersect, but they are distinct in cognitive terms (see also Gianno, 1986).

We argue that the Ka'apor exhibit a classification of latexes, resins, and saps that is separate from the classification of the organisms that produce them. At the same time, the special classification of latexes, resins, and saps appears to resemble the classification of plants based on stem habit.

The Semantic Field of Hɨk

The term hɨk covers the botanical range of resins, saps, and latexes. This term labels a se-

mantic domain different from that of "water" in the plant stem and/or "floral nectar," which are labeled by the word tikwer. Like Ka'apor words for roots, leaves, flowers, fruits, and the like, hɨk "crosscuts" (Berlin et al., 1974: 154) life form classes. One can speak, therefore, of hɨk from vines, trees, and herbs. Hɨk is distinctively a plant morphological term. It labels the parts and wholes of no other organisms; this usage implies the existence of the covert taxon, "plant" (see above).

Cognates with hɨk in other Tupi-Guarani languages are shown in Table II. These cognates appear to represent lexical retention and not lexical diffusion, since by the comparative method of linguistics, these terms are reconstructable in Proto-Tupi-Guarani. In other words, the concept of lumping resins, saps, and latexes under a single term is ancient in the Tupi-Guarani family.

In the Ka'apor language, hɨk is incorporated in a folk generic name. The resin, sap, or latex of any botanical species may be thus denoted. For example, the latex of Manilkara huberi (Ducke) Standl. (Sapotaceae), irikɨwa-'ɨ, is denoted as irikɨwa-hɨk. In terms of its post-positioning to generic plant names, hɨk functions syntactically very much like the suffix 'ɨ. Lemle (1971: 116, n. 16) has reconstructed this in Proto-Tupi-Guarani as *ɨb and assigned it the gloss of "tree." The term 'ɨ (in Ka'apor), is, however, a bound morpheme (unlike its presumably English gloss from Lemle). As such, 'ɨ is syntactically more like the Portuguese -zeiro (as in cupuaçuzeiro— the 'cupuaçu tree') than English "tree," which is a free morpheme. When incorporated in a folk generic name, 'ɨ denotes a tall plant (usually greater than 2 m in height) with an orthotropic (erect-growing) stem; alternatively, it may be used to denote the stem of any erect growing plant regardless of height. Unlike the term *mira* ("tree"), *'i* is not polysemous with "wood." Thus, the herbaceous stem of "manioc" (*mani*) is denoted by *mani'i*. Moreover, although palms are not considered to be "trees," since by and large they lack the character of "woodiness," tall, erect growing palms such as *Oenocarpus distichus* Mart. (*piniwa-'i*) are invariably named with a folk generic head and the postposed *'i*. In other contexts, the term *'i* is applied to a folk generic head which denotes a "tree." In this usage, *'i* refers to the whole organism. To refer to *Manilkara huberi* as an organism, for example, one employs the head *irikiwa* with the postposed adjunct *'i*, hence, *irikiwa'i*. This implies that obligatory life form-like suffixes for certain folk generic heads are grammatically no more primary than other suffixes labeling other properties and conditions, such as that for flammable, oxidized sap. Just as *hik* is a special taxon of an ethnobiological rank comparable to that of life form, it, too, is further subdivided into special generic-like taxa.

The Semantic Field of *Kanei*[1]

The most accurate gloss for the Ka'apor word *kanei* appears to be "any combustible resin, latex, or sap." The taxon *kanei* compares favorably to a covert complex (cf. Berlin et al., 1974), since some resins, saps, and latexes are not named with the head *kanei,* but nevertheless are implicitly considered to be constituents of this taxon. As such, the term *kanei* covers substances which occur in numerous morphologically distinct taxa

of organisms. The term is limited, however, to substances from trees, although this may be a simple artifact of plant morphological discontinuity. In other words, there are no herbs or vines whose saps are used in the one way that the Ka'apor use all *kanei,* namely as a source of light. It is suggestive, nevertheless, that the term *kanei* functions at a rank comparable to that of the folk generic, although only in a special purpose sense.

The head *kanei* together with the suffix *'i* are used, moreover, to refer to numerous trees of the genus *Protium* (Burseraceae). These species exude resins which upon oxidization are useful in the illuminating of interiors. Five folk specifics are included in generic taxon *kanei-i.* These are *ara-kanei-i* ("macaw's *kanei* tree"), *kanei-aka'i* ("fat *kanei* tree"), *kanei-ape-i* ("*Apeiba* spp. *kanei* tree"), *kanei-'i-pitag̃* ("red *kanei* tree"), and *kanei'i-tuwir* ("white *kanei* tree"). Each of these folk specifics possesses a unique activity signature (Hunn, 1982), here defined in terms of cultural uses and the place of the plant in relation to non-botanical organisms (such as animals that feed on its fruits, seeds, and other parts—see Table III). The only common facet of the activity signatures of all these *kanei* folk specifics is edibility of the fruits by human beings and flammability of the resin, which is by far the most important property in a special purpose sense. The fruits of most *Protium* species are small (less than 2.5 cm in diameter), and their white pulps are considered to be only 'children's food' (*ta'i-ta-mi'u*) and/or 'starvation food' (*u'u-awa-sorok-hu-rahã,* lit., "one eats them when very hungry"). *Protium* fruits are never the objects of intensive gathering trips, in contrast to the highly prized fruits of, for example, *Theobroma grandiflorum* K. Sch. (Sterculiaceae; *cupuaçu* in Portuguese) and *Platonia insignis* Mart. (Clusiaceae; *bacuri* in Portuguese). On the other hand, Ka'apor gathering groups do, on occasion, collect basketsful of *Protium* resins to be used in illuminating the interiors of their houses at night. The light produced by only small quantities of these burning resins is comparable to that of a small bonfire and far brighter than the light given off by a typical Ka'apor hearth. There are other species not of the folk generic-like taxon *kanei* which also possess saps that are combustible and useful for the purpose of illumination (see Table III). These include *Trattinnickia burserifolia* Mart. (Burseraceae), *Symphonia globulifera* L.f.

---

[1] Curiously, the Ka'apor term *kanei* does not appear to be of indigenous origin (J. Jangoux and A. D. Rodrigues, pers. comm.). There seem to be no cognates for *kanei* in other Tupi-Guarani languages. The term *kanei,* in fact, seems to have been borrowed from an old Portuguese word for "candle of beeswax" and/or "small utensil used for illumination"–*candeia* (Holanda Ferreira, n.d.: 267). The word *candeia* would have been introduced into the Ka'apor language in colonial times, prior to about the one hundred years of warfare between Ka'apor society and Brazilian society, which ended in 1928 (Balée, 1988). The fact that *kanei* itself is probably not of Ka'apor origin, however, does not affect the Ka'apor system of resin classification.

(Clusiaceae), *Hymenaea* spp. (Caesalpiniaceae), and *Manilkara huberi*. Each of these tree species is also assigned a unique activity signature (see Hunn, 1982). For example, only resin from *T. burserifolia* (*kirihu'i*) is employed by the Ka'apor shaman as a tobacco sweetener and as an aid in achieving trance (it is described as a *hesak-ha*, lit., "vision generator"). It is not a true hallucinogen, however, since it appears to be devoid of any psychoactive compounds (Guilherme Maia, pers. comm., 1987). Only resin from *S. globulifera* (*irati*) is used for dyeing arrow shafts black, as a glue for arrow fletching, and for inducing abortion and/or achieving contraception when taken orally (this latter use is under study to determine whether etically this species is effective). Resins from all three *Hymenaea* species occurring in the region (*trapai-pihun, trapai-pitag,* and *yeta'i*) are used by the Ka'apor as pottery slips, but only resin from *yeta'i* (*Hymenaea parvifolia* Huber) serves in treating injured eyes (drops of the burning resin are mixed with water at room temperature; the resulting solution is dropped into the eyes). Only *trapai-pihun* fruits are consumed by collared peccaries, according to informants, thus setting off its activity signature from that of *trapai-pitag*, among other differences (see Table III).

The resin, latex, and sap of all these species are denoted with the postposed term *hik*. Thus, *irikiwa-hik, kanei-hik, kirihu-hik, yeta-hik,* and so on are compound names denoting the resins produced by *irikiwa* (*Manilkara*), *kanei* (*Protium*), *kirihu* (*Trattinnickia*), and *yeta* (*Hymenaea*).

All these folk taxa seem to be included within a suprageneric, covert complex which is best called *kanei*. This complex is covert insofar as not all of its members are labeled by *kanei*. The resin of several species of *Protium* is called simply *kanei,* or less commonly, *kanei-hik*. The other volatile resins, from species that are not labeled with the head *kanei*, are, in fact, also included in this suprageneric complex, because of their high flammability and usefulness in illuminating interiors. This inclusion seems to be cognitively true, as is readily suggested in a Ka'apor origin myth, recounted by the shaman of the settlement of Urutawi:

Long ago, no one possessed any kanei. *The settlement, therefore, was dark at night and people had to eat game meat in the dark. There was a man named* Kanei-yar *(lit., "kanei-owner"). Another man,* Sarakur *(lit., "wood rail") wanted light, so that he would need not fear snakes while dining at night. So* Sarakur *killed* Kanei-yar *by scoring his neck with a machete.* Sarakur *then reached into the opened nape of* Kanei-yar *and pulled out pieces of bone, which were really pieces of* irikiwa-hik *(latex from* Manilkara huberi*). Then,* Sarakur *ignited these pieces and the settlement, henceforth, was well-illuminated at night.*

That this hardened latex from *Manilkara huberi* is identified as a kind of *kanei* in this origin myth supports the contention that *kanei* is a suprageneric, special-purpose taxon. Folk generic names pertaining to the suprageneric complex *kanei* but which are not formed with the head *kanei* may partly result from the uses of their sap other than as material for illumination. Some are used as pottery slips, glues, dye, and so on (see Table III). These disjunct uses are, no doubt, partly artifacts of morphological discontinuities between the trees themselves.

Within the complex *kanei*, properties of the resins produced by trees may supersede morphological characters in the Ka'apor classification of them at the botanical species level. For this reason, Ka'apor classification and Western classification of the Burseraceae (e.g., Daly, 1987) are not isomorphic. In some cases, closely related taxa may have distinctive folk specifics, while in the others, phylogenetically less related taxa are lumped under the same folk specific. Morphologically, *Protium decandrum* (Aublet) Marchand and *P. giganteum* Engl. are distinguished only with difficulty, but they nonetheless have different folk specific names (*kanei'i-pitag* and *kanei'i-tuwir*, respectively). On the other hand, *P. giganteum* Engl., *P. pallidum* Cuatrec., and *P. spruceanum* (Benth.) Engl. are not closely related, but they are assigned the same folk specific name (*kanei'i-tuwir*). *Protium polybotryum* (Turcz.) Engl. and *P. tenuifolium* (Engl.) Engl. belong to different sections of the genus, but they have the same folk specific name (*kanei-ape'i*). It is interesting to observe that *P. trifoliolatum* Engl., which was placed in the genus *Tetragastris* by Cuatrecasas (1961), is labeled *sekãtã'i*, although it still belongs to the covert complex *kanei*. On the other hand, *P. aracouchini* (Aublet) Marchand is also given a name other than *kanei* (*yawa-mira*), but its placement in *Protium* has not been questioned in this century. Although this species produces small quantities of resin, it

Table III

Useul resin-producing trees and their Ka'apor activity signatures

Ka'apor name	Latin name	Voucher[a]	Activity signature[b]
ara-kanei-'ɨ	*Protium altsonii*	272	2, 23, 25
ɨratɨ	*Symphonia globulifera*	1023	1, 6, 12, 14, 20, 24, 25
ɨrɨkɨwa-'ɨ	*Manilkara huberi*		1, 8, 13, 25
kanei-aka'ɨ	*Protium sagotianum*	855	1, 3, 23, 25
kanei-ape-'ɨ	*P. polybotryum,*	98	
	P. tenuifolium	157	1, 3, 17, 23, 25, 27
kanei'-ɨ-pitã	*P. decandrum*	34	1, 2, 8, 23, 25
kanei'-ɨ-tuwɨr	*P. giganteum*	1061	
	P. pallidum	265	
	P. spruceanum	843	1, 10, 18, 23, 25, 26
kɨrɨhu-'ɨ	*Trattinnickia burserifolia*	876	6, 7, 21, 22, 24, 25
sekãtã-'ɨ	*Protium trifoliolatum*	413	1, 2, 3, 15, 23
trapaɨ-pihun	*Hymenaea intermedia*	1044	1, 5, 8, 9, 11, 25
trapaɨ-pitã	*H. courbaril*	1000	1, 11, 25
yawa-mɨra	*Protium aracouchini*	969	1, 3, 4, 5, 23
yeta-'ɨ	*Hymenaea parvifolia*	880	1, 11, 16, 19, 24, 25

[a] Vouchers are in the collection series of Balée.

[b] Activity signatures are coded as follows:

Human food

    1. Edible pulp of fruit (for the Ka'apor)

Game food

    2. Edible fruit for macaws
    3. Edible fruit for parrots in general
    4. Edible fruit for Dusky Parrot
    5. Edible fruit for Red-fan Parrot
    6. Edible fruit and/or flowers for Golden Parakeet
    7. Edible fruit for toucans
    8. Edible fruit for Brocket Deer
    9. Edible fruit for Collared Peccary
  10. Edible fruit for Yellow-footed Tortoise

Technological uses

  11. Resin used as pottery slip
  12. Sap used to glue arrow point to arrow shaft and to glue on arrow feathers to shaft
  13. Sap used as a glue in feather art
  14. Sap used as dye to blacken bow and arrow
  15. Wood used to make posts for temporary shelter in forest

Medicinal uses

  16. Resin used as remedy for conjunctivitis
  17. Resin used as remedy for sinusitis
  18. Resin used as remedy for toothache
  19. Decoction of resin droplets and water taken orally by women to reduce excessive menstrual discharge
  20. Decoction of sap and water taken orally by women to prevent menstruation (and hence ovulation)

Uses in magic

  21. Resin smoked as an emic aid to achieving shamanistic trance and in sweetening tobacco
  22. Resin pellets wrapped in small cloth and tied to infants' wrists to prevent sickness

Uses in combustion

  23. Firewood (especially used in toasting manioc flour)
  24. Firewood for cooking and heat at night
  25. Resin, sap, or latex used to illuminate household interiors at night (i.e., it is a kind of *kanei*)

Miscellaneous uses

  26. Stem shavings mixed with tobacco to sweeten taste of smoke
  27. Resin used as a deodorant (for men only)

is not used by the Ka'apor for illuminating interiors, as is the case with all species designated with the head *kanei*.

Discussion

Figure 1 shows a hierarchical tree of Ka'apor resin classification. Six ranks can be distinguished. At rank 1 is the covert class, since it is unnamed, of "plant parts" (cf. Gianno, 1986). One encounters at rank 2 the major Ka'apor classes of plant parts, including the covert class "plant fluids," along with named taxa such as "leaf" (*ho*) and "root" (*hapo*). At rank 3 the class "plant fluids" is subdivided into "resin, sap, latex" (*hɨk*) and "watery plant fluid" (*tikwer*). At rank 4, *hɨk* may be subdivided further into two covert suprageneric complexes. We label these "*kanei*" and "*non-kanei*." "*Non-kanei*" kinds of *hɨk* are those resins, saps, and latexes which the Ka'apor do not use in illuminating their houses. These would include resins, saps, and latexes from species such as *Ficus* spp. (strangler figs), *Clusia* spp. (strangler vines), and *Manihot esculenta* (manioc), the Ka'apor names for which (not shown in Fig. 1) would occur at rank 5. The suprageneric-like complex *kanei* is subdivided into names for resins, saps, and latexes from several botanical genera and species which do possess the property of flammability upon ignition (see Fig. 1, rank 5). These can be further subdivided into names which resemble folk specifics for resins, saps, and latexes, at rank 6 (see Fig. 1, subdivisions of *kanei-hɨk*). As such, Ka'apor resin classification resembles a general plant classification based on stem habit. Rank 1 ("plant parts") corresponds to the covert unique beginner "plant"; rank 2 corresponds to the life form rank; rank 4 corresponds to a suprageneric complex of plants; rank 5 corresponds to the folk generic rank; and rank 6 corresponds to the folk specific rank of general ethnobiological classification (see Berlin et al., 1973, 1974; Berlin, 1976). There is no rank comparable to the folk varietal in Ka'apor resin classification.

Despite the relative depth of Ka'apor resin classification, with six hierarchical ranks, it cannot be correlated with a classification of the organisms that produce these resins, saps, and latexes. This is abundantly clear at rank 5, which resembles the folk generic rank of general purpose ethnobiological classification. In a test, whereby informants were asked separately to group verbally various folk generic names for trees according to their affinity, it was found that the organisms which produce *kanei* are classified into different groups, whereby some species which produce no *kanei* are grouped with those that do. This test was based on a persistent Ka'apor theme with regard to folk botanical genera: some are *anam* (lit., "relative, akin") to a given taxon, while others are *anam-'ɨm* ("not relative, not akin") to the same taxon. This exercise is similar to that used by Berlin et al. (1974) in discovering Tzeltal Mayan suprageneric plant complexes, the chief difference being that Ka'apor informants used were illiterate and they were only asked to associate freely.

This test showed that all folk specific names formed partly with the head *kanei* were considered to be akin, which is not surprising. Yet *sekãtã-'ɨ* (*Protium trifoliolatum*) and *kanei-'ɨ* (*Protium* spp.) were also considered to be "akin" (*anam*), even though *sekãtã-'ɨ* produces no flammable resin. Likewise, *yawa-mɨra* (*P. aracouchini*) was also classed with *kanei-'ɨ*, although *yawa-mɨra* produces no resin used for illumination. The *Hymenaea* spp. (*yeta-'ɨ, trapaɨ-pitag̃,* and *trapaɨ-pihun*) were all believed to be *anam* to each other and not akin to the *kanei* group, even though resins from all *Hymenaea* spp. are used for illuminating interiors. In addition, *kumaru'ɨ* (*Dipteryx odorata* Willd.—Fabaceae), which like *Hymenaea* is a large leguminous tree, but which produces no combustible resin in its stem (even though oil from its seeds have medicinal uses in Ka'apor culture), was grouped as being *anam* to the *Hymenaea* species. Finally, *Manilkara huberi* (*ɨrɨkɨwa'ɨ*) was considered to be *anam* to *Manilkara amazonica* (*ɨrɨkɨwa-yu-'ɨ*; lit., "yellow *ɨrɨkɨwa'ɨ*"), yet the latex of *M. amazonica* is not used by the Ka'apor, and hence is not a kind of *kanei*.

Therefore, it is clear that although in terms of its hierarchical structure and its depth Ka'apor resin classification resembles the Ka'apor general classification of non-cultivated plants and those which are not ambiguous morphologically (such as palms), the two systems are classifying phenomena for different purposes, using different criteria of contrast. Ka'apor resin classification is distinct from their general botanical classification in that resins are classified according to

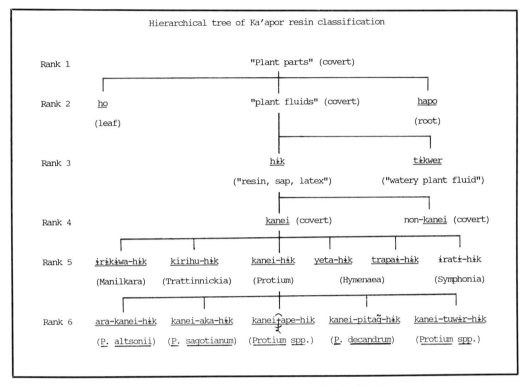

FIG. 1.   Hierarchical tree of Ka'apor resin classification.

their usefulness, mainly, in illuminating interiors; the general classification is based, principally, on stem habit and other morphological characters regardless of usefulness or type of use of the species in question. As such, Ka'apor resin classification is clearly of the special purpose sort (Gilmour, 1961).

Conclusions

The Ka'apor system of resin classification based on resin utility parallels the Ka'apor system of plant classification based on habit. This parallelism is seen in syntax, the lexicon, in myth, and in the hierarchical aspect of both systems. In syntactic terms, an equivalence of the suffixes indicating resins, saps, latexes (-*hɨk*) and plants of erect and tall habit (-*ɨ*) implies that *hɨk* parallels a life form class, such as 'tree.' In lexical terms, the head *kanei,* on the other hand, labels a group of combustible resins, saps, and latexes, and, on the other, it labels a group of trees of the genus *Protium.* As such, *kanei* insofar as it denotes combustible resins, saps and latexes, resembles a suprageneric complex in folk biological classification. The reality of this suprageneric-like complex, here labeled *kanei,* is affirmed in a Ka'apor origin myth, whereby the combustible latex of *Manilkara huberi* (*ɨrɨkɨwa-hɨk*) is revealed as being a kind of *kanei,* although its name is not formed with the head, *kanei.* The term *kanei* is best glossed as "combustible resin, sap, or latex used in illuminating interiors." The special factor of usefulness as an illuminating agent, therefore, appears to be the basis of Ka'apor resin classification.

With six identifiable ranks, Ka'apor resin classification possesses depth, which is not usually accorded to special purpose classifications (e.g., Hays, 1979). In addition, there appears to be no overlapping in class membership. For example, *kanei-ape-hɨk* (from *Protium polybotryum* and *P. tenuifolium*) is a kind of *kanei-hɨk* which, in turn, is a kind of *kanei*; but *kanei-ape-hɨk* is not considered to be a kind of *yeta-hɨk,* which is itself also a kind of *kanei* (see Fig. 1). In other words, genuine contrast sets exist in Ka'apor resin classification, just as these do in general ethnobiological classification (cf. Berlin et al., 1974).

Special purpose classifications have been described as "secondary" (Berlin et al., 1974: 154) and "quasi-taxonomic" (Hays, 1979: 257). The data presented here suggest that Ka'apor resin classification is neither secondary nor primary to the Ka'apor taxonomic system embracing the organisms themselves that produce resins, saps, and latexes. Rather, the two systems are cognitively distinct. Ka'apor resin classification is as "taxonomic," in terms of its hierarchical depth, as is general Ka'apor botanical classification. The chief difference between the two systems is associated with the semantic range of phenomena being classified, namely useful plant substances vs. the plants (in particular, non-domesticated plants) themselves. In Ka'apor ethnobotany, plant resins, saps, and latexes are hierarchically classified (as are plants themselves) probably because of the common, abiding utility of some of these in supplying light.

Acknowledgments

We are grateful to Jean Langenheim, Dennis Moore, and David Williams for comments on earlier versions of this paper. Ethnobotanical field work in Brazil (1984–1987) was generously funded by the Edward John Noble Foundation in a series of grants to the Institute of Economic Botany of the New York Botanical Garden. This is publication number 97 in the series of the Institute of Economic Botany of The New York Botanical Garden.

Literature Cited

Atran, S. 1983. Covert fragmenta and the origins of the botanical family. Man 18(1): 51–71.
——. 1985. The nature of folk botanical life-forms. Amer. Anthropol. 87(2): 298–315.
Balée, W. 1984. The persistence of Ka'apor culture. Unpubl. Ph.D. Dissertation. Columbia University, New York.
——. 1986. Análise preliminar de inventário florestal e a etnobotânica Ka'apor (Maranhão). Bol. Mus. Paraense Hist. Nat. N. Sér. (Bot.) 2(2): 151–167.
——. 1988. The Ka'apor Indian wars of lower Amazonia, ca. 1825–1928. Pages 155–169 in R. R. Randolph, D. M. Schneider & M. N. Diaz (eds.), Dialectics and gender. Westview Press, Boulder.
——. (In press). Nomenclatural patterns in Ka'apor ethnobotany. J. Ethnobiol.
——— & A. Gély. 1989. Managed forest succession in Amazonia: The Ka'apor case. In Resource management in Amazonia: Indigenous and folk strategies. Adv. Econ. Bot. 7: 129–158.
Berlin, B. 1973. Folk systematics in relation to biological classification and nomenclature. Ann. Rev. Ecol. Syst. 4: 259–271.
——. 1976. The concept of rank in ethnobiological classification. Amer. Ethnol. 3: 381–399.
———, D. E. Breedlove, & P. H. Raven. 1973. General principles of classification and nomenclature in folk biology. Amer. Anthropol. 75(1): 214–242.
———, ——— & ———. 1974. Principles of Tzeltal plant classification. Academic Press, New York.
Brown, C. H. 1977. Folk botanical life-forms: Their universality and growth. Amer. Anthropol. 79: 317–342.
——. 1984. Language and living things. Rutgers University Press, New Brunswick, New Jersey.
Conklin, H. 1954. The relation of Hanunóo culture to the plant world. Unpubl. Ph.D. Dissertation. Yale University.
Cuatrecasas, J. 1961. Burseraceae brasiliae Novae. Bol. Mus. Hist. Nat., Nov. Sér., (Bot.) 11: 1–10.
Daly, D. C. 1987. A taxonomic revision of Protium (Burseraceae) in eastern Amazonia and the Guianas. Ph.D. Dissertation. City University of New York. 469.
Friedberg, C. 1974. Les processus classificatoires appliqués aux objets naturels et leur mises en évidence: Quelques principes méthologiques. J. Agric. Trop. Bot. Appl. 21: 313–334.
Gianno, R. 1986. Resin classification among the Semelai of Tasek Bera, Pahang, Malaysia. Econ. Bot. 40(2): 186–200.
Gilmour, J. S. L. 1961. Taxonomy. Pages 27–45 in A. M. Macleod & L. S. Cobley (eds.), Contemporary botanical thought. Oliver & Boyd, Edinburgh.
Grenand, F. & C. Haxaire. 1977. Monographie d'un abattis Wayãpi. J. Agric. Trop. Bot. Appl. 24: 285–310.
Grenand, P. 1980. Introduction à l'étude de l'univers Wayãpi. SELAF, Paris.
Hays, T. 1979. Plant classification and nomenclature in Ndumba, Papua New Guinea Highlands. Ethnology 18(3): 253–270.
——. 1983. Ndumba folk biology and general principles of ethnobotanical classification and nomenclature. Amer. Anthropol. 85(3): 592–611.
Holanda Ferreira, A. B. de. (N.d.). Novo dicionário da língua portuguesa. Editora Nova Fronteira, Rio de Janeiro.
Hunn, E. 1982. The utilitarian factor in folk biological classification. Amer. Anthropol. 84(4): 830–847.
Huxley, F. 1957. Affable savages. Viking Press, New York.
Lemle, M. 1971. Internal classification of the Tupi-Guarani linguistic family. Pages 107–129 in D. Bendor-Samuel (ed.), Tupi Studies I. Summer Institute of Linguistics, Norman, Oklahoma.

Pires, J. M. & G. T. Prance. 1985. The vegetation types of the Brazilian Amazon. Pages 109–145 *in* G. T. Prance & T. E. Lovejoy (eds.), Key environments: Amazonia. Pergamon Press, New York.

Prance, G. T., W. Balée, B. Boom & R. Carneiro. 1987. Quantitative ethnobotany and the case for conservation in Amazonia. Conserv. Biol. **1(4):** 296–310.

Ribeiro, D. 1970. Os índios e a civilização. Editora Civilização Brasileira, Rio de Janeiro.

———. 1976. Os índios Urubus: Ciclo anual das atividades de subsistência de uma tribo da floresta tropical. Pages 31–59 *in* Uirá sai à procura de Deus. Paz e Terra, Rio de Janeiro.

A Nutritional Study of *Aiphanes caryotifolia* (Kunth) Wendl. (Palmae) Fruit: An Exceptional Source of Vitamin A and High Quality Protein from Tropical America

Michael J. Balick and Stanley N. Gershoff

Table of Contents

Abstract

BALICK, M. J. (Institute of Economic Botany, New York Botanical Garden, Bronx, New York 10458-5126, U.S.A.) and S. N. GERSHOFF (School of Nutrition, Tufts University, Medford, Massachusetts 02155, U.S.A.). A nutritional study of *Aiphanes caryotifolia* (Kunth) Wendl. (Palmae) fruit: An exceptional source of vitamin A and high quality protein from tropical America. Advances in Economic Botany 8: 35–40. 1990. *Aiphanes caryotifolia* (Kunth) Wendl. is a palm native to northern areas of South America, primarily occurring in the lowlands, but found at elevations to 1800 m. Its fruits are a common food in the areas where it grows. The mesocarp of the fruit contains a pulp rich in vitamin A, with 16,000 IU/100 g (wet weight basis). Fatty acid analysis shows that the endosperm is rich in lauric acid. While the palm is apparently not a major component of the local diet, it is still a rich source of nutrition.

Key words: *Aiphanes*; palm; nutrition; vitamin A; beta carotene; protein; oil.

Resumen

Aiphanes caryotifolia (Kunth.) Wendl. es una palmera oriunda a las áreas del norte de Sur América, encontrándose principalmente en las tierras bajas a elevaciones menores de 1800 m. Sus frutos son frecuentemente usados como alimento en las áreas donde crece. El mesocarpio contiene una pulpa de alto contenido de vitamina A, con 16,000 IU/100 g (en base de peso húmedo). El endospermo contiene 4.3% de una proteina de alto valor nutricional (en base de peso húmedo). El análisis de los ácidos grasos del endospermo revela que éste es rico en ácido láurico. Aunque la palmera no es un componente principal de la dieta en éstas áreas, si es una fuente de buena nutrición.

Introduction

Aiphanes Wendl., a neotropical genus, is primarily known for its horticultural value. The most recent taxonomic revision of the genus was by Burret (1932), who reported 28 species. Moore (1973) recognized 38 species, of which seven were included in the checklist of cultivated palms he published in 1963. Dransfield and Uhl (1986) placed *Aiphanes* in the subfamily Arecoideae, Cocoeae: Bactridinae. The genus *Martinezia* is considered a synonym of *Aiphanes,* although the name is still occasionally found in the literature.

Aiphanes caryotifolia (Kunth) Wendl., is distributed primarily in the lowlands of northern South America, but also occurs at elevations up to 1800 m (Pérez-Arbeláez, 1978). In Venezuela it is found in the partially open dense forests of the upper Orinoco region (Braun, 1968). Galeano and Bernal (1987) noted that it was originally a plant of the Inter-Andean valleys of Colombia, but is now frequently cultivated in Tropical America. Patiño (1963) considered it to probably have been a species domesticated since pre-Colombian times, although no conclusive proof was cited.

Aiphanes caryotifolia grows to 8 m or more, and is solitary in habit. Its stem is covered with black spines, and the sheath, petiole and leaf rachis are also covered with prickles. The leaves are pinnate with wedge-shaped pinnae in groups of four to six along the rachis. The fruits are borne in interfoliar panicles ca. 0.6–0.75 m long. Mature fruits are globose, ca. 1.5–2.0 cm in diameter, and bright red in color. The endocarp is bony, often pitted, with three pores at or above the middle. The endosperm is white, oily, and homogeneous.

Aiphanes caryotifolia is known by various local names throughout its distribution (Table I). Aside from its value as an ornamental plant, several other uses have been reported. The Island Caribs of Dominica eat the mesocarp and endosperm (Hodge & Taylor, 1957). Pérez-Arbeláez (1978) noted that the tasty endosperm is used in confections as a substitute for almonds and filberts. The leaves of *Martinezia* sp. (=*Aiphanes* sp.) are apparently edible (Martin & Ruberté, 1979) although we have never observed this practice. In general, the consumption of this palm appears limited to the fruits.

In July of 1984, one of us (MJB) travelled to Colombia to participate in a palm germplasm collecting expedition in the Chocó region of that country. Due to a limitation on internal air travel, the team was forced to take an overland route from Medellín to Calí, where our trip was to begin. At various points on the Medellín–Calí road, people were selling panicles of *Aiphanes caryotifolia* fruit (Figs. 1, 3). We stopped to speak to the vendors, and they reported that both the mesocarp and endocarp were edible, and commonly eaten in the area. A few kilograms of fruits were purchased and stored for later analysis. Upon our arrival in Calí, economic botanist Dr. Victor Manuel Patiño confirmed the identification and use of this local food.

Analysis of Fruit

Because the bright pigmentation of the mesocarp and oil-rich endosperm suggested high levels of carotenes and Vitamin A, we decided to analyse the composition of this fruit. Analyses were carried out by Hazleton Laboratories America, Inc., in Madison, Wisconsin. A single collection was submitted for analysis; the results are presented in Tables II–V.

Most striking is the content of Vitamin A from carotene, 16,000 IU/100 g (wet weight basis; dry weight basis is 73,392 IU/100 g). This fruit has a remarkable content of Vitamin A, superior to the best sources known in the plant kingdom such as carrots, which have a maximum of 12,000 IU/100 g (range 2000–12,000 IU/100 g). Sweet potatoes, another excellent source of Vitamin A,

Table I

Common names for *Aiphanes caryotifolia*

Name	Country (region, group or language)	Reference
charascal	Colombia (Antioquia)	Peréz-Arbeláez, 1978
chontaruro	Ecuador	Peréz-Arbeláez, 1978
corozo	Colombia (Antioquia)	Galeano & Bernal, 1978
corozo anchame	Venezuela (Bolívar)	Peréz-Arbeláez, 1978
corozo chiquito	Colombia (Antioquia)	Galeano & Bernal, 1987
corozo colorado	Colombia (Antioquia)	Peréz-Arbeláez, 1978
corozo de chascará	Colombia (Valle de Cauca)	Patiño, 1963
corozo del Orinoco	Venezuela (Bolívar)	Peréz-Arbeláez, 1978
fish tail palm	Zanzibar	Williams, 1949
gri-gri	Trinidad and Tobago; Dominica (French)	Williams & Williams, 1951; Hodge & Taylor, 1957
majerona	Brazil	Patiño, 1963
mararabe	Colombia (Llanos)	Peréz-Arbeláez, 1978
mararai	Colombia (Magdalena; Llanos)	Patiño, 1963
palma de corozo	Venezuela	Patiño, 1963
paxiuba mangerona	Brazil	Patiño, 1963
pujamo	Colombia (Pujamo)	Peréz-Arbeláez, 1978
qualte	Colombia (Nariño)	Peréz-Arbeláez, 1978
quindío	Colombia; Ecuador; Peru	Patiño, 1963
rókri	Dominica (Carib)	Hodge & Taylor, 1957
ruffle palm	U.S.	Staff, L. H. Bailey Hortorium 1976
spine palm	U.S.	Staff, L. H. Bailey Hortorium 1976

contain 8800 IU/100 g of raw portion (Bogert et al., 1973).

Vitamin A in the Diet

Vitamin A is an essential component of the diet. Unfortunately, Vitamin A deficiency remains one of the major public health problems in many developing countries, particularly in Asia and the Middle East and some parts of Africa. Even in countries such as the United States and Canada there has been recent concern about the Vitamin A nutriture of many of their citizens. It has been estimated that at least half a million Asian children develop potentially blinding corneal involvement every year as a result of Vitamin A deficiency (Sommer, 1982). Except for the northeastern part of Brazil and Haiti, most Latin American countries do not report a major xerophthalmia problem despite low intakes and blood levels of Vitamin A (WHO, 1976). Severe Vitamin A deficiency often results in death from infection. A recent report (Committee on Diet, Nutrition and Cancer, 1982) states: "A growing accumulation of epidemiological evidence indicates that there is an inverse relationship between the risk of cancer and the consumption of foods that contain Vitamin A or its precursors."

The precursors are carotenoids, particularly beta carotene which are found in plant foods. Because they must be converted to Vitamin A, the carotenoids show less Vitamin A activity when fed than Vitamin A (retinol) itself. The Vitamin A of a food is expressed in retinol equivalents or international units (IU). One retinol equivalent equals 1 μg retinol, 6 μg beta carotene or 12 μg of other provitamin carotenoids. A retinol equivalent equals 3.33 IU from retinol or 10 IU from beta carotene.

It has been frustrating to nutritionists that even in countries where edible plant sources of beta carotene exist in quantity, it has been frequently difficult to get people to eat them. For example, in many parts of the world red palm oil, an excellent source for the carotenoids which give it its color, is consumed as a dietary component. However, in Indonesia it has been impossible to get people to consume this readily available oil because it has been traditionally used for nonfood purposes. Programs which enrich foods with Vitamin A or provide vitamins to individuals in capsules or by injection are expensive and have had limited success. Clearly, the availability of

Table II

Proximate composition of *Aiphanes caryotifolia* mesocarp

	Dry weight	Wet weight
Protein	6.0 g/100 g	1.3 g/100 g
Moisture (70° vac. oven)	—	78.2 g/100 g
Fat	1.8 g/100 g	0.4 g/100 g
Ash	6.9 g/100 g	1.5 g/100 g
Crude fiber	15.6 g/100 g	3.4 g/100 g
Carbohydrates	69.7 g/100 g	15.2 g/100 g
Beta carotene	44.1 mg/100 g	9.61 mg/100 g
Vitamin A from carotene	73,392 IU/100 g	16,000 IU/100 g
Calories	319 cal/100 g	69.6 cal/100 g

palatable foods high in Vitamin A activity would have major public health value in many parts of the world.

Oil and Protein in *Aiphanes* fruit

Another interesting observation is that the endosperm of this species (Table III) contains 37% fat (dry weight basis). It is, then, a good source of calories and oil in the local diet. The oil is primarily (63%) lauric acid (a saturated fat, and thus relatively similar to coconut oil).

The protein of *Aiphanes caryotifolia* is of a high biological value (Table IV), containing as its most limiting amino acid threonine, at 77% of the ideal protein value (FAO/WHO, 1973). With its total protein composition of 37%, the endosperm of this palm is a good source of nutrition.

Aiphanes sp. fruit is also reported to be consumed in small amounts by the Oilbird, *Stea-*

tornis caripensis (Snow & Snow, 1978), and presumably is a valuable source of nutrition for this and other animals.

Cultivation and Use of *Aiphanes*

During a recent trip to Honduras, one of us (MJB) observed this palm in cultivation in the palm garden of Dr. William Plowden in Peña Blanca, Department of Cortés (Fig. 2). Plowden reported that he had collected seed of this palm in the vicinity of Medellín, Colombia, and transported it to his farm in Honduras. Interestingly, the fruits of this palm are also collected and eaten by the people of the area in the same manner as

Table III

Proximate composition of *Aiphanes caryotifolia* endosperm

	Dry weight	Wet weight
Protein	7.9 g/100 g	4.3 g/100 g
Moisture (70° vac. oven)	—	45.9 g/100 g
Fat	37.0 g/100 g	20.0 g/100 g
Ash	2.0 g/100 g	1.1 g/100 g
Crude fiber	34.68 g/100 g	18.7 g/100 g
Carbohydrates	18.5 g/100 g	10.0 g/100 g
Calories	438 cal/100g	237 cal/100 g

Table IV

Fatty acid composition of *Aiphanes caryotifolia* endosperm

Fatty acid	Percent composition
Saturated	
Arachidic (20:0)	0.07%
Capric (10:0)	2.1
Caprylic (8:0)	1.7
Lauric (12:0)	62.8
Myristic (14:0)	17.9
Palmitic (16:0)	5.4
Pentadecanoic (15:0)	0.08
Stearic (18:0)	3.0
Unsaturated	
Linoleic (18:2)	2.2
Oleic (18:1)	4.7

←

FIGS. 1–3. **1.** Panicle of ripe *Aiphanes caryotifolia* fruit for sale along the Medellín–Calí road. Note bag of loose fruit also for sale. **2.** Mature specimen of *Aiphanes caryotifolia,* grown from seed collected in Colombia. **3.** View of fruit stand where *Aiphanes caryotifolia* is being sold (far right and bag on left).

Table V

Analysis of essential amino acids in *Aiphanes caryotifolia* endosperm protein

Amino acid component	mg/g	FAO/ WHO amino acid scoring pattern	% of scoring pattern
Isoleucine	31.87	40	80
Leucine	58.62	70	84
Lysine	55.82	55	101
Methionine & cystine	28	35	80
Phenylalanine & tyrosine	56.52	60	94
Threonine	30.7	40	77
Tryptophan	10	10	100
Valine	50.47	50	101

the fruits are consumed in Colombia as well as other areas where the palm is native.

While *Aiphanes caryotifolia* is a palm with great nutritional value, it is likely that it will remain a minor food plant throughout its distribution, eaten on an opportunistic basis and occasionally sold by local farmers, as in the instance along the Medellín–Calí road. However limited in importance, this species should be recognized as an important source of both Vitamin A and protein and therefore a valuable supplement to the diet of those who consume it. As long as economic and social dislocations continue to result in a decrease in the quality of the diet of both rural and urban dwellers in many Third World countries, little-known plant species such as this palm will continue to play a useful role in human nutrition.

Acknowledgments

The work discussed in this study was supported by grants from the Armand G. Erpf Fund, the U.S. Agency for International Development, and the Joyce Mertz-Gilmore Foundation. Luis E. Forero P., Steven King and Nelly Hernández were participants in the germplasm expedition to Colombia during which this palm was collected, and their participation is gratefully acknowledged. Steven King kindly provided comments on an earlier version of this paper. This is publication number 117 in the series of the Institute of Economic Botany of The New York Botanical Garden.

Literature Cited

Bogert, L. J., G. M. Briggs & D. H. Calloway. 1973. Nutrition and physical fitness, 9th ed. W. B. Saunders Co., Philadelphia.

Braun, A. 1968. Cultivated palms of Venezuela. Principes 12: 39–103.

Burret, M. 1932. Die palmengattungen *Martinezia* und *Aiphanes*. Notizblatt Botanischen Gartens und Museums zu Berlin-Dahlem 11: 557–577.

Committee on Diet, Nutrition and Cancer. 1982. Diet, Nutrition and Cancer, Assembly of Life Sciences, National Research Council, National Academy Press, Washington, D.C.

Dransfield, J. & N. W. Uhl. 1986. An outline of a classification of palms. Principes 30: 3–11.

FAO/WHO. 1973. Energy and protein requirements. Food and Agricultural Organization of the United Nations/World Health Organization Technical Report Series No. 522, Geneva.

Galeano, G. & R. Bernal. 1987. Palmas del Departamento de Antioquia, región occidental. Universidad Nacional de Colombia-Centro Editorial, Bogotá.

Hodge, W. H. & D. Taylor. 1957. The ethnobotany of the island Caribs of Dominica. Webbia 12: 513–644.

Martin, F. W. & R. B. Ruberté. 1979. Edible leaves of the tropics, 2nd ed. Antillian College Press, Mayagüez, Puerto Rico.

Moore, H. E., Jr. 1963. An annotated checklist of cultivated palms. Principes 7: 119–182.

———. 1973. The major groups of palms and their distribution. Gentes Herbarum 11: 27–141.

Patiño, V. M. 1963. Plantas cultivadas y animales domesticos en América equinoccial. Imprenta Departmental, Calí.

Pérez-Arbeláez, E. 1978. Plantas útiles de Colombia. Litografía Arco, Bogotá.

Snow, D. W. & B. K. Snow. 1978. Palm fruits in the diet of the Oilbird, *Steatornis caripensis*. Principes 22: 107–109.

Sommer, A. 1982. Nutritional blindness: Xerophthalmia and keratomalacia. Oxford University Press, New York.

Staff, L. H. Bailey Hortorium. 1976. Hortus Third. Macmillan Publishing Co., Inc., New York.

Williams, R. O. 1949. The useful and ornamental plants in Zanzibar and Pemba. Zanzibar Protectorate, Zanzibar.

——— & R. O. Williams, Jr. 1951. The useful and ornamental plants in Trinidad and Tobago. Guardian Commercial Printery, Port-of-Spain.

WHO. 1976. Vitamin A deficiency and xerophthalmia. World Health Organization Technical Report Series, No. 590, Geneva.

A Survey of the Useful Species of *Paullinia* L. (Sapindaceae)

Hans T. Beck

Table of Contents

Abstract

BECK, H. T. (Institute of Economic Botany, New York Botanical Garden, Bronx, New York 10458). A survey of the useful species of *Paullinia* L. (Sapindaceae). Advances in Economic Botany **8**: 41–56. 1990. From information gleaned from a survey of the botanical, medical and ethnobotanical literature, 39 species of *Paullinia* are reported to be useful. The species are used primarily in tropical America and tropical Africa. Fresh plant material or preparations are used for medicines, fish poisons, caffeine-rich beverages, binding material for huts, weaving material, food, wood for barrel staves and hoops, human poisons and admixtures for arrow poisons. Some species are believed to have special magical and charm properties. Detailed information on the uses is provided, along with scientific names (and taxonomic synonyms), vernacular names, geographical distribution and reported chemical compositions.

Key words: *Paullinia*; fish poisons; beverages; construction; arrow poisons; human poisons; food.

Resumo

Com base em pesquisas de literatura botânica, médica e etnobotânica, 39 espécies de *Paullinia* são consideradas de importância econômica. As espécies são utilizadas principalmente na América e Africa tropicais. Material fresco ou preparados são utilizados como remédios, ictiotóxicos, bebidas com alto teor de cafeína, material de construção para diversos artigos em habitações e artesanato, alimentção e como substâncias tóxicas. Acredita-se que algumas espécies possuem propriedades especiais de magia e encantamento. Informações detalhadas sobre os diversos usos são fornecidas com nomes científicos (e sinônimos taxonômicos), nomes vulgares, distribuição geográfica e composição química.

Introduction

The documentation of the past and present uses of plants by people in the tropics has been an important aspect of the research done by ethnobotanists, plant systematists, anthropologists, pharmacognosists and physicians. It is their research and observations, published over the last 150 years, that has been drawn upon for this survey of the uses of the species of *Paullinia*.

The genus *Paullinia* L. is a mostly neotropical group of woody lianas and small rain forest trees having approximately 180 species with one species extending out to tropical Africa (Radlkofer, 1933; Willis, 1966); of these species, 39 are reportedly used in subsistence or commercial preparations. The primary uses of these species are for fish poisons, medicines and caffeine-rich beverages; minor uses include admixtures for arrow poisons, human poisons, food from edible arils, fiber for weaving and wood for construction. Much use of these species of *Paullinia* occurs at a local level by indigenous peoples in tropical America and tropical Africa, but some species are cultivated and marketed regionally, and one species, *P. cupana* var. *sorbilis,* has even become an international export commodity for Brazil (Erickson et al., 1984; Herman, 1982; Ypiranga Monteiro, 1965). These uses stem from the high percentages of alkaloids, tannins and saponins present in various parts of the plants.

Summary of Uses

A summary of the uses of the species of *Paullinia* is given here to provide a brief overview of the diversity; consequently, specific references are withheld from this section. Full details and references are provided in the next section entitled Synopsis of Taxonomic, Geographical, Utilitarian and Phytochemical Information for *Paullinia* Species, and uses are categorized in Table I.

The uses of species of *Paullinia* reported in the literature involve different parts of the plants. The seeds and bark are used for stimulating beverages. The whole plant, as well as parts thereof, is used as a fish poison. The stems and twigs are used for fiber and construction materials. Medicines are derived from all parts of the plants. Stem sap and seeds are used for human poisons and as admixtures in arrow poisons. For some

of the species, the aril is edible and is consumed as food.

Paullinia preparations are used at local, regional and, rarely, international levels. Locally, the indigenous use of *Paullinia* species has been extensive. Regionally and internationally, the use of *Paullinia* has been limited to two species, *P. cupana* var. *sorbilis* and *P. pinnata*. The main areas for the use of the species of *Paullinia* have been in the neotropics, from Mexico through Central America, the West Indies and South America; however, one widespread species, *P. pinnata,* is also found in the African tropics, from west tropical Africa to Madagascar.

The most commonly reported use of *Paullinia* species has been as a fish poison. The plant parts employed for this purpose include the leaves, fruits, stems, roots, flowers, or the entire plant. The plant material is commonly beaten and then thrown into the water, the effects being transitory but sufficient to immobilize the fish for easy catching[1]. Species reported as fish poisons are *P. acuminata, P. alata, P. aspera, P. barbadensis, P. carpopodea, P. costata, P. cupana, P. cururu, P. dasyphylla, P. echinata, P. elegans, P. fuscescens, P. imberbis, P. jamaicensis, P. macrophylla, P. meliaefolia, P. neglecta, P. pachycarpa, P. pinnata, P. rhizantha, P. rhomboidea, P. rubiginosa, P. scarlatina, P. seminuda, P. spicata, P. subnuda, P. thalictrifolia, P. trigonia* and *P. xestophylla.*

In contrast to the extensive use of the many species of *Paullinia* as a fish poison, only two species, *P. cururu* and *P. pinnata,* are reported as being admixtures in arrow and dart poisons. The stem sap is the plant part most reported for this use.

Human poisons are reportedly derived from *Paullinia.* Consumed internally, the seeds of *P. cururu, P. fuscescens* and the seeds and roots of *P. pinnata* are reported to be slow-acting but deadly poisons; furthermore, it is commonly reported that the seeds are employed as a method of criminal poisoning, especially in the West Indies.

Medicinal remedies and oral hygiene are another primary use of the species of *Paullinia*; many are reported as having medicinal proper-

---

[1] For a more complete discussion of fish poisons and their uses, see P. Acevedo Rodríguez in this volume.

Table I

Uses of the species of *Paullinia* according to category

Species	Fish poison	Medicine	Food	Handi-crafts/construc-tion	Human poison	Beverage	Other[a]
acuminata	X						
alata	X						
aspera	X						
barbadensis	X	X		X			
bracteosa			X				
carpopodea	X			X			
costata	X						
costaricensis		X					
cupana	X	X				X	Xm
cururu	X			X	X		Xa
dasyphylla	X						
echinata	X						
elegans	X		X				Xo
emetica		X					
fuscescens		X	X	X	X		
imberbis	X	X					
jamaicensis	X	X					
macrophylla	X		X				
meliaefolia	X	X					
neglecta	X						
pachycarpa	X						
pinnata	X	X	X	X	X		Xa, m
pterophylla		X				X	
rhizantha	X						
rhomboidea	X						
rubiginosa	X						
scarlatina	X						
seminuda	X						
sphaerocarpa			X				
spicata	X		X				
subnuda	X						
subrotunda			X				
tarapotensis			X				
tetragona				X			
thalictrifolia	X						Xo
tomentosa		X					
tricornis			X				
trigonia	X	X					
uloptera							Xu
xestophylla	X						
yoco		X				X	
Total	28	13	10	6	3	3	7

[a] Other refers to one or more of the following uses: (o) ornamental, (a) arrow poison, (m) magic/charm, or (u) unspecified use.

ties and purposes: *Paullinia barbadensis, P. costaricensis, P. cupana, P. emetica, P. fuscescens, P. imberbis, P. jamaicensis, P. meliaefolia, P. pinnata, P. pterophylla, P. tomentosa, P. trigonia* and *P. yoco.* These species are reportedly used in treating a range of illnesses and afflictions affecting the nervous system (i.e., neuralgia, headaches, mental illness), and the gastrointestinal system (i.e., dysentery, diarrhea), the urogenital system (i.e., menstrual cramps and excessive menstrual bleeding, inflammation of the ovaries, postpartum pain), the circulation system (i.e., cardiac derangements, arteriosclerosis), and general body ailments (i.e., rheumatism, bronchitis). Some species are reportedly used as contraceptives. Weight control and loss preparations employ *guaraná* powder from *P. cupana* var. *sorbilis* because of the high caffeine content. The use of the roots of *P. jamaicensis* and *P. pinnata* as chew sticks for cleaning teeth and maintaining good oral hygiene has been reported.

One of the most well-known uses of *Paullinia* is for the preparation of stimulating, caffeine-rich beverages. The species used in the preparation of such beverages are *P. cupana* var. *sorbilis* and *P. yoco.* A third species, *P. pterophylla,* is possibly employed for the same purpose.

Paullinia cupana var. *sorbilis* is the source of a caffeine-rich beverage called *guaraná.* The seeds contain from 2.3 to 5% caffeine as well as other important chemical substances such as theobromine, tannins and saponins. In the preparation of *guaraná,* the seeds are processed into either powder, syrup or extract. *Guaraná* is used by native peoples, especially in Brazil, in the form of a *bastão* (a small brown cylindrical stick). The *bastões* are made for home use or for sale in local markets. A *bastão* is turned into powder by grating, commonly against the dried tongue of a *pirarucu* fish; this powdered *guaraná* is added directly to hot or cold water and is consumed as a stimulating, refreshing beverage. Ready-to-use powdered *guaraná* can be purchased in bags or jars, the preparation with hot or cold water being basically the same except for the occasional addition of sugar. The syrup and extract forms of *guaraná* are used more commonly in commercial beverage preparations. *Guaraná* syrup, sold in bottles, is diluted with water to make an iced-tea-like beverage. *Guaraná* extract is employed as a flavoring agent and a source of caffeine for the carbonated and noncarbonated soda and soft drink industry.

The other species of *Paullinia* used for a caffeine-rich beverage is *P. yoco.* Lowland forest Indian tribes of Colombia and Ecuador scrape the bark and squeeze the sap from the stems to make a cold-water infusion, which is consumed commonly in the morning on a daily basis for staying hunger and relieving fatigue.

Several species of *Paullinia* are reportedly used as sources of fiber for ropes and plaited objects, as sources of wood for special construction material, and as sources of tying and binding material. The tough, flexible branches of *P. barbadensis* are made into riding switches and walking sticks. The wood of *P. carpopodea* is employed in the manufacture of barrel staves and barrel hoops. *Paullinia cururu* stems are macerated and used for binding material. *Paullinia fuscescens* stems provide fiber used for twine and rope substitutes in the construction of hut frameworks and fences. *Paullinia pinnata* stems and twigs are employed for binding materials and the wood for barrel staves. The stems of *P. tetragona* are reported to be useful for weaving and plaiting material for baskets and hats.

Certain species of *Paullinia* have been reported to be used for food or at least have edible fruit. The arils from *P. bracteosa, P. elegans, P. fuscescens, P. macrophylla, P. pinnata, P. sphaerocarpa, P. spicata, P. subrotunda, P. tarapotensis* and *P. tricornis* are reported as edible. Additionally, edible flowers are reported for *P. pinnata* and edible seeds for *P. subrotunda.*

Two species of *Paullinia* are reported to have magical properties or charms. *Paullinia pinnata* is used to provide immunity from wounds, to prevent kidney troubles and other illnesses, and to cure female sterility. *Paullinia cupana* is reported to bring good luck, bring good rains and protect against storms.

Horticulturally, two species of *Paullinia* are grown. *Paullinia elegans* is reportedly cultivated as an ornamental, and *P. thalictrifolia* is cited as an ornamental plant cultivated in Rio de Janeiro and São Paulo.

The uses of the species of *Paullinia* are summarized according to use category in Table 1. Fish poisons are by far the most reported use, occurring in 28 species. Medicinal uses and food are the next most reported uses with 13 and 10 species, respectively. Six species are reported for handicrafts and construction materials. Human poisons, arrow poisons, beverages, magic/charms, and ornamentals are represented by three or fewer species. The remainder of this paper presents species-specific information for the scientific name, taxonomic synonyms, geographic distribution, vernacular names, detailed references for the uses, and a listing of the reported chemical compositions.

Taxonomic Note

There are some reports in the literature of species with ethnobotanical and economic uses that have been taxonomically assigned to *Paullinia* in the past. The most noteworthy examples of this are *Paullinia australis* St. Hil. now held to be *Serjania herteri* Ferrucci, *Paullinia curassavica* Jacq. now held to be *Serjania mexicana* (L.) Willd., *Paullinia grandiflora* St. Hil. now held to be *Serjania erecta* Radlk., and *Paullinia mexicana* Jacq. now held to be *Serjania mexicana* (L.) Willd. A discussion of these species uses is not included here, but many of the references, particularly the older ones in the Literature Cited do list their uses (1, 22, 24, 32, 33, 40, 49, 61, 65, 70, 77, 83, 84, 87, 95, 106, 113, 119, 133)[2].

Synopsis of Taxonomic, Geographical, Utilitarian and Phytochemical Information for *Paullinia* Species

Paullinia acuminata Uitt. GEOGRAPHIC DISTRIBUTION: Guianas. USES: Reported as a fish poison in the Guianas (1, 86). CHEMICAL CONSTITUENTS: Saponins (1).

Paullinia alata (Ruiz & Pavón) G. Don. GEOGRAPHIC DISTRIBUTION: Brazil, Ecuador, Peru, Colombia and Panama. TAXONOMIC SYNONYMS: *Semarillaria alata* Ruiz & Pavón. VERNACULAR NAMES: Peru: macote (111). Brazil: caferana, cipó de tres quinas (106), cofferana (61, 110), timbó, tingui (87, 113), urarirana (61, 106, 110). Ecuador: vejuco de tres esquinas (106, 110). USES: The roots are reported as a fish poison in tropical America (1, 61, 87, 111, 113). CHEMICAL CONSTITUENTS: Saponins (1).

Paullinia aspera Radlk. GEOGRAPHIC DISTRIBUTION: Central Brazil. VERNACULAR NAMES: Brazil: timbó (106). USES: Reported as a fish poison in Brazil (1, 106). CHEMICAL CONSTITUENTS: Alkaloids: timbonine (106). Saponins (1).

Paullinia barbadensis Jacq. GEOGRAPHIC DISTRIBUTION: West Indies and Mexico. VERNACULAR NAMES: Mexico: guarana (24). Jamaica: supple jacks (34). USES: In Jamaica, the flexible and tough branches are employed as riding switches and walking sticks, and the seeds are reported as a fish poison (34). This plant is reported to be indigenous to the West Indies (i.e., not introduced by slaves) and is reported to be used formerly as a medicinal by slaves (44).

Paullinia bracteosa Radlk. GEOGRAPHIC DISTRIBUTION: Central and northern South America. VERNACULAR NAMES: Ecuador: okwe yoko, 'oko yoko (139). USES: The Siona Indians of Ecuador are reported to suck on the edible flesh (aril) around the seed (139).

Paullinia carpopodea Camb. GEOGRAPHIC DISTRIBUTION: Southern Brazil. TAXONOMIC SYNONYMS: *Paullinia timbo* Vell. (61). VERNACULAR NAMES: Brazil: timbó (61, 106). USES: Reported as a fish poison in Brazil (61, 106). The wood has been used for barrel staves and barrel hoops in Brazil as well (106). CHEMICAL CONSTITUENTS: Alkaloids: timbonine (106). Saponins (1).

Paullinia costata Schlect. & Cham. GEOGRAPHIC DISTRIBUTION: Mexico and Central America. VERNACULAR NAMES: Mexico: bejuco de agua (127), guarana (24). USES: Reported as a fish poison in Mexico (1, 95). CHEMICAL CONSTITUENTS: Saponins (95).

Paullinia costaricensis Radlk. GEOGRAPHIC DISTRIBUTION: Mexico and Central America. VERNACULAR NAMES: Honduras: pate (128). USES: Reported to be a medicinal remedy in Guatemala, the common name "pate" is derived from the Nahuatl word "patl," which refers to medicine (128).

Paullinia cupana H.B.K. var. *sorbilis* (Mart.) Ducke & var. *cupana* GEOGRAPHIC DISRIBUTION: var. *sorbilis*: northern Brazil. var. *cupana*: Venezuela. TAXONOMIC SYNONYMS: *Paullinia sorbilis* Mart. VERNACULAR NAMES: Brazil: cupana (28, 54, 79), guaraná (7, 9, 10,

---

[2] The references are numbered for use in the Synopsis that follows, to save space.

28, 46, 49, 54, 61, 79, 87, 98, 99, 137); gisipó (61), guarana-iiva (101), guaraná-ripó (46), guaraná-sipo (101), guaraná-uva (99), guarana-üva (61, 77, 79), guaranauva (46), naranzeiro (99), uarana (61), uaraná (54, 99), uraná (79, 101). Colombia: cupana (7, 106), cupána (27), cupania, guarana (100), guaraná (122), yocco (106). England/United States: Brazilian cocoa (29, 83, 90), guarana shrub (129). France: paullinia (37). Germany: guarana-kletterstrauch (129), schwimfwort (37). Netherlands: guaranstuik (129). Panama: palo cuadrado (3). Peru: guaraná (122). Venezuela: cupana (46, 61, 101, 106), cupána (27), guaraná (122), yocco (106). Other: guaraná (12, 15, 16, 17, 18, 19, 20, 22, 24, 26, 27, 29, 30, 31, 32, 36, 40, 50, 51, 53, 55, 63, 64, 65, 66, 67, 68, 69, 71, 72, 74, 80, 81, 82, 84, 85, 88, 89, 91, 92, 93, 95, 96, 97, 102, 103, 104, 106, 108, 109, 110, 116, 119, 121, 123, 124, 125, 127, 128, 132, 135, 140, 142), uabano (29). USES: Taxonomic recognition of two varieties of guaraná (27), one the typical or wild variety and the other the cultivated or domesticated variety, has only happened in this century; consequently, older historical references to guaraná are potentially difficult to interpret with respect to variety. It can probably be safely assumed that the majority of references to guaraná relate to the cultivated variety *sorbilis* and not the more rare wild variety *cupana* (*typica*).

Guaraná-producing regions occur in Amazonas, Brazil, and are concentrated around Maués and Manaus (31). Traditional guaraná agriculture (family-based guaraná orchards) is practised by the Sateré-Maué Indians in the Maués region (50, 144). Commercially, peasant farmers and agroindustrial firms grow this plantation crop on newly cleared forest land (31, 50). Present total production ranges from 250 to 900 tons per annum, and more than 75% is used in the soft drink industry (31, 50).

A refreshing, stimulating, caffeine-rich beverage is made from the seeds (9, 40, 41, 63, 64, 65, 81, 98, 100, 110, 116, 119, 121) and is often served as a coffee or tea substitute (9, 54, 72, 89, 95, 119, 124, 125). This stimulating beverage has been reportedly used as a diet drink (29, 72). Guaraná pills are sold both as diet aids and exotic natural stimulants (7, 50). The seeds of this species supply a flavor ingredient for use in unfermented, carbonated and noncarbonated soft drinks (12, 29, 67, 74, 82, 142); the guaraná ingredient may be in the form of an extract (67, 74) or syrup (41). There are reports of guaraná being made into alcoholic beverages (29, 69, 135), such as liqueurs and cordials (29, 67). A guaraná candy has been reported (67).

The seeds of guaraná are reported to be a preferred food for birds (67). This species has been reported as being employed as a fish poison in northern Brazil and Venezuela (1, 46, 49, 61, 95). Guaraná is reported as a source of a tannin dye or stain (124, 132); the Indians of Brazil stain their faces with it (124).

Guaraná has many cited medicinal applications. With respect to the nervous system, guaraná has been reported to be a stimulant (3, 9, 10, 27, 29, 41, 54, 65, 67, 72, 81, 82, 97, 99, 100, 109, 124, 129, 132), useful for treating headaches (9, 29, 82, 97, 124), hangovers (29, 116), menstrual headaches (116), migraines (29, 51, 129), paralysis (9), and neuralgia (7, 9, 29, 65, 84, 91, 93, 96, 99, 102, 129, 136); furthermore, it is reported to produce wakefulness (80, 97, 116, 124), quicken perception (116), and relax nervous tension (109). With respect to the circulation system, guaraná has been reported to combat arteriosclerosis (22, 65, 99), dilate blood vessels (99), stimulate circulation (80), slow the pulse (116), and serve as a cardiotonic (10, 41, 65) for cardiac derangements (29). Guaraná and *Nasturtium officinale* (Cruciferae) are consumed as an infusion for catarrhal jaundice as prescribed by Kallawaya herbalists in Bolivia (8). Guaraná is also cited as a geriatric tonic (22). With respect to the gastrointestinal and urogenital systems, guaraná has been reported as a digestive or stomachic (71, 77, 80, 91, 93, 99, 132), antiflatulent (22, 99), purgative (124), and diuretic (22, 29, 54, 99); furthermore, it has been cited as useful in treating diarrhea (7, 9, 22, 26, 29, 37, 41, 51, 65, 67, 74, 84, 91, 93, 96, 97, 104, 109, 125, 129, 135, 136), dysentery (37, 51, 65, 102, 104), dyspepsia (37, 51, 79), intestinal disorders and colic (29, 99, 109), malaria (91), urinary tract maladies (8, 9, 26, 97), menstrual cramps and painful menstruation (82, 91), uterine hemorrhages (93), and blennorrhagia (37, 51, 104). Other reported medicinal uses of guaraná include treatment for pleuritis, pneumonia (93), general debility (37, 51), and lumbago (29); guaraná is reportedly used as a curative, prophylactic (9), tonic (3, 9, 29, 37, 65, 79, 82, 84, 93, 96, 99, 109, 122, 129), restorative (9, 65, 72, 124), nutritive (9, 122), astringent (37, 51, 67, 71, 129), febrifuge (9, 68, 71, 74, 82, 84, 93, 99, 100, 102), and appetite suppressant (74, 82, 99, 116). Other medicinal properties reported for guaraná include curing impotence (97) and use as an aphrodisiac (7, 9, 29, 65, 71, 77, 80, 82, 99).

Guaraná is reported to hold luck, magic and special charms: the Sateré-Maué Indians believe that good luck in transactions is brought by guaraná, as well as giving joy and a stimulus to work; it provides protection or charm for them in the form of bringing rains and protecting against storms, protecting their farms, curing certain diseases and preventing others, and bringing success in war and in love (92).

CHEMICAL CONSTITUENTS: Alkaloids: caffeine (12, 18, 19, 29, 41, 80, 82, 88, 89, 90, 95, 116, 119, 134); guaranine (29, 77, 80, 124, 128, 135); theobromine (29, 88, 134); theophylline; timbonine (29, 134). Saponins (18, 95). Tannins (18, 29, 82, 119). Fixed oils (18, 95). Others (29, 122).

Paullinia cururu L. GEOGRAPHIC DISTRIBUTION: Mexico, West Indies, Central and South America. TAXONOMIC SYNONYMS: *Paullinia riparia* H.B.K. *Semarillaria cururu* Ruiz & Pavón. *Serjania nodosa* Radlk. VERNACULAR NAMES: Brazil: arary (106), cipó cruapé branco (49, 106), cipó cururú (79), curucú (46), cururú (49, 106), timbó (46, 79), timbó cururú (61). Colombia: azucarito (11), barbasco (100), bejuco mulato (11), bejuco de San Pedro (46, 61, 100), cupania (100), perita de monte (46, 100). Dominican Republic: bejuco tres filhos (76). Nicaragua: chilmecate (106, 125). Mexico: colorín (100, 106, 125), pahuch-ac, pajuj-ac (128), p'ahuch'ak, p'ahuh-ak' (78). Venezuela: azucarito (46, 61, 100, 106, 125), bejuco mulato (46, 61, 100, 106). USES: Toxic principles are reported for the seeds (26, 40, 100, 106). Numerous reports cited the use of this species as a fish poison (1, 11, 46, 49, 61, 76, 78, 106, 126) and as a curare admixture (11, 26, 69, 100, 106, 119) in the preparation of arrow poisons (83, 106) in French Guiana and other countries in South America. The seeds are re-

ported harmful to domesticated birds (76). The alkaloid timbonine extracted from this species has been cited as harmful to humans (76). Small stems and macerated large stems are useful binding material (106). CHEMICAL CONSTITUENTS: Alkaloids: timbonine (76). Saponins (1).

Paullinia dasyphylla Radlk. GEOGRAPHIC DISTRIBUTION: Central Brazil. VERNACULAR NAMES: Brazil: timbó (106). USES: Reported as a fish poison in Brazil (1, 106). CHEMICAL CONSTITUENTS: Alkaloids: timbonine (106). Saponins (1).

Paullinia echinata Hub. GEOGRAPHIC DISTRIBUTION: Northern Brazil and Peru. VERNACULAR NAMES: Brazil: cipó timbó (106). USES: Reported as being employed by sertanejos (peasants) as a fish poison in Brazil (106).

Paullinia elegans Camb. GEOGRAPHIC DISTRIBUTION: Colombia, Bolivia, Brazil, Uruguay, Paraguay, Argentina. VERNACULAR NAMES: Argentina: caí-escalera-rá, isipó-moroti (57). Brazil: cipó de timbó (46, 106), cipó timbó (13, 79), timbó (13). USES: Reported as having an edible aril (13) and as a fish poison in Brazil (1, 13, 46, 61, 79, 106). The plant is reportedly cultivated as an ornamental in Argentina (57). CHEMICAL CONSTITUENTS: Saponins (1).

Paullinia emetica R. E. Schultes. GEOGRAPHIC DISTRIBUTION: Colombia. USES: An infusion of this plant is cited as an emetic (39, 120); specifically, the Carijona Indians of Miraflores (Colombia) are reported to use leaf infusions as strong emetics (120).

Paullinia fuscescens H.B.K. GEOGRAPHIC DISTRIBUTION: Mexico, West Indies, Central and northern South America. TAXONOMIC SYNONYMS: *Paullinia curassavica* Millsp. *Paullinia velutina* DC. VERNACULAR NAMES: Central America: barbasco (142), bejuco de barbasco (46, 61), bejuco colorado, bejuco cuadrado, bejuco de barbasco, bejuquillo de gusano (142), campalaca (46, 61), chilmecate (142), nistamal (46, 61, 142), nistamalillo (46), palo de mimbre (46, 61, 142), sebo de pollo (142). Colombia: peronilla (46). El Salvador: barbasco (125, 128), bejuco cuadro (111), bejuco cuadrado (125, 128), nistamal (111, 125,

128), nistamalillo (125, 128). Guatemala: barbasco (60, 125), bejuco barbasco, bejuco colorado, bejuquillo de gusano, chimecate (128), luruche (52), punche (60), sebo de pollo (128). Honduras: cainpalaca (125), campalaca (66, 128), pate (128). Mexico: aquiztli (24), bejuco costillón (24, 46, 61, 66, 125), bejuquillo (38), bix-chemac (128), chilmecate (38), kexac (128), kex-ak (24, 66, 117, 126), nixtamalillo (38), ojo de perro (46, 612), panoquera (46, 61, 66, 125), pico de quiloche (125). Panama: hierba de alacrán (125). Venezuela: bejuco mulato (46, 61, 66, 125, 138). USES: Reported as a venomous insect bite remedy (125), an antivenereal agent (138) and a fish poison (1, 46, 52, 60, 61, 111, 128, 142). The tough, flexible stems are employed as twine and rope substitute (126) in Mexico for binding frameworks for houses, huts and fences (66, 117). The Conquistadores are reported to have used this plant to alleviate "contamination and suffering of the groin" (38). The seeds are cited as poisonous (64, 111, 128, 142) and the fleshy aril as edible (64, 111, 125, 128, 142). CHEMICAL CONSTITUENTS: Saponins (1).

Paullinia imberbis Radlk. GEOGRAPHIC DISTRIBUTION: Peru, Guianas and Brazil. TAXONOMIC SYNONYMS: *Paullinia macrophylla* Sagot. VERNACULAR NAMES: Brazil: cipó timbó (106). South America: cipó timbó, cipo grande arbustivo (65), timbó (32, 65). USES: Reported as a fish poison in Brazil (32, 65, 106). An alcohol tincture of the seeds is cited as a rub used for rheumatism and partial paralysis (65).

Paullinia jamaicensis Macf. GEOGRAPHIC DISTRIBUTION: West Indies. TAXONOMIC SYNONYMS: *Paullinia sarmentosa* Browne. USES: Reported as a fish poison in Jamaica (1). Jamaican blacks reportedly used the ends of roots for cleaning their teeth (44).

Paullinia macrophylla H.B.K. GEOGRAPHIC DISTRIBUTION: Colombia. VERNACULAR NAMES: Colombia: bejuco prieto, ojito de nene (114), ojo de nene (46, 114). USES: Reported as a fish poison in Colombia (1, 46, 61, 95). The aril is cited as a food snack for native Colombians (114). CHEMICAL CONSTITUENTS: Saponins (1, 95).

Paullinia meliaefolia Juss. GEOGRAPHIC DISTRIBUTION: Southern Brazil, Paraguay and Argentina. TAXONOMIC SYNONYMS: *Paullinia falcata* Gardn. *Paullinia maritima* Vell. *Paullinia sericea* Camb. VERNACULAR NAMES: Brazil: timbó de cipó (47), timbó peba (49, 61, 106), timbó vermelho, tingui, tingui da folha grande (61, 106). USES: Reported as a fish poison in Brazil (1, 49, 95, 106). An alcohol extract of the stem is cited for treating tumors in Brazil (47). CHEMICAL CONSTITUENTS: Saponins (1, 95).

Paullinia neglecta Radlk. GEOGRAPHIC DISTRIBUTION: Peru and Bolivia. VERNACULAR NAMES: Brazil: timbó (106). USES: Reported as a fish poison in Brazil (1, 106). CHEMICAL CONSTITUENTS: Alkaloids: timbonine (106). Saponins (1).

Paullinia pachycarpa Benth. GEOGRAPHIC DISTRIBUTION: Colombia, central and northern Brazil. VERNACULAR NAMES: Brazil: timbo preto (112). USES: The stems and wood are reported as a fish poison in Brazil (1, 112).

Paullinia pinnata L. GEOGRAPHIC DISTRIBUTION: Tropical America and tropical Africa. TAXONOMIC SYNONYMS: *Paullinia africana* G. Don. *Paullinia grandiflora* Camb. *Paullinia senegalensis* Juss. *Paullinia timbo* Vell. *Paullinia uvata* Schum. & Thonn. VERNACULAR NAMES: Argentina: sachahuasca (106). Belize: macalte-ic, tietie (128). Brazil: cipó cruapé vermelho (65, 106), cipó cumaru-apé (106), cipó cururu (13), cipó cururú (79), cipó timbó (22, 106), cipo tingui (46, 61), cruape vermelho (49), cumaru ape (79), curuapé (65), cururú (79), cururú-apé (22, 46, 49, 61, 79, 86, 99, 106), de cruape vermelho (79), guaratimbo (22, 99), mafome (46, 61, 79, 106), mata fome (65), mata porco (46, 61), matta porco (49), sipo timbo (86), timbo (108), timbó (13, 46, 61, 74, 84, 99, 113, 133), timbó cipó (22, 46, 49, 61, 65, 99, 106), timbó de peixe (32, 65), timbó liane (46, 61), timbo sipo (83), timbó-sipo (133), tingui (113), tururú ape (79). British Antilles: bread-and-cheese (106, 125, 127). Central America: barbasco (61, 142), bejuco de Costilla, bejuco prieto, bejuco vaquero, nistamal, nistamalillo, pozolillo (61), pate, pietie

(142). Colombia: bejuco de zarcillo (11), morolico (46). Congo: lopasi, lusambo, tchinkolokosso (106). Cuba: azucarito (76, 106, 127), bejuco de costilla (106), bejuco de vieja (76). East Africa: mgogote (62). French Guiana: feifi-finga (115), kutupu (86), laurier du Sénégal (106), liane à scie (46, 61), liane carrée (65, 86), liane quarré (46, 61, 106). Gambia: bambaco, bissako, jumba lulu (6). Germany: gefiederte Paullinie (40). Ghana: adfiehotsi, akodwen-akodwen, akonkyerew-akonbyerew, akoron-koronduawa, akplokinakpa, berekueyene, gbadzajeka, gbobiloi amamu, gbobilomamu, gbolantshere, gboloi, obomofo-bese, toa-ntini, tolondi, tuentin, tumardiahbah (6). Gold Coast: adifiẽ-hotsi, akronkrŏnduawa (23), aplokinakpa (56), aplukniakpa, berekuĕkyĕne, gbadzafeka, gbŏbilŏi-amamu (23), gbobiloi amamu (56), gbŏbilōmamu (23), gbobilo-mamn (56), tolondi, tumardiahbah (23), twentin (23, 56). Guatemala: bejuco de barbasco (52). Hispaniola (Dominican Republic & Haiti): bejuco costilla (73), bejuco de tres filos, liane carrée (73, 76), tres filos, trefilo (73). Honduras: nistamal (5, 106, 125, 12, 135), pate (5, 106, 125, 128). Ivory Coast/Upper Volta: amoralia, bissagbibro, follo, gemoun haablou, guéguébro, kabéra, kakala, kanguélépéssé, kokoli kokolou, kotokwatra, ména nomo, korondi, kourounowvo, koutoulou, kpéfaka, kpéfanga, maci, mlanovo, moudembé, mounoudingbé, nagnan, namia, pidiakou, piendatriké, pinia guédré, samênéba, tchouebieban, tienkolé, torondi, trondi, twendini (6). Jamaica: bread-and-cheese, supple jack (5). Liberia: bge-se (6), gbe-se (23), gbĕsĕ-kŏ (6, 23). Malagasy Republic (Madagascar): varimarinhanga (106). Mexico: barbasco (5, 46, 61, 106, 125, 126, 127, 128, 135), bejuco de Costilla, bejuco prieto (46, 61), bejuco vaquero (5, 46, 61, 106, 125, 127), bejuquillo (106, 128), cuamecate (106), cuaumecate (5, 125), guarana (24), nistamal, nistamalillo (46, 61), pozolillo (61), pozonillo (46). Nigeria: aza (6), ènù kàkànchelà (6, 23), furen amariya (6), furin amariya (23), goron dorina, hannu biyar (6, 23), kakanchela (6), kàkànchelò (23), kaka shenla (6, 23), kana kana (23), ogbe-okeye (6), ŏgbé-okùje (23), yalsa biyar (6), yutsa biyar (23), zaga rafi (6), zaria zaga rafi (23). Panama: barbasco (5). Puerto Rico: bejuco de Costilla (76, 127). Salvador:

chilmecate, nistamalillo, pozolillo (5, 106, 125, 128). Senegal: fungo, kep i gney (6, 23). Sierra Leone: ĕ-funt, ĕ-bonka, ĕ-funtere-bonka, ndogbŏ-la-gule, ndgobŏ-lawule, firi-tembe, filifire-tembe (6, 23). Surinam: tondin (61), tondín (46). Togo: adioke-hotschi, gorogadam, tolundi (6, 23). Venezuela: bejuco de zarcillo (61), bejuco zarcillo (46). West Africa: gbese ko (6).

USES: This species has numerous references to the whole plant, leaves, seeds, fruit, bark, and roots being used as a fish poison (1, 5, 11, 13, 22, 23, 26, 28, 40, 46, 49, 52, 56, 60, 61, 65, 74, 76, 82, 86, 87, 105, 106, 110, 113, 119, 126, 127, 128, 135, 140, 141, 142); however, the dried leaves are reported as having no piscicidal action (141). The seeds (64, 125) and bark (5) are reported as poisonous. The roots and seeds have been reported as being used in the Antilles for criminal poisoning (5, 11, 40, 106, 125, 127, 128, 135, 142). The sap from the stems is reportedly used by Amerindians to poison arrows and darts (5, 11, 73, 106, 127, 135). Cattle are reported to become sterile after ingesting water tainted with this plant (76).

Stems and twigs of this species are employed in rural economies for cordage (106, 141). Fibers are obtained from the stems (23, 32) and are used in the manufacture of hats and other plaited objects (13, 79, 106) as well as serving as a rope substitute (125, 127, 128) for house construction, fence ties and thatch ties (23, 56, 128, 135, 142). The wood from the flexible stem is used in the manufacture of barrel staves (32, 106).

The Congo Indians are reported to use powdered flowers from this species mixed with salt in their diet (106). An edible aril is reported (13, 14, 23, 56, 64, 115). The roots are useful for oral hygiene and reportedly serve as a chewstick or brushstick (6, 23, 56, 69, 106). African children reportedly use the fruit gum for sticking papers together (56).

There are many reported medicinal applications, including treatment for hydropsy (106), hydrophobia (77, 106), melancholy (40, 77, 106), mental disease and insanity (6, 77, 99, 106, 141), stomach and intestinal pains (6, 106), gland tumors (106), lumbago and rheumatism (6, 23, 60), respiratory paralysis (76), forced respiration and rickets in children (6), leprosy (69), jaundice (5, 69), paralysis (62), bronchitis

(22), yellow fever (6), hypochondria (40, 99, 141), hysteria (22, 99), constipation (99), asthenia (6), neuralgia, ovary inflammation (22, 99), amaurosis (6, 23, 141), ophthalmia (6, 23, 60, 106, 141), urogenital afflictions (6, 106), and toothache (6, 23). Fractures, abcesses, wounds and bruises are reportedly attended to with various decoctions of the roots, leaves and seeds along with other plant admixtures (6, 23, 56, 141). Snakebites are administered with leaves from this plant (6, 62, 69). Afflictions of the liver and spleen are reportedly treated with leaf poultices (127) or other measures (22, 73, 99, 106, 141). Gonorrhea is treated with the roots or leaves usually placed in palm wine (6, 62, 127). This species has been reported for use as an anodyne liniment (127), emolient (106), emmenagogue (13), sedative (79, 84, 106), narcotic (22, 106, 133), slow and potent poison (6, 11, 22, 23, 56, 69, 71, 83, 84, 119), emetic (6, 76), purgative (6, 60), hemostatic (6, 23, 26, 56, 62, 94), febrifuge (6, 23, 58, 60), diuretic (22), abortifacient (94). Dysentery is treated usually with the leaves and roots (6, 23, 56, 58, 60, 69, 94). The prevention of abortion or miscarriage by root or leaf infusions has been reported (6, 23, 56, 60). Leaves and sap have been employed for treating coughs (6, 22, 56, 99) and colic (6, 22, 56). A topical application of the root is reportedly used for eczema (141). Leaves are reported to provide relief from painful menstruation (63). Twigs of this species are mixed with *Phyllanthus floribundus* (Euphorbiaceae), salt and palm oil as an internal remedy for the relief of postpartum pain (141). Headaches and migraines are reportedly treated with a root infusion (22, 99), which is sometimes snuffed through the nostrils (56).

This species is reported to be an aphrodisiac (69). Other reports indicate that certain peoples believe that a mixture of gunpowder and the dried and pulverized leaves, when rubbed on the body, will provide immunity from wounds (23); lumbago and kidney troubles are cured or prevented by tying this plant to the waist; and women are reported to add pounded roots to rice as a cure for sterility (6, 23, 56), the liquid being snuffed through the nostrils (6). This species is cited to serve as a medicinal for Bacaïri shamans used in healing rites as a diagnostic or divination for illness (2).

This species is used medicinally in the United States (118), and it is encountered in commercial pharmaceuticals in Brazil as fluid extracts and tinctures, commonly recommended by physicians (106).

CHEMICAL CONSTITUENTS: Alkaloids: timbonine (bark) (76, 127, 141). Saponins: cardiotonic steroid heterosides (94).

Paullinia pterophylla Tr. & Planch. GEOGRAPHIC DISTRIBUTION: Central America and northern South America. USES: Closely related to *Paullinia yoco* R. E. Schultes, it is possibly employed as a stimulating beverage (20, 102), similar to yoco, by Indians in Ecuador and Colombia (20). Indians on the upper Rio Caquetá (Colombia) reportedly used it medicinally as a febrifugal tonic (119).

Paullinia rhizantha Poepp. GEOGRAPHIC DISTRIBUTION: Peru, Colombia and northern Brazil. VERNACULAR NAMES: Brazil: cipó timbó (106). USES: Reported to be used by sertanejos (peasants) as a fish poison in Brazil (106).

Paullinia rhomboidea Radlk. GEOGRAPHIC DISTRIBUTION: Southeastern Brazil. VERNACULAR NAMES: Brazil: cipó timbó (106). USES: Reported to be used by sertanejos (peasants) as a fish poison in Brazil (106).

Paullinia rubiginosa Camb. GEOGRAPHIC DISTRIBUTION: Eastern and southeastern Brazil. VERNACULAR NAMES: Brazil: cipó cruapé vermelho, cruapé vermelho, timbó cabelludo (61). USES: Reported as a fish poison in Brazil (1, 61). CHEMICAL CONSTITUENTS: Saponins (1).

Paullinia scarlatina Radlk. GEOGRAPHIC DISTRIBUTION: Central America. VERNACULAR NAMES: Central America: barbasco, pate (142). Honduras: pate (128). USES: Reported as a fish poison in Belize, Guatemala and Honduras (128, 142). CHEMICAL CONSTITUENTS: Alkaloids: caffeine (in bark) (134).

Paullinia seminuda Radlk. GEOGRAPHIC DISTRIBUTION: Southern Brazil. VERNACULAR NAMES: Brazil: cururú, tingui (61). USES: Reported as a fish poison in Brazil (1). CHEMICAL CONSTITUENTS: Saponins (1).

Paullinia sphaerocarpa L. C. Rich ex Juss. GEOGRAPHIC DISTRIBUTION: Northern South America. USES: Reported as having a white, mealy-tasting aril, spider monkeys are cited as known dispersal agents (endozoochory) in the Guianas (115).

Paullinia spicata Benth. GEOGRAPHIC DISTRIBUTION: Ecuador, Colombia, Guianas and Brazil. VERNACULAR NAMES: Brazil: timbó, tingui (87, 113). French Guiana: taitetu lapi'a (86). USES: Reported as a fish poison in Brazil and French Guiana (1, 86, 87, 113). The white, mealy-tasting arils are reportedly edible, and spider monkeys are cited as known dispersal agents (endozoochory) in the Guianas (115). CHEMICAL CONSTITUENTS: Saponins (1).

Paullinia subnuda Radlk. GEOGRAPHIC DISTRIBUTION: Central America, Guianas and western Brazil. VERNACULAR NAMES: Brazil: cipó timbó (106). USES: Reported as being employed by sertanejos (peasants) as a fish poison in Brazil (106).

Paullinia subrotunda (Ruiz & Pavón) Pers. GEOGRAPHIC DISTRIBUTION: Peru. USES: Reported to have edible arils (64) and seeds (64, 80).

Paullinia tarapotensis Radlk. GEOGRAPHIC DISTRIBUTION: Peru. VERNACULAR NAMES: Peru: lúcumia, ycánchem (112). USES: Reported as having an edible mature fruit pulp (aril) (112).

Paullinia tetragona Aubl. GEOGRAPHIC DISTRIBUTION: Northern South America. USES: Reported to be used as an indigenous weaving material for the manufacture of household products, the stems are soaked in water and macerated into four parts to provide material for making baskets and coarse hats (44).

Paullinia thalictrifolia A. L. Juss. GEOGRAPHIC DISTRIBUTION: Central and southern Brazil. VERNACULAR NAMES: Brazil: camaihua cipó (106), camihua cipó (61), jacatupé (49), quanacai cipó (61). USES: Reported as a fish poison in Brazil (49, 95, 141) and grown in Rio de Janeiro and São Paulo as an ornamental plant (106). CHEMICAL CONSTITUENTS: Saponins (95, 141).

Paullinia tomentosa Jacq. GEOGRAPHIC DISTRIBUTION: Mexico and Central America. TAXONOMIC SYNONYMS: *Paullinia pteropoda* DC. VERNACULAR NAMES: Mexico: barbasco, barbasquillo (127), t'in kamab, t'in kamaab, tu' kamaab (4). USES: The Teenek (Huastec) Indians of Mexico are reported as having many medicinal uses for this species. It is used in dermatological medicine for falling hair and grey hair. For gastrointestinal medicinal applications, it is used alone or in mixes for dysentery and gastrointestinal pain. In gynecological medicine, the boiled roots serve as a contraceptive (this infusion is consumed early in the morning on an empty stomach once a day for three days at the end of a menstruation period, being repeated each month) and as a remedy for halting or reducing profuse menstrual bleeding (root infusion consumed three times a day for two days when period begins) (4).

Paullinia tricornis Radlk. GEOGRAPHIC DISTRIBUTION: Guianas. USES: Reported to have a white, mealy-tasting aril, spider monkeys are cited as known seed dispersers (endozoochory) in the Guianas (115).

Paullinia trigonia Vell. GEOGRAPHIC DISTRIBUTION: Eastern and southern Brazil. VERNACULAR NAMES: Brazil: timbó (61, 106), timbo aitica (61), timbó de cipó (47), tingui cipó (61, 106). USES: The stems, leaves and fruit are reported as a fish poison in Brazil (1, 96, 106), are cited as an insecticide and are toxic to cattle (106). An alcoholic extract of the stem is reportedly used to treat tumors in Brazil (47). CHEMICAL CONSTITUENTS: Saponins (1, 95).

Paullinia uloptera Radlk. GEOGRAPHIC DISTRIBUTION: Southeastern Brazil. VERNACULAR NAMES: Brazil: cipó raxa (106). USES: Reported as a useful plant of Brazil, but no specific details were provided as to its uses (106).

Paullinia xestophylla Radlk. GEOGRAPHIC DISTRIBUTION: Southeastern Brazil. VERNACULAR NAMES: Brazil: timbó (106). USES: Reported as a fish poison in Brazil (1, 106). CHEMICAL CONSTITUENTS: Alkaloids: timbonine (106). Saponins (1).

Paullinia yoco Schultes & Killip. GEOGRAPHIC DISTRIBUTION: Colombia and Ecuador. VERNACULAR NAMES: Colombia: bejuco amazonico (122), bejuco mulato, bejuco San Pedro, boyo mohoso (101), turuca yoco, yagé-yoco (39), yoco (39, 111, 135), yocó (39, 102), yocoó (102), yopo (30). Ecuador: yoko (139).

USES: Reports indicate the plant is used as a stimulating beverage (50, 63, 64, 97, 119, 135), the beverage being prepared from the bark (50, 63, 64, 135) and the expressed sap from twigs and branches (63, 111). This beverage has been reportedly used by some Ecuadorian (139) and Colombian (135) Indians. The Indians of the Colombian Putamayo use this beverage for a stimulating morning sustenance to stay hunger and fatigue (30), the beverage being prepared by scraping bark off stems into cold water (39). The Siona and Secoya Indians of Ecuador are reported to scrape the bark into cold water and squeeze the bark to make a bitter-tasting, caffeine-rich beverage; this beverage is usually consumed in the early morning by men before beginning the task of weaving hammocks (139). These Indians also consume the stimulating beverage in the morning during yahé (*Banisteriopsis*: Malphigiaceae) ceremonies, the observers to these curing rituals being maintained by the beverage (139).

The beverage is reported as a tonic (122), stimulant (20, 39, 74, 102), intestinal disinfectant (39), antimalarial (39, 119), antipyretic (20, 39, 119), purgative (39), antibilious (20, 119) and emetic (130). The bark infusion beverage is reportedly consumed for allaying hunger and fatigue (119, 130) and for providing agility and endurance for long expeditions (103).

CHEMICAL CONSTITUENTS: Alkaloids: caffeine (30, 63, 69, 103, 119, 134, 135).

Acknowledgments

The author gratefully acknowledges the useful comments of P. Acevedo Rodríguez, D. E. Williams, the late T. Plowman and H. D. Hammond in the preparation of this manuscript. This is publication number 133 in the series of the Institute of Economic Botany of the New York Botanical Garden.

Literature Cited

1. **Acevedo Rodríguez, P.** 1989. The systematic occurrence of fish poison plants. Adv. Econ. Bot. **8**: 1–23.
2. **Ackerknecht, E. H.** 1949. Medical practices. Pages 621–643 *in* J. H. Steward (ed.), Handbook of South American Indians. Vol. 5. The comparative ethnology of South American Indians. Smithsonian Institution Bureau of American Ethnology Bulletin 143. U.S. Government Printing Office, Washington, D.C.
3. **Alba, A. F.** 1936. Propiedades curatives de algunas plantas de Panama. The Star & Herald Co., Panama.
4. **Alcorn, J. B.** 1984. Huastec Mayan ethnobotany. University of Texas Press, Austin.
5. **Allen, P. H.** 1943. Poisonous and injurious plants of Panama. Suppl., Amer. J. Trop Med. **23(1)**: 1–30.
6. **Ayensu, E. S.** 1978. Medicinal plants of West Africa. Reference Publications, Inc., Algonac, Michigan.
7. **Balick, M. J.** 1986. Useful plants of Amazonia: A resource of global importance. Ch. **19**: 339–368 *in* G. T. Prance & T. E. Lovejoy (eds.), Key environments: Amazonia. Pergamon Press, Oxford, England.
8. **Bastien, J. W.** 1987. Healers of the Andes: Kallawaya herbalists and their medicinal plants. University of Utah Press, Salt Lake City.
9. **Bentley, R. & H. Trimen.** 1880. Medicinal plants: Being descriptions with original figures of the principal plants employed in medicine and an account of the characters, properties, and uses of their parts and products of medicinal value. Vol. 1. J. & A. Churchill, Oxford, England.
10. **Berg, M. E. van den.** 1982. Plantas medicinais na Amazônia: Contribução ao seu conhecimento sistemático. Conselho Nacional de Desenvolvimento Cientifico e Tecnológico, Programa Trópico Umido/MPEG, Belém.
11. **Blohm, H.** 1962. Poisonous plants of Venezuela. Wissenschaftliche Verlagsgesellschaft M.B.H., Stuttgart.
12. **Brady, G. S. & H. R. Clauser.** 1979. Materials handbook, 11th ed. McGraw-Hill Book Company, New York.
13. **Braga, R.** 1960. Plantas do Nordeste, especialmente do Ceará, 2a edição. Imprensa Oficial, Fortaleza, Ceará, Brasil.
14. **Busson, F.** 1965. Plantes alimentaires de l'ouest Africain, étude botanique, biologique et chimique. L'Imprimerie Lecointe, Marseille.
15. **Cavalcante, P. B.** 1976. Frutas comestíveis da Amazônia. Vol. I, 3a edição. INPA, Manaus, Brasil.
16. ———. 1979. Frutas comestíveis da Amazônia. Vol. III. Publições Avulsas do Museu Goeldi, Belém, Brasil.
17. **Cheney, R. H.** 1946. The biology and economics of the beverage industry. Econ. Bot. **1**: 243–275.
18. **Chopra, R. N., R. L. Badwahr & S. Ghosh.** 1949. Poisonous plants of India. Vol. 1. (Scientific Monograph No. 17, 1940, The Indian Council of Agricultural Research.) Manager of Publications, Delhi.
19. ———, **I. C. Chopra, K. L. Handa & L. D. Kapur.** 1958. Chopra's indigenous drugs of India, 2nd

ed. U. N. Dhur & Sons Private Limited, Calcutta.

20. **Cooper, J. M.** 1949. Stimulants and narcotics. Pages 525–558 *in* J. H. Steward (ed.), Handbook of South American Indians. Vol. 5. The comparative ethnology of South American Indians. Smithsonian Institution Bureau of American Ethnology Bulletin 143. U.S. Government Printing Office, Washington, D.C.

21. **Cordero, A. B.** 1978. Manual de medicina doméstica (Plantas medicinales Dominicanas). Taller, Santo Domingo, República Dominicana.

22. **Cruz, G. L.** 1982. Dicionário das plantas úteis do Brasil, 2a edição. Editora Civilização Brasilieira S.A., Rio de Janeiro.

23. **Dalziel, J. M.** 1937. The useful plants of West Tropical Africa. Crown Agents for the Colonies, London.

24. **Díaz, J. L.** 1976. Indice y sinonimia de las plantas medicinales de México. (Monografías Científicas 1.) Instituto Mexicano para el Estudio del las Plantas Medicinales (IMEPLAM), México.

25. **Dragendorff, G.** 1898. Die Heilpflanzen der verschiedenen Völker und Zeiten. Verlag von Ferdinand Enke, Stuttgart.

26. **Duchesne, E. A.** 1836. Répertoire de plantes utiles et de plantes vénéneuses du globe. Jules Renouard, Libraire-Editeur, Paris.

27. **Ducke, A.** 1937. Diversidade dos guaranás. Rodriguesia **10**: 155–156.

28. ———. 1946. Plantas de cultura precolombiana na Amazônia Brasileira. Notas sôbre as espécies ou formas espontâneas que supostamente lhes teriam dado origem. Bol. Tecn. No. 8. Inst. Agron. do Norte, Belém, Brasil.

29. **Duke, J. A.** 1985. CRC handbook of medicinal herbs. CRC Press, Inc., Boca Raton, Florida.

30. **Emboden, W.** 1979. Narcotic plants. Macmillan Publishing Co., Inc., New York.

31. **Erickson, H. T., M. P. F. Corrêa & J. R. Escobar.** 1984. Guaraná (*Paullinia cupana*) as a commercial crop in Brazilian Amazonia. Econ. Bot. **38**(3): 273–286.

32. **Farquhar, D. P. & B. J. Siegel.** 1944. A glossary of useful Amazonian flora, compiled from the Strategic Index of the Americas. Office of the Coordinator of Interamerican Affairs, Research Division, Washington, D.C.

33. **Fawcett, W.** 1891. Economic plants: An index to economic products of the vegetable kingdom in Jamaica. Government Printing Establishment, Kingston, Jamaica.

34. ——— **& A. B. Rendle.** 1926. Flora of Jamaica, containing descriptions of the flowering plants known from the island. Vol. V. Trustees of the British Museum, London.

35. **Fearnside, P. M.** 1985. Agriculture in Amazonia. Ch. **21**: 393–418 *in* G. T. Prance & T. E. Lovejoy (eds.), Key environments: Amazonia. Pergamon Press, Oxford, England.

36. ———. 1986. Human carrying capacity of the Brazilian rainforest. Columbia University Press, New York.

37. **Ferrándiz, V. L.** 1967. Guia de medicina vegetal. Imprenta J. Bilbeny, San Celoni.

38. **Figuero Marroquin, H.** 1955. Enfermedades de los Conquistadores. Ministerio de Cultura Departamento Editorial, San Salvador, El Salvador.

39. **García Barriga, H.** 1974. Flora medicinal de Colombia, botánica médica. Tomo Segundo. Instituto de Ciencias Naturales, Universidad Nacional, Bogotá, D.E., Colombia.

40. **Geiger, P. L.** 1840. Handbuch der Pharmacie. Zweiter Band. Zweite Auflage. Zweite Abtheilung. Pharmaceutische Botanik. Zweite Hälfte. C. F. Winter, Heidelberg.

41. **Gonzalez Torres, D. M.** 1980. Catálogo de plantas medicinales (y alimenticias y útiles) usadas en Paraguay. Editorial Comuneros, Asunción, Paraguay.

42. **Gottlieb, O. R.** 1985. The chemical uses and chemical geography of Amazon plants. Ch. **12**: 218–238 *in* G. T. Prance & T. E. Lovejoy (eds.), Key environments: Amazonia. Pergamon Press, Oxford, England.

43. **Grieve, M.** 1971. A modern herbal. Vol. 1. Dover Publications, Inc., New York.

44. **Grimé, W. E.** 1979. Ethno-botany of the Black Americans. Reference Publications, Inc., Algonac, Michigan.

45. **Grosourdy, D. R. de.** 1864. El médico botánico criollo. Tomo I. Librería de Francisco Brachet, Paris.

46. **Gutierrez V., G.** 1947. Estudio sobre los principales barbascos colombianos. Revist. Fac. Nac. Agron. **7**(25): 77–93.

47. **Hartwell, J. L.** 1982. Plants used against cancer: A survey. Quarterman Publications, Inc., Lawrence, Massachusetts.

48. **Hedrick, U. P. (ed.).** 1919. Sturtevant's notes on edible plants. State of New York—Department of Agriculture, 27th Annual Report, Vol. 2, Part II. J. B. Lyon Company, State Printers, Albany, New York.

49. **Heizer, R. F.** 1949. Fish poisons. Pages 277–281 *in* J. H. Steward (ed.), Handbook of South American Indians. Vol. 5. The comparative ethnology of South American Indians. Smithsonian Institution Bureau of American Ethnology Bulletin 143. U.S. Government Printing Office, Washington, D.C.

50. **Henman, A. R.** 1982. Guaraná (*Paullinia cupana* var. *sorbilis*): Ecological and social perspectives on an economic plant of the central Amazon basin. J. Ethnopharmacol. **6**: 311–338.

51. **Héraud, A.** 1875. Nouveau dictionnaire des plants médicinales. Librairie J.-B. Baillière et Fils, Paris.

52. **Herrarte, M. P.** 1933. Plantes employées comme "barbasco" au Guatémala. Rev. Bot. Appl. Agr. Trop. **13**: 351–353.

53. **Hill, A. F.** 1937. Economic botany: A textbook

of useful plants and plant products. McGraw-Hill Book Company, Inc., New York.

54. **Hoehne, F. C.** 1920. O que vendem os hervanarios da cidade de S. Paulo. Serviço Sanitaria do estado de São Paulo, N.S., N. 14. Casa Duprat, S. Paulo, Brasil.

55. ———. 1939. Plantas es substâncias vegetais tóxicas e medicinais. Graphicars, S. Paulo, Brasil.

56. **Irvine, F. R.** 1930. Plants of the Gold Coast. Oxford University Press, London.

57. **Jozami, J. M. & J. de Dios Muñoz.** 1982. Arboles y arbustos indígenas de la Prov. de Entre Ríos. IPNAYS (CONICET-UNL), Sante Fé, Argentina.

58. **Kerharo, J., F. Guichard & A. Bouquet.** 1959. Les végétaux ichtyotoxiques (poisons de pêche). 1re Partie. Introduction à l'étude de poison de pêche. Bull. Mem. Fac. Med. Pharm. Dakar. **8**: 313–329.

59. ———, ——— & ———. 1960. Les végétaux ichtyotoxiques (poisons de pêche). 2e Partie. Inventaire de poison de pêche. Bull. Mem. Fac. Med. Pharm. Dakar. **9**: 355–386.

60. ———, ——— & ———. 1961. Les végétaux ichtyotoxiques (poisons de pêche). III. 2e Partie. Inventaire de poison de pêche. Bull. Mem. Fac. Med. Pharm. Dakar. **10**: 223–242.

61. **Killip, E. P. & A. C. Smith.** 1935. Some American plants used as fish poisons. Bureau of Plant Industry, U.S. Department of Agriculture.

62. **Kokwaro, J. O.** 1976. Medicinal plants of East Africa. East African Literature Bureau, Kampala, Uganda.

63. **Krieg, M. B.** 1964. Green medicine, the search for plants that heal. . . . Rand McNally & Co., Chicago.

64. **Kunkel, G.** 1984. Plants for human consumption: An annotated checklist of the edible phanerogams and ferns. Koeltz Scientific Books, Koenigstein.

65. **LeCointe, P.** 1947. Arvores e plantas úteis (indigenas e aclimadas), 2a edição. Companhia Editor Nacional, São Paulo, Brasil.

66. **Leon, J.** 1968. Fundamentos botánicos de los cultivos tropicales. Instituto Interamericano de Ciencias Agrícolas de la O.E.A., San José, Costa Rica.

67. **Leung, A. Y.** 1980. Encyclopedia of common natural ingredients, used in food, drugs, and cosmetics. John Wiley & Sons, New York.

68. **Lévi-Strauss, C.** 1950. The use of wild plants in tropical South America. Pages 465–486 in J. H. Steward (ed.), Handbook of South American Indians. Vol. 6. Physical anthropology, linguistics and cultural geography of South American Indians. Smithsonian Institution Bureau of American Ethnology Bulletin 143. U.S. Government Printing Office, Washington, D.C.

69. **Lewis, W. H. & M. P. F. Lewis.** 1977. Medical botany: Plants affecting man's health. John Wiley & Sons, New York.

70. **Lindley, J.** 1838. Flora medica; a botanical account of all the more important plants used in medicine, in different parts of the world. Longman, Orme, Brown, Green, & Longmans, London.

71. ———. 1849. Medical and oeconomical botany. Bradbury & Evans, London.

72. ——— & T. Moore (eds.). 1870. The treasury of botany: A popular dictionary of the vegetable kingdom. Part II. Longmans, Green, & Co., London.

73. **Liogier, A. H.** 1974. Diccionario botánico de nombres vulgares de la Española. Imprensa UNPHY, Santo Domingo, República Dominicana.

74. **Lowie, R. H.** 1948. The tropical forests: An introduction. Pages 1–56 in J. H. Steward (ed.), Handbook of South American Indians. Vol. 3. The tropical forest tribes. Smithsonian Institution Bureau of American Ethnology Bulletin 143. U.S. Government Printing Office, Washington, D.C.

75. **Luerssen, C.** 1882. Medicinisch-Pharmaceutische Botanik zugleich als Handbuch der systematischen Botanik für Botaniker, Ärzte, und Apotheker. Band II. Verlag von H. Haessel. Leipzig.

76. **Marcano F., E. De Js.** 1977. Plantas venenosas en la República Dominicana. Publicaciones de la Asociación Médica Dominicana, Vol. 1, No. 1. Editorial Gremio, Santo Domingo, República Dominicana.

77. **Martius, C. F. P. de** 1843. Systema materiae medicae vegetabilis brasiliensis. Frid. Fleischer, Lipsiae.

78. **Mendieta, R. M. & S. del Amo R.** 1981. Plantas medicinales del estado de Yucatan. Instituto Nacional de Investigaciones sobre Recursos Bioticos (INIREB), Xalapa, Vera Cruz. Compañia Editorial Continental, S.A. de C.V., Mexico.

79. **Menezes, A. I. de.** 1949. Flóra da Bahia. Companhia Editora Nacional, São Paulo, Brasil.

80. **Menninger, E. A.** 1977. Edible nuts of the world. Horticultural Books, Inc., Stuart, Florida.

81. **Métraux, A.** 1948. The Tupinamba. Pages 95–111 in J. H. Steward (ed.), Handbook of South American Indians. Vol. 3. The tropical forest tribes. Smithsonian Institution Bureau of American Ethnology Bulletin 143. U.S. Government Printing Office, Washington, D.C.

82. ———. 1948. Tribes of eastern Bolivia and the Madeira headwaters. Pages 381–414 in J. H. Steward (ed.), Handbook of South American Indians. Vol. 3. The tropical forest tribes. Smithsonian Institution Bureau of American Ethnology Bulletin 143. U.S. Government Printing Office, Washington, D.C.

83. **Millspaugh, C. F.** 1974. American medicinal plants. Dover Publications, Inc., New York.

84. **Montesano, A.** 1913. Plantas medicinales (extranjeras e indígenas). Imprenta Suiza de Imsand y Cia., Buenos Aires, Argentina.

85. **Moreira Filho, H.** 1972. Plantas medicinais I. Imprensa da Universidade Federal do Paraná, Curitiba, Brasil.

86. **Moretti, C. & P. Grenand.** 1982. Les nivrées ou plants ichtyotoxiques de la Guyane Française. J. Ethnopharmacol. **6:** 139–160.

87. **Mors, W. B. & C. T. Rizzini.** 1966. Useful plants of Brazil. Holden-Day, Inc., San Francisco, California.

88. **Mosig, A.** 1961. Kurze Systematik der Arzneipflanzen. Fünfte Auflage. Verlag von Theodor Steinkopf, Dresden.

89. **Nair, P. K. R.** 1980. Agroforestry species: A crop sheets manual. International Council for Research in Agroforestry (ICRAF), Nairobi, Kenya.

90. **Nelson, A.** 1951. Medical botany. E. & S. Livingstone, Ltd., Edinburgh.

91. **New York World's Fair. 1939.** 1939. Brazil: Medicinal plants and herbs. Official Publication. Carioca de Artes Graphicas, Rio de Janeiro, Brasil.

92. **Nimuendajú, C.** 1948. The Maué and Arapium. Pages 245–254 *in* J. H. Steward (ed.), Handbook of South American Indians. Vol. 3. The tropical forest tribes. Smithsonian Institution Bureau of American Ethnology Bulletin 143. U.S. Government Printing Office, Washington, D.C.

93. **Oblitas Poblete, E.** 1969. Plantas medicinales de Bolivia. Editorial "Los Amigos del Libro," Chochabamba, Bolivia.

94. **Oliver-Bever, B.** 1986. Medicinal plants in tropical West Africa. Cambridge University Press, Cambridge, England.

95. **Pammel, L. H.** 1911. A manual of poisonous plants. Part. II. The Torch Press, Cedar Rapids, Iowa.

96. **Parker, J.** 1964. Mil plantas medicinales. Editorial Cayari, Buenos Aires, Argentina.

97. **Patiño, V. M.** 1967. Plantas cultivadas y animales domésticos en América Equinoccial. Tomo III. Fibras, medicinas, miscelaneas, la edicíon. Imprenta Departmental, Cali, Colombia.

98. **Passarinho, J. G.** 1971. Amazônia: O deadio dos trópicos. Primor, Rio de Janeiro, Brasil.

99. **Penna, M.** 1946. Dicionario brasilero de plantas medicinais, 3a edição. Livraria Kosmos Editora Erich Eichner & Cia., Ltda., Rio de Janeiro, Brasil.

100. **Perez Arbelaez, E.** 1937. Plantas medicinales de Colombia. Editorial Cromos, Bogotá, Colombia.

101. ———. 1978. Plantas útiles de Colombia, 4a edición. Litografia Arco, Bogotá, Colombia.

102. **Perez de Barradas, J.** 1957. Plantas mágicas Americanas. Consejo Superior de Investigaciones Científicas, Instituto "Bernardino de Sahagun," Madrid.

103. **Perrot, E.** 1943–1944. Matières premières usuelles du rène végétal: thérapeutique, hygiéne, industrie. Tome Second. Masson et Cie.,

Editeurs, Libraires de l'Académie de Médicine, Paris.

104. **Pinto, J. de A.** 1873. Diccionario de botânica brasileira ou compendio dos vegetaes do Brasil, tanto indigenas como aclimados. Tipografia PERSEVERANÇA, Rio de Janeiro, Brasil.

105. **Pires, J. M.** 1978. Plantas ichthyotoxicas: aspecto do botânica sistemática. Pages 37–41 *in* Ciencia e Cultura Supplemento. V Simposio de Plantas Medicinais do Brasil, São Paulo, 4 a 6 de Setembro. [Some specific names lacking in the article were later provided by J. M. Pires, pers. comm.]

106. **Pio Corrêa, M.** 1926–1978. Diccionario das plantas úteis do Brasil e das exoticas cultivadas. 6 Vols. Imprensa Nacional e Ministério da Agricultura, Rio de Janeiro.

107. **Pittier, H.** 1978. Plantas usuales de Costa Rica. Editorial Costa Rica, San José, Costa Rica.

108. **Prance, G. T.** 1985. The increased importance of ethnobotany and underexploited plants in a changing Amazon. Pages 129–136 *in* J. Hemming (ed.), Change in the Amazon Basin. Vol. I. Man's impact on the forests and rivers. Manchester University Press, Manchester, England.

109. **Quiros Calvo, M.** 1945. Botánica aplicada a la farmacia. Universidad de Costa Rica y la Secretaría de Educación Pública, San José, Costa Rica.

110. **Radlkofer, L. T.** 1933. Sapindaceae. Vols. 1 & 2. *In* A. Engler (ed.), Das Pflanzenreich. IV 165. Verlag von W. Englermann, Leipzig.

111. **Reis Altschul, S. von.** 1973. Drugs and foods from little-known plants: Notes in Harvard University herbaria. Harvard University Press, Cambridge, Massachusetts.

112. **Reis, S. von & F. J. Lipp, Jr.** 1982. New plant sources for drugs and foods from the New York Botanical Garden Herbarium. Harvard University Press, Cambridge, Massachusetts.

113. **Rizzini, C. T. & W. B. Mors.** 1976. Botânica econômica brasileira. Editora da Universidade de São Paulo (EDUSP), São Paulo, Brasil.

114. **Romero Castañeda, R.** 1969. Frutas silvestres de Colombia. Vol. II. Universidad Nacional de Colombia, Instituto de Ciencias Naturales, Bogotá, D.E.

115. **Roosmalen, M. G. M. van.** 1985. Fruits of the Guianan flora. Institute of Systematic Botany, Utrecht University, Netherlands.

116. **Rose, J.** 1972. Herbs and things: Jeanne Rose's herbal. Workman Publishing Company, New York.

117. **Roys, R. L.** 1972. The ethno-botany of the Maya. Publication No. 2, Middle American Research Series. Department of Middle American Research, Tulane University, New Orleans, Louisiana.

118. **Schneider, A.** 1912. Pharmacal plants and their culture. Bulletin No. 2. California State Board of Forestry, Sacramento, California.

119. **Schultes, R. E.** 1942. Plantae colombianae II.

Yoco: A stimulant of southern Colombia. Bot. Mus. Leafl. Harvard Univ. **10(10):** 301–324.

120. ———. 1944. Plantae colombianae, VII: Novae notiones generis *Paullinia.* Caldisia **10:** 419–423.

121. ———. 1979. The Amazon as a source of new economic plants. Econ. Bot. **33(3):** 259–266.

122. **Secretaria Ejecutiva Permanete del Convenio "Andrés Bello" (SECAB).** 1983. Especies vegetales promisorias de los países del convenio "Andrés Bello." SECAB y COLCIENCIAS, Bogotá, D.E., Colombia.

123. **Sievers, A. F. & E. C. Higbee.** 1948. Plantas medicinales de regiones tropicales y subtropicales. Publicacíon Agrícola, Nos. 154–158. Union Panamericana, Washington, D.C.

124. **Simmonds, P. L.** 1889. Tropical agriculture: A treatise on the culture, preparation, commerce and consumption of principal products of the vegetable kingdom. E. & F. N. Spon, New York.

125. **Standley, P. C.** 1928. Flora of the Panama Canal Zone. Contributions from the United States National Herbarium, Vol. 27.

126. ———. 1930. Flora of Yucatan. Publication 279, Botanical Series, Vol. 3, No. 3, Field Museum of Natural History, Chicago, Illinois.

127. ———. 1961. Trees and shrubs of Mexico. Contributions from the United States Herbarium, Vol. 23.

128. ——— & J. A. Steyermark. 1949. Flora of Guatemala. Fieldiana: Botany, Vol. 24, Part VI.

129. **Steinmetz, E. F.** 1957. Codex vegetabilis, 2nd ed. E. F. Steinmetz, Amsterdam.

130. **Steward, J. H.** 1948. Tribes of the montana: An introduction. Pages 507–531 *in* J. H. Steward (ed.), Handbook of South American Indians. Vol. 3. The tropical forest tribes. Smithsonian Institution Bureau of American Ethnology Bulletin 143. U.S. Government Printing Office, Washington, D.C.

131. **Taylor, N.** 1944. Guarana (*Paullinia cupana*): Preliminary report. Cinchona Products Institute, Inc. [3-page manuscript in New York Botanical Garden Library.]

132. **Teixeira da Fonseca, E.** 1922. Indicador de madeiras e plantas úteis do Brasil. Officinas Graphicas Villas-Boas & Co., Rio de Janeiro.

133. **Toursarkissian, M.** 1980. Plantas medicinales de la Argentina: sus nombres botánicos, vulgares, usos y distribución geográfica. Editorial Hemisferio Sur S.A., Buenos Aires, Argentina.

134. **United States Department of Agriculture.** N.D. Alkaloid bearing plants and their contained alkaloids. Agricultural Research Service Technical Bulletin, No. 1234.

135. **Uphof, J. C. T.** 1968. Dictionary of economic plants, 2nd ed. Verlag von J. Cramer, New York.

136. **Vander, A.** 1967. Plantas medicinales: Las enfermedades y su tratamiento por las plantas. Adrian Van der Put, Barcelona, Spain.

137. **Vasconcelos, A., J. C. Nascimento & A. Lemos Maia.** 1976. A cultura do guaraná. Pages 61–71 *in* Simposio internacional sobre plantas de interés económico de la flora Amazónica. Belém, Brasil, May 29–Junio 2. IICA, Unidad de Documentación, Turrialba, Costa Rica.

138. **Velez Salas, F.** 1982. Plantas medicinales de Venezuela. INAGRO, Caracas, Venezuela.

139. **Vickers, W. T. & T. Plowman.** 1984. Useful plants of the Siona and Secoya Indians of eastern Ecuador. Fieldiana: Bot. New Series No. **15:** 1–63.

140. **Wagley, C. & E. Galvão.** 1948. The Tapirapé. Pages 167–170 *in* J. H. Steward (ed.), Handbook of South American Indians. Vol. 3. The tropical forest tribes. Smithsonian Institution Bureau of American Ethnology Bulletin 143. U.S. Government Printing Office, Washington, D.C.

141. **Watt, J. M. & M. G. Breyer-Brandwijk.** 1962. The medicinal and poisonous plants of southern and eastern Africa. E. & S. Livingstone Ltd., Edinburgh.

142. **Williams, L. O.** 1981. The useful plants of Central America. Ceiba **24(1 & 2):** 1–297.

143. **Willis, J. C.** 1966. A dictionary of the flowering plants and ferns, 7th ed. Cambridge University Press, Cambridge, England.

144. **Ypiranga Monteiro, M.** 1965. Antropogeografia do guaraná. Cadernos da Amazônia 6. Instituto Nacional de Pesquisas da Amazônia (INPA), Manaus, Brasil.

Useful Plants of the Panare Indians of the Venezuelan Guayana

Brian M. Boom

Table of Contents

Abstract

BOOM, B. M. (The New York Botanical Garden, Bronx, New York 10458, U.S.A.). Useful plants of the Panare Indians of the Venezuelan Guayana. Adv. Econ. Bot. **8**: 57–76. 1990. Results are presented for an ethnobotanical study of the Panare Indians, an indigenous Carib-speaking group of the Guayana Highland region in southern Venezuela. A total of 376 species of angiosperms are accounted for, based on nearly five months of intensive collecting and interviewing in the Panare village of Corozal. Of this total, 112 species (30%) are recorded as being utilized by the Panare. In order to quantify forest utilization per unit area, a one-hectare ethnobotanical tree inventory was conducted in each of the Panare's two principal resource areas: mountain forest at 400–500 m elevation, and tall woody savanna at 90–100 m elevation. All trees with a dbh of 10 cm and greater were marked and vouchers were collected. In the mountain forest plot, 72 species were found of which the Panare indicated uses for 36 (50.0%). In the tall woody savanna inventory, 44 species were recorded of which 23 (52.3%) were regarded as useful.

Key words: Panare; Guayana Highland; Venezuela; mountain forest; woody savanna; ethnobotanical tree inventory.

Resumen

Se presentan los resultados de un estudio etnobotánico de los indígenas Panare, de la Guayana Venezolana. En este estudio, que comprendió cinco meses de intensa colección y entrevistas en la aldea de Corozal, se registró un total de 376 especies de angiospermas de las cuales 112 especies son utilizadas (30%). Para cuantificar la utilización del bosque por área de unidad, se hizo un inventario etnobotánico de los árboles en un tramo de una hectárea en cada una de las dos áreas principales de recursos naturales: el bosque del montaña (400–500 m altura) y el bosque claro en la sabana (90–100 m altura). Se anotaron y recolectaron los árboles que alcanzaron por lo menos 10 cm de diámetro. En el tramo de la montaña, se encontraron 72 especies de las cuales 36 son utilizadas (50%). En el tramo del bosque claro en la sabana, se encontraron 44 especies de las cuales 23 son utilizadas (52.3%).

Introduction

The Guayana Highland—the "Lost World" of the Conan Doyle novels—is a phytogeographic province of some one million km located just to the north of the Amazon Basin in South America. The region, which is characterized by several hundred sandstone table mountains overlying a layer of Precambrian igneous rock, exhibits a diversity of habitats and a flora that is legendary for its species richness and extraordinary endemism. Partly because of its biotic richness and its relative inaccessibility, the Guayana Highland remains today one of the least-known floristic provinces in the world.

As with the flora, the indigenous peoples of the Guayana Highland have not, with very few exceptions, received the attention from ethnobotanists that they deserve. At present, a certain urgency for such studies exists due to the increasing threats the region is facing from forestry, mining, and agricultural concerns. In southern Venezuela, the heart of the Guayana, a road linking the city of Puerto Ayacucho (capital of the Territorio Federal Amazonas) with the northern part of the country was blacktopped in 1986. This road passes along the western flank of the territory of the Panare, a Carib-speaking indigenous group of some 2000 individuals whose ethnobotany comprises the subject of this chapter. One can expect that this road will undermine the cultural integrity of the Panare in the coming years. The precarious situation of the Panare, and the prospects for the loss of much of the Panare's knowledge about plants in the near future, provided the impetus for the present study.

Perhaps the best overview of regional ethnobotany in southern Venezuela is provided by Colchester and Lister (unpubl.) who report on the ethnobotany of the Orinoco-Ventuari region, located slightly to the south of Panare territory. These authors conducted preliminary investigations among the Piaroa, Maco, Yabarana, Guahibo, Hoti, Makiritare, Ye'Kuana, and Kuripaku. Although the fieldwork for this project was carried out between October 1975 and August 1976, their report, unfortunately, has not yet been published.

The Panare themselves have not been the subjects of an ethnobotanical study, but two anthropologists recently have made some contributions in this area. Dumont (1976, 1978) carried out ethnological investigations with the Panare at Turiba. Henley (1982) studied the large Panare settlement in the valley of the Caño Colorado. In both cases, a number of plants are mentioned, but, regrettably, voucher specimens were either not collected or not saved. Therefore, a well-documented ethnobotanical study of the Panare was still lacking. The Dumont and Henley works, however, provide an excellent background for putting any such ethnobotanical study in its proper cultural context.

In this regard, chapters 2 ("Ecology and subsistence") and 3 ("The economic system") in Henley (1982) are the most relevant. The traditional territory of the Panare is the forested mountains of the upper Río Cuchivero in Bolívar State. Over the past several decades, however, there has been an expansion to the north and west (i.e., towards the Orinoco River savannas). Presently, nearly all Panare communities, of which Henley (1982: 20) estimates there are 38, are situated on the savanna, but near the base of the mountains. Apparently, the Panare have moved as they have in order to increase trading possibilities with *criollos* (a term for Venezuelan non-Indians in general). Interestingly, trading provides about as much interaction with *criollos* as the Panare desire. For the most part, the Panare try to maintain their traditional subsistence and economic system. This is done despite intensive Protestant missionary activity in the region. The Panare's success in taking only what they wanted from *criollo* culture, and resisting what they did not want, has led Henley to describe the Panare as unique among lowland South American Indians in the extent to which they have resisted acculturation (despite the first Hispanic settlers having arrived in the middle Orinoco region in the 1600's and contact of some sort has existed since then).

The Panare derive most of their subsistence from the forest. Although the savanna offers fishing, and, to a minor extent, hunting and gathering opportunities, it is in the forest that the Panare practice swidden agriculture, do most of their hunting, and gather a wide range of plants for food, construction, technology, medicine, and commerce. To quote Henley (1982: 32), ". . . the ecological adaptation of the Panare is principally silvine, secondarily riverine and only marginally savanna-oriented. Thus when the Panare settle on the plains, it is the pockets of forest and the rivers that they exploit to meet their subsistence needs rather than the savanna itself."

Henley terms the Panare's attitude towards the exploitation of nature as "prodigal." While their traditional semi-nomadic settlement pattern resulted in a conservation of natural resources, it was conservation by default. Henley (1982: 50) states that "the principle that appears to underlie their appropriation of nature is that of exploiting a given resource site until a significant decline in the return given for the labour invested makes it less arduous to start anew elsewhere."

The Panare economic system must be understood in light of their attitude toward natural resources and their subsistence pattern. Henley (1982: 84) summarizes the relationship: ". . . the Panare economic system as a whole is antithetical to the planned use of resources. In the sphere of production, each hearth group works as a more or less independent entity. In the sphere of distribution and consumption, the food-sharing ideology militates against the careful husbanding of the food supply. There is nothing integral to the structure of the economic organization that serves to regulate and coordinate the productive efforts of independent hearth groups. Thus the system of economic relations also provides no social mechanism for regulating the way in which the settlement group as a whole appropriates the natural environment. Nor . . . does the kinship system or the political system, such as it is, provide any such mechanism. This lack of planning

and coordination in the use of labour resources and the Panare's 'prodigal' attitude towards the use of natural resources go hand in hand, complementing and reinforcing one another."

It is of great importance in this context that we view the Panare's present tendency of living in greater densities on the plains where natural resources are less abundant than in the mountains. "The collective system of distribution and consumption, which holds the settlement group together as an economic entity, depends on the willingness of individual hearth groups to surrender the greater part of the food that they have produced independently to the common pool. This they are willing to do under normal circumstances since there are certain obvious advantages to be gained from participation in the system of collective consumption and, moreover, such participation is strongly reinforced ideologically. Yet, under conditions of food scarcity, there is a tendency for this collective system to break down, as each hearth group reverts to feeding itself. As a result, the settlement group becomes nothing more than a conglomeration of economically independent entities held together only by kinship ties and the physical constraints of the communal house" (Henley, 1982: 85). The situation is even more serious in those communities where cash-cropping is practiced.

While the above discussion explains the interaction of ecology, subsistence and economics, it does not account for the Panare's selective resistance to acculturation. This must be due only in part to inherent characteristics of their culture. Of greater importance is that, at least until recently, the Panare and *criollos* did not compete for the same resources. The *criollos* raised cattle on the plains, but not in the mountains. As the Panare moved to the plains to live, land disputes arose. Furthermore, diamond and bauxite mining is on the increase in the region and is starting to have an impact on the Panare. These developments, along with the road paving I mentioned above, no doubt mean that the Panare's ability to resist acculturation will be put to a serious test in upcoming years. The reader is referred to Henley (1982) for an excellent discussion on the current status and prospects for the Panare; nearly one-half of his book deals with the issues of Panare-*criollo* relations.

The specific goal of the present project was to quantify the importance of the forest to the Panare. The idea of quantifying plant use can be traced back as far as 1920. In that year, Kroeber (1920) published a major critique of ethnobotany and stated, among other things, that quantitative information about plants that are regarded as important by informants would be invaluable for understanding the role of plants in the culture being studied. No one seems to have really attempted this, however, until Carneiro (1978) described the knowledge and use of trees by the Kuikuru Indians of central Brazil. The flaw in Carneiro's work was that he did not make plant collections, so we can never really know how many species or the identities of trees that the Kuikuru used. Furthermore, the area of the forest that he sampled was very small—only one-sixth of an acre. Campbell et al. (1986) have shown that at least one hectare of *terra firme* forest in Amazonia needs to be sampled in order to determine local tree diversity adequately.

Setting and Methods

Although I visited the large Panare community in the Colorado valley (ca. 230 people, vide Henley 1982: 28) several times, the majority of the fieldwork was conducted in the much smaller village of Corozal (ca. 30 people) located along the Puerto Ayacucho/Caicara road some 6 km north of the Venezuelan National Guard post at Maniapure (6°55′N, 66°30′W). Figures 1–3 show some Panare of the Colorado community. Although there would have been more potential informants at the Colorado community than at Corozal, the situation in the former community was too tense due to the presence of a New Tribes Mission station at Colorado; the missionaries did not encourage my research plans in their area. Therefore, it was decided to work in a smaller village without the presence of missionaries (although there was a regularly attended church at Corozal). Figures 4–5 show two residents of Corozal in the process of making baskets; one of these, Mendoza, was the principal informant in the present study, while the younger man in Figure 4 is one of Mendoza's sons.

The terrain around Corozal ranges from the Orinoco floodplains in the west at about 90 m elevation to the mountains in the east at about 500 m elevation. Granitic outcrops are abundant both on the savanna and up in the mountains. The average annual rainfall in Panare territory is about 2400 mm (Henley, 1982: 32). There is a dry season from about December to May.

Two methods were employed to study the ethnobotany of the Panare. One of these was simply the traditional approach to the subject: interviews were conducted with informants to elicit names of useful plants and observations on plant uses were made around the village that led to the collection of the species involved.

The second method was the reverse of the first approach; it involved, first, the collection of plant specimens that were later shown to informants to obtain indigenous names and uses. This is the approach that has come to be called "quantitative ethnobotany," and it formed the basis for a recent comparison of four Amazonian indigenous groups, including the Panare (Prance et al., 1987).

Quantitative ethnobotany is the method by which the question "How important is the forest to the Panare?" could be at least partially answered. I conducted two inventories, one in each of the two major resource areas of the Panare—mountain forest and tall woody savanna. (The plants of the tree and shrub savanna are scarcely used by the Panare, as is evident from the list of plants presented herein.) The physiognomic savanna terminology of Sarmiento (1984) is adopted here. A tall woody savanna is characterized by having trees over 8 m tall that comprise more than 30% of total plant cover. A tree and shrub savanna is marked by having a low (less than 8 m tall) woody cover that comprises no more than 2%. Patches of tall woody savanna, of up to several hectares in size, occur sporadically in the tree and shrub savanna that dominates this portion of the Orinoco lowlands.

During August–November 1985, I conducted the tall woody savanna inventory on a plot about 1 km from Corozal at an elevation of 90–100 m. The second inventory, that of mountain forest, was conducted during April–May 1986 about 1 km to the east of the first plot, but at an elevation of 400–500 m. In each case an area of vegetation 20 m wide and 500 m long was delimited. All trees at least 10 cm dbh were marked and specimens were collected. These specimens were shown to Panare informants in order to determine indigenous names and uses for the plants. The interviews were tape-recorded and the tapes were kindly transcribed by linguist Marie-Claude Müller. In addition to the specific tree invento-

FIG. 1. Panare boys on savanna in Colorado valley, the principal tribal village. (Photograph by B. Boom.)
FIG. 2. Panare fishing with spears in the Caño Colorado. (Photograph by M. de Cuevas.)

5

FIG. 5. Mendoza, headman of village of Corozal, weaving a burden basket from leaves of *Maximiliana maripa* (Arecaceae). The tumpline (unattached, by his left foot) is from the fibrous inner bark of *Lecythis corrugata* subsp. *rosea* (Lecythidaceae). (Photograph by B. Boom.)

←

FIG. 3. Panare from Colorado valley relaxing, reading Bible stories, and making decorative baskets for sale to tourists. (Photograph by M. de Cuevas.)

FIG. 4. Young man from Corozal village making decorative baskets. Basket material is from *Ischnosiphon arouma* (Marantaceae). Red lacquer (lighter colored in photo) comes from *Byrsonima crassifolia* (Malpighiaceae). Black lacquer comes from *Casearia sylvestris* (Flacourtiaceae). (Photograph by B. Boom.)

ries mentioned, general collections were made all around the area of all plants, native and cultivated. These results are presented in the list of useful plants that follows. Because it is of interest to know of plants that are present, but not being used, all plants collected in the project—useful or not—are included in the list.

The first set of specimens from the August–November 1985 work is deposited at MYF; the first set from the April–May 1986 work is at PORT. The second set in all cases is at VEN. The third set is at NY, the fourth set was sent to relevant taxonomic specialists as gifts for determination, and the fifth set is at MO. Additional duplicates have been sent to various botanical institutions having research interests in the Guayana flora.

Results

In total, 376 species of plants were collected. Of these, 112 (30%) were indicated as being useful by Panare informants. Of the inventories, it was seen that a higher percentage of the species were useful than was the case in the overall general collections. This is because the general collections included many species of graminoids and other herbs of the savanna that are hardly given folk specific names by the Panare, much less regarded as useful. In the mountain forest inventory, 72 species of trees were collected; of these 36 (50.0%) were used. In the tall woody savanna inventory, 44 tree species were collected; of these 23 (52.3%) were used. The methods in which these species were employed, along with indigenous name(s) and voucher numbers (all on Boom's series) are indicated below. A general, semi-popular overview of these data was presented in Boom (1987). The ethnobotanical results of all collecting activity are presented in Table I. The ecological results of the inventories will be presented elsewhere (Boom, in prep.).

Seventeen species, out of the total of 376 collected, are cultivated in dooryard gardens or agricultural fields. With the exception of cotton (*Gossypium barbadense*, Malvaceae), which is grown for fiber, and achiote (*Bixa orellana*, Bixaceae), which is grown for the red dye extracted from the seeds, all other cultigens are used as food. Three of these are trees that are grown in dooryard gardens: cashew (*Anacardium occidentale*, Anacardiaceae), mango (*Mangifera indica*,

Anacardiaceae), and papaya (*Carica papaya*, Caricaceae). In the agricultural fields, usually located several kilometers from the village, most of the food crops are grown. There are six species of tuber crops; of these, manioc (*Manihot esculenta*, Euphorbiaceae) is the most important; I only encountered the sweet variety at Corozal. The other tuber crops are sweet potato (*Ipomoea batatas*, Convolvulaceae), yams (*Dioscorea alata* and *D. trifida*, Dioscoreaceae), and *yautia* (*Xanthosoma sagittifolium* and *X. violaceum*, Araceae). The only grain cultivated is maize (*Zea mays*, Poaceae) and it is nearly as important to the Panare as is manioc. Sweet bananas (*Musa ×paradisiaca*, Musaceae) are commonly cultivated. Two species of Cucurbitaceae are grown: melon (*Cucumis melo*) and pumpkin (*Cucurbita moschata*), but neither seems especially important. Other minor food crops include peanut (*Arachis hypogaea*, Fabaceae) and chili pepper (*Capsicum frutescens*, Solanaceae).

This accounting of Panare ethnobotany should be regarded as a preliminary report on the subject. There remains much research to be done even at Corozal, and Henley (1982) lists at least 38 communities throughout Panare territory. Much more effort needs to be made, especially as regards medicinal plants. In the Panare culture, detailed knowledge of medicinals is restricted to shamans. There are very few shamans still living, however, and I never had the opportunity to meet one. This accounts for the rather sparse listing of medicinal plants in Table I. Often, the Panare did not give a specific name to a specimen shown them in an interview. For example, the term *kë' ñëtë* (=liana) was applied to any number of species in various families; such generic or plant form terms are not given in Table I. In other cases the "name" reported is really a descriptive phrase. For example, *o'mohchahcho'*, which means "the plant of a special ant named *o'mohcha*," is used a number of times. A more analytical approach to these data, emphasizing the linguistic aspects, is planned in collaboration with Marie-Claude Müller.

Acknowledgments

I would like to acknowledge the generous funding by the Charles A. Lindbergh Fund that made this project possible. In Venezuela numerous people have aided the effort, including the staffs at MYF, PORT, and VEN, the various governmental officials who issued permits for the project, and friends who helped with moral support and free lodging during stays in Caracas. The identification of a number of specimens was facilitated by specialists at NY (S. Mori, T. Koyama, R. Barneby, D. Johnson, W. Thomas, M. Nee, D. Daly, J. Mitchell, J. Pruski, A. Henderson, P. Acevedo, C. Vincent); at VDB (R. Kral); at MO (G. Davidse, T. Croat, A. Gentry, J. Steyermark, B. Holst, H. van der Werff, E. Zardini); at U (P. Maas); at VEN (G. Carnevalli); at A (R. Howard); at USF (B. Hansen); at WAG (A. Leeuwenberg); at BG (C. Berg); at K (T. Pennington, R. Harley, G. Prance, P. Taylor); at LEA (J. Kuijt); at MICH (W. Anderson); at FAU (D. Austin); at UBC (H. Kennedy); at GB (L. Andersson); at TAES (P. Fryxell); at F (T. Plowman, M. Huft); and at US (J. Wurdack, R. Faden, H. Robinson, L. Smith, D. Wasshausen). I would particularly like to thank Margot Grillo for her field assistance with the tall woody savanna inventory and Sondra Wentzel for her field assistance with the mountain forest inventory. Marie-Claude Müller provided transcriptions of Panare plant names along with general encouragement for the project. This paper is dedicated to the Panare Indians of Corozal in the hope that it will contribute in some small way toward their future well-being. This is publication number 124 in the series of the Institute of Economic Botany of the New York Botanical Garden.

Literature Cited

Boom, B. M. 1987. The Panare Indians and their forest: Survival of a Venezuelan culture. Jour. Wash. Acad. Sci. **77(4):** 178–182.

———. In prep. Composition and structure of tall woody savanna and adjacent moist tropical forest in the Venezuelan Guayana.

Campbell, D. G., D. C. Daly, G. T. Prance & U. N. Maciel. 1986. Quantitative ecological inventory of *terra firme* and *várzea* forest on the Rio Xingu, Brazilian Amazon. Brittonia 38(4): 369–393.

Carneiro, R. L. 1978. The knowledge and use of rain forest trees by the Kuikuru Indians of central Brazil. Pages 201–216 *in* R. I. Ford (ed.), The nature and status of ethnobotany. Anthropological Paper No. 67, Museum of Anthropology, University of Michigan, Ann Arbor.

Colchester, M. & J. R. Lister. Unpubl. The ethnobotany of the Orinoco-Ventuari region. London

Dumont, J.-P. 1976. Under the rainbow: Nature and supernature among the Panare Indians. University of Texas Press, Austin.

———. 1978. The headman and I: Ambiguity and ambivalence in the fieldworking experience. University of Texas Press, Austin.

Henley, P. 1982. The Panare: Tradition and change on the Amazonian frontier. Yale University Press, New Haven.

Kroeber, A. L. 1920. Review of uses of plants by the Indians of the Missouri River region by Melvin Randolph Gilmore. Amer. Anthropol. 22: 384–385.

Prance, G. T., W. Balée, B. M. Boom & R. L. Carneiro. 1987. Quantitative ethnobotany and the case for conservation in Amazonia. Conservation Biology **1(4):** 296–310.

Sarmiento, G. 1984. The ecology of neotropical savannas. Harvard University Press, Cambridge.

Table I

List of plant species collected at Corozal. The tree species at least 10 cm dbh that were encountered in the one-hectare inventory plots are indicated by asterisks; species with one asterisk (*) occur in the tall woody savanna plot and those with two asterisks (**) occur in the mountain forest plot. Species with no asterisk either occur outside the aforementioned plots or else are not trees of at least 10 cm dbh. The Panare plant names and uses, listed last in each entry, if known, were transcribed by Marie-Claude Müller from tape recordings made by the author. Voucher numbers are on Boom's series. All specimens are deposited at NY and VEN; additionally, some duplicates are located at MYF and PORT.

ACANTHACEAE

Aphelandra deppeana Schl. & Cham. Shrub of tall woody savanna, 90 m elev. No name given (*6079*).

Chaetochlamys wurdackii Leonard. Weak shrub in agricultural field on tree and shrub savanna, 90 m elev. No name given (*6318*).

Justicia sp. Weak shrub in margin of agricultural field on tree and shrub savanna, 90 m elev. No name given (*6315*).

Ruellia pterocaulon Leonard. Herb of mountain slopes, 200 m elev. No name given (*6468*).

ANACARDIACEAE

Anacardium occidentale L. Tree cultivated in village, 90 m elev. *mere* (no voucher): edible fruit.

Astronium lecointei Ducke*.**. Tree of mountain forest, 400 m elev. and tall woody savanna, 90 m elev. *marana kai' mën* (*6229*).

Astronium ulei Mattick**. Tree of mountain forest, 400 m elev. *kiña* (*6637*): edible fruit.

Mangifera indica L. Tree cultivated in village, 90 m elev. *mankoyo'* (*mankayo'*) (no voucher): edible fruit.

Spondias mombin L.* Tree of tall woody savanna, 90 m elev. *mopiyo'* (*mapiyo'*) (*6257*): edible fruit (Henley, 1982).

Tapirira guianensis Aubl.**. Tree of mountain forest, 400 m elev. and tall woody savanna, 90 m elev. *tamu'ñe mën* (*6601*): edible fruit.

Tapirira velutinifolia (Cowan) Marcano Berti*. Tree of tall woody savanna, 90 m elev. and granite outcrops, 200 m elev. *mapiyo'* (*mopiyo'*) *kai' mën* (*6219*): wood used as support beams in houses.

Thyrsodium schomburgkianum Benth. Tree of mountain forest, 400 m elev. *che'ke* (*6619*).

ANNONACEAE

Annona densicoma Mart. Shrub of tall woody savanna, 60–100 m elev. *ya'ra payayo'* (*6569*).

Annona montana Macf. Sapling collected in tall woody savanna, 90 m elev. *ya'ra payayo'* (*6092*): edible fruit.

Annona purpurea Dunal. Shrub cultivated in village, 90 m elev. *ya'ra payayo'* (*6371*): edible fruit?

Rollinia cf. *exsucca* (Dunal) A. DC.*. Tree of tall woody savanna, 90 m elev. *tuwaeyo'* (*6140*): fibers from inner bark used to make handles on decorative baskets.

Xylopia aromatica (Lam.) Mart.*. Tree of tall woody savanna, 90 m elev. *chirinyo'* (*6279*): edible fruit.

Xylopia sericea St. Hil.**. Tree of mountain forest, 400 m elev. *chirinyo'* (*6624*).

APOCYNACEAE

Himatanthus articulatus (Vahl) Woodson*.**. Tree of mountain forest, 400 m elev. and tall woody savanna, 90 m elev. *parunkayo'* (*6698*). *aru'ru'* (*6285*).

Mandevilla javitensis (H.B.K.) K. Schum. Vining shrub of mountain granite outcrops, 370 m elev. *mᵻtë tëpu'kë mën* (*6383*).

Mandevilla steyermarkii Woodson. Vine of mountain granite outcrops, 370 m elev. No name given (*6362*).

Mandevilla scabra (R. & P.) K. Schum. Vine of mountain granite outcrops, 370 m elev. *kë'ñatë to'pa mën* (*6388*).

Parahancornia oblongata (Muell. Arg.) Monachino. Tree of tall woody savanna, 90 m elev. *aru'ru'* (*6561*): edible fruit.

Plumeria inodora Jacq. Shrub of mountain granite outcrops, 370 m elev. *to'pa mën; tëpu'kë mën* (*6389*).

Prestonia acutifolia (Benth.) K. Schum. Vine of tall woody savanna, 90–100 m elev. *kë'ñatë tëpu'kë mën* (*6155*). *o'mochahcho'* (*6284*).

Genus indet.**. Tree of mountain forest, 400 m elev. *kahka' kai'* (*6611*): edible fruit.

ARACEAE

Caladium bicolor (Ait.) Vent. Terrestrial herb of tall woody savanna, 90 m elev. *ko'ya* (*6110*).

Philodendron brevispathum Schott. Climbing herb on rocks of lower mountain slopes, 100 m elev. *kaya* (*6306*).

Philodendron hederaceum (Jacq.) Schott. Epiphyte of tall woody savanna, 90 m elev. *kaya* (*6132, 6142*).

Xanthosoma sp. Herb cultivated in agricultural fields. *ocumo* (no voucher): edible tubers.

ARALIACEAE

Schefflera morototoni (Aubl.) Maguire, Steyermark & Frodin**. Tree of mountain forest, 400 m elev. *anku'ka'* (*6635*).

Table I
Continued

ARECACEAE

Astrocaryum aculeatum Meyer*. Tree of tall woody savanna, 90 m elev. *amankayo'* (*6213*): edible fruits.
Bactris guineensis (L.) Moore. Acaulescent palm of tall woody savanna, 90 m elev. *pinkicho'* (*6144*): edible fruits.
Euterpe precatoria Mart.**. Tree of mountain forest, 400 m elev. *anku'* (*6720*).
Mauritia flexuosa L.f.**. Tree of mountain forest, 400 m elev. and tree and shrub savanna, 60–100 m elev. in wet areas. *ankayano* (no voucher): edible fruits; leaves used for thatch.
Maximiliana maripa (Correa de Serra) Drude**. Tree of mountain forest, 400 m elev. *we'sae* (no voucher): edible fruits; leaves used for burden baskets.
Syagrus orinocensis (Spruce) Burret**. Tree of mountain forest. 400 m elev. *kopa* (*6616*): edible fruits.

ARISTOLOCHIACEAE

Aristolochia argentina Griseb. Vine of tall woody savanna, 100 m elev. *tumuru' kichañe mën* (*6295*).
Aristolochia nummularifolia H.B.K. Vine of tall woody savanna, 90 m elev. *kowi'nëtë* (*6130*).

ASCLEPIADACEAE

Blepharodon nitidum (Vell.) Macbr. Vine of lower slopes of mountain forest, 100 m elev. *o'mochahcho'* (*6283*).
Matelea trianae (Dcne. ex Triana) Spellman. Vine of tall woody savanna, 90 m elev. *o'mochahcho'* (*6113*).
Vincetoxicum lasiostomum (Dcne.) Blake. Vine of mountains in abandoned agricultural field, 300 m elev. *tëpu'kë mën* (*6333*).

ASTERACEAE

Calea berteriana DC. Shrub of tall woody savanna, 100 m elev. No name given (*6300*).
Bidens riparia Kunth. Herb of tall woody savanna, 90 m elev. *siihpen kai' mën* (*6138*).
Chromolaena squalida (DC.) King & Robinson. Shrub of tree and shrub savanna, 100 m elev. No name given (*6308*).
Guayania cerasifolia (Sch. Bip.) King & Robinson. Herb on rocks along streams of mountain forest, 400 m elev. *tepapa mën* (*6344*).
Mikania micrantha Kunth. Vine on rocks of tall woody savanna along stream, 90 m elev. *o'mochahcho'* (*6461*).
Mikania tillettii King & Robinson. Vine on mountain forest slopes, 370 m elev. *mankowa kai' mën* (*6366*).
Pollalesta schomburgkii (Sch. Bip.) Arteg. Shrub of mountain forest, 400 m elev. *tamu'ñe mën* (*6335*).
Praxelis pauciflora (H.B.K.) King & Robinson. Herb

of tall woody savanna, 90 m elev. No name given (*6125*).
Synedrella nodiflora (L.) Gaertn. Herb of mountain forest, 400 m elev. and tree and shrub savanna, 90 m elev. *mitë* (*6271*). *kayen kai' mën* (*6541*).
Unxia camphorata L.f. Herb of granite outcrops, 370 m elev. No name given (*6398*).
Wedelia calycina L. C. Rich. Shrub of granite outcrops, 370 m elev. *kayen kai' mën* (*6387*).
Wulffia baccata (L.f.) Kuntze. Shrub of tall woody savanna, 90–100 m elev. *kayewankae* (*6066*). *peya wankae* (*6294*).

BEGONIACEAE

Begonia aff. *humillima* Smith & Wasshausen. Herb on rocks of streambed of mountain slopes, 150 m elev. *wëntë* (*6265*).

BIGNONIACEAE

Anemopaegma karstenii Bur. & K. Schum. Liana of mountain forest, 400 m elev. and tall woody savanna, 90 m elev. No name given (*6095, 6513*): used as a lashing material in dwellings.
Arrabidaea inaequalis (DC. ex Splitg.) K. Schum. Liana of mountain forest, 400 m elev. No name given (*6597*): used as a lashing material.
Arrabidaea mollissima (H.B.K.) Bur. & K. Schum. Liana of tall woody savanna, 90 m elev. No name given (*6122*).
Cydista aequinoctialis (L.) Miers. Liana of mountain forest, 400 m elev. No name given (*6545*): used as a lashing material.
Jacaranda rhombifolia G. F. W. Mey. Tree of tall woody savanna, 90 m elev. *tamu'ñe mën* (*6202*). *tayeche mën tamu'ñe mën* (*6120*).
Melloa quadrivalis (Jacq.) A. Gentry. Liana of tall woody savanna, 60–100 m elev. No name given (*6576*): used as a lashing material.
Tabebuia ochracea (Cham.) Standl. s.l. Tree of tall woody savanna, 60–100 m elev. *kooñe* (*6573*).
Tabebuia rosea (Bertol.) DC.*. Tree of tall woody savanna, 90 m elev. *kawañe'* (*6254*): unspecified medicinal.
Tabebuia serratifolia (Vahl) Nichols.*.**. Tree of mountain forest, 400 m elev. and tall woody savanna, 90 m elev. *kawañe'* (*6209*): used as a medicinal for stomachache. *kooñe* (*6704*). *kërë'këre'mayo'* (*6057*): unspecified medicinal.

BIXACEAE

Bixa orellana L. Shrub cultivated in village, 90 m elev. *wace* (no voucher): red dye extracted from seeds.

Table I
Continued

Cochlospermum orinocense (H.B.K.) Steud.*.**. Tree of mountain forest, 400 m elev. and tall woody savanna, 90 m elev. *wanayo' (6511, 6244, 6654)*: fibrous inner bark used to lash burden baskets together.

Cochlospermum vitifolium (Willd.) Spreng.*. Tree of tall woody savanna, 90 m elev. *wanayo' (6086, 6116, 6216)*: fibrous inner bark used to form handles for decorative baskets.

BORAGINACEAE

Cordia alliodora (R. & P.) Cham. ex DC.*. Tree of tall woody savanna, 90 m elev. *utaatiyo' (6158, 6128)*. *mataatiyo (6223)*.

Cordia bicolor A. DC.*.**. Tree of mountain forest, 400 m elev. and tall woody savanna, 90 m elev. *untaatiyo' (6252)*: edible fruit. *mataatiyo' (6076, 6625, 6555)*: edible fruit.

Cordia nodosa Lam. Shrub of mountain forest, 400 m elev. *o'mochahcho' (6669)*: edible fruit.

BROMELIACEAE

Tillandsia flexuosa Sw. Epiphyte of mountain forest, 400 m elev. *onkeyo' kai' mën (6393)*.

BURMANNIACEAE

Burmannia capitata (Walt. ex Gmel.) Mart. Herb of tree and shrub savanna, 90 m elev. *mara'po (6176)*.

BURSERACEAE

Protium heptaphyllum (Aubl.) March. s.l.*.**. Tree of mountain forest, 400 m elev. and tall woody savanna, 90 m elev. (and cultivated in village?). *marahkwa (6236)*: edible fruit.

Protium sagotianum March.**. Tree of mountain forest, 400 m elev. *marahkwa (6475)*: edible fruit. *kiña (6655)*: edible fruit.

Tetragastris altissima (Aubl.) Swart**. Tree of mountain forest, 400 m elev. *marahkwa (6714)*: edible fruit. *kiña (6467)*: edible fruit.

CACTACEAE

Cephalocereus smithianus Britton & Rose. Shrub on granite outcrops, 100 m elev. *këmëhke (6469)*.

CARICACEAE

Carica papaya L. Tree cultivated in village, 90 m elev. *paya* (no voucher): edible fruit.

CECROPIACEAE

Cecropia riparia Snethl.**. Tree of mountain forest, 400 m elev. *kasanka' (6493)*.

Coussapoa asperifolia Trec.**. Tree of mountain forest, 400 m elev. *kanwae (6713)*.

CHRYSOBALANACEAE

Hirtella glandulosa Spreng. Tree of mountain forest, 400 m elev. *kañmare kai' mën (6542)*: edible fruit.

Hirtella racemosa Lam. var. **hexandra** (Willd. ex R. & S.) Prance. Shrub of tall woody savanna, 60–100 m elev. *kañmare kae'ñ' (6574)*: edible fruit.

Licania canescens R. Ben.**. Tree of mountain forest, 400 m elev. *weyo' (weeyo') (6636)*.

Licania cruegeriana Urb.**. Tree of mountain forest, 200–400 m elev. *weyo' (weeyo') (6359)*: used as firewood.

Licania hypoleuca Benth.**. Tree of mountain forest, 400 m elev. and tall woody savanna, 90–100 m elev. *weyo' (weeyo') picha mën (6677)*: edible fruit. *weyo' (weeyo') (6534)*: edible fruit.

Licania leucosepala Griseb.**. Tree of mountain forest, 400 m elev. *weyo' (weeyo') (6592)*.

Licania octandra (Hoffmgg. ex R. & S.) Kuntze. Tree of mountain forest, 400 m elev. *usiipayo' (wosiipayo') (6536)*.

Parinari excelsa Sabine**. Tree of mountain forest, 400 m elev. and (cultivated?) in village, 90 m elev. *inkeewe (6514)*: edible fruit.

CLUSIACEAE

Clusia palmicida L. C. Rich. Shrub of granite outcrops in mountains, 370 m elev. *parata (6404)*.

Oedematopus sp.**. Tree of mountain forest, 400 m elev. *soweyo' (6690)*: edible fruit.

Vismia cayenensis (Jacq.) Pers. Tree of mountain forest, 400–500 m elev. *mahcho' kai' mën (6721)*.

Vismia japurensis Reichardt. Tree of mountain forest, 400 m elev. *mahcho' kai' mën (6546)*.

COMBRETACEAE

Combretum frangulifolium Kunth. Liana of tall woody savanna, 60–100 m elev. *arkon nëka mukon (6577)*.

Terminalia amazonia (J. Gmel.) Exell**. Tree of mountain forest, 400 m elev. *chimin (chemin) (6550)*.

COMMELINACEAE

Dichorisandra hexandra (Aubl.) Standl. s.l. Weak shrub of mountain forest slopes, 200 m elev. *korono kai' mën (6421)*.

Table I
Continued

CONNARACEAE

Rourea grosourdyana Baill. Shrub of tree and shrub savanna, 90 m elev. No name given *(6312)*.

CONVOLVULACEAE

Ipomoea batatas (L.) Lam. Vine cultivated in agricultural fields, 90 m elev. *cho'* *(6327)*: edible tubers.

Ipomoea discolor (H.B.K.) G. Don. Vine of granite outcrops, 370 m elev. *chokun kae* *(6399)*.

Ipomoea hederifolia L. Vine in agricultural fields, 90 m elev. *chokun kae* *(6319)*.

Ipomoea trifida (H.B.K.) G. Don. Vine of tall woody savanna, 90 m elev. *chokun kae* *(6439)*.

Jacquemontia pentantha (Jacq.) G. Don. Vine of tall woody savanna, 90 m elev. *ke'ñatë o'mochahcho'* *(6207)*.

Merremia macrocalyx (R. & P.) O'Donell. Vine of granite outcrops, 400 m elev. No name given *(6504)*.

CUCURBITACEAE

Cayaponia racemosa (Mill.) Cogn. Vine of tall woody savanna, 90 m elev. *kaso'kana* *(6441)*.

Cucumis melo L. Vine cultivated in agricultural fields, 90 m elev. *miron* *(6321)*: edible fruit.

Cucurbita moschata (Lam.) Poir. Vine cultivated in village, 90 m elev. *kayama* *(auyama)* *(6376)*: edible fruit.

CYPERACEAE

Bulbostylis hirta (Thunb.) Svenson. Herb of tree and shrub savanna, 90 m elev. *kurawae'* *(6185)*.

Bulbostylis junciformis Liebm. Herb of tree and shrub savanna, 90 m elev. *wana'ipun kawe mën* *(6450)*.

Calyptrocarya glomerulata Urb. Herb of streambanks in mountain forest, 400 m elev. and tall woody savanna, 90 m elev. *kuruwañ* *(6498)*. *soowo* *(6463)*.

Cyperus diffusus Vahl. Herb of tall woody savanna, 90 m elev. *soowo* *(6133)*.

Cyperus haspan L. Herb of wet sand and gravel along road in tree and shrub savanna, 90 m elev. *kuruwoi'* *(6429)*.

Cyperus simplex H.B.K. Herb of tall woody savanna, 90 m elev. *kuruwae'* *(6159)*.

Diplacrum longifolium C. B. Clarke. Herb of granite outcrops of mountains, 370 m elev. *kuruwae'* *(6413)*.

Fuirena umbellata Roteb. Herb of flooded, tall woody savanna, 90 m elev. *soowo* *(6443)*.

Hypolytrum pulchrum (Rudge) Pfeiff. Herb of mountain forest, 400 m elev. *soowo* *(6519)*.

Mariscus hermaphroditus Vahl. Herb at margin of agricultural fields, 90 m elev. *kuruwai'* *(6317)*.

Rhynchospora armerioides Presl. Herb of wet sand and gravel along road in tree and shrub savanna, 90 m elev. *soowo* *(6425)*.

Rhynchospora cephalotes (L.) Vahl. Herb of mountain forest, 400 m elev. and tall woody savanna, 90 m elev. *soowo* *(6520)*.

Rhynchospora curvula Griseb. Herb of granite outcrops, 370 m elev. *kuruwai'* *(6407)*.

Rhynchospora eximia (Nees) Bockl. Herb of tree and shrub savanna, 90 m elev. *kuruwaeñ* *(6175)*.

Rhynchospora junciformis (Kunth) Bockl. Herb of granite outcrops, 370 m elev. *mïtë* *(6382)*.

Rhynchospora nervosa Vahl. Herb of tall woody savanna, 90 m elev. *kuruwa* *(6166)*.

Rhynchospora pubera (Vahl) Bockl. Herb of wet sand and gravel along road in tree and shrub savanna, 90 m elev. *kuruwoi'* *(6433)*. *kurawae* *(6182)*.

Rhynchospora tenerrima Bockl. Herb of tree and shrub savanna, 90 m elev. *wana'ipun* *(6184)*.

Scleria hirtella Sw. Herb of tree and shrub savanna, 90 m elev. *mïtë* *(6183)*.

DILLENIACEAE

Curatella americana L.* Dominant tree of tree and shrub savanna and common in tall woody savanna, 90 m elev. *takucho* (no voucher).

DIOSCOREACEAE

Dioscorea alata L. Vine cultivated in agricultural fields, 90 m elev. *wanka tamuñe mën* *(6326)*: edible tubers.

Dioscorea trifida L.f. Vine cultivated in agricultural fields, 90 m elev. *wanka* *(6322)*: edible tubers.

Dioscorea sp. Vine at margin of agricultural fields, 90 m elev. *wankayan kai' mën* *(6296)*.

ERIOCAULACEAE

Paepalanthus lamarckii Kunth. Herb of wet sand and gravel along road in tree and shrub savanna, 90 m elev. *to'pa mën* *(6424)*.

Syngonanthus gracilis (Bong.) Ruhl. Herb of wet sand and gravel along road in tree and shrub savanna, 90 m elev. *mënkën* *(6423)*.

Syngonanthus sp. Herb of wet sand and gravel along road in savanna. *tënahkwa mën* *(6428)*.

ERYTHROXYLACEAE

Erythroxylum impressum O. E. Schulz. Shrub of granite outcrops, 370 m elev. *tëpapa mën* *(6402)*.

EUPHORBIACEAE

Acalypha cuspidata Jacq. Herb of wet, shaded area of mountain forest, 400 m elev. *tayeeche mën* *(6303)*.

Table I

Continued

Alcornia schomburgkii Kl.**. Tree of mountain forest, 400 m elev. *tëkaman si̇ mën (6719).*

Croton sp. Tree of mountain forest, 400 m elev. No name given *(6339).*

Jatropha gossypiifolia L. Shrub of tree and shrub savanna, collected along road near *criollo* habitation, 90 m elev. *pinonyo' (6329).*

Mabea occidentalis Benth.*. Shrub or small tree of mountain forest, 400 m elev. and tall woody savanna, 90 m elev. *tëpu'kë mën (tepoku mën) (6422).*

Manihot esculenta Crantz. Stout herb cultivated in agricultural fields, 90 m elev. *amaka* (no voucher): edible tubers; sweet variety.

Manihot tristis Muell. Arg. Shrub of granite outcrops, 370 m elev. *poorara utonyo' (6384).*

Margaritaria nobilis L.f.*.**. Tree of mountain forest, 400 m elev. and tall woody savanna, 90 m elev. *mayon (6699). waayayo' kai' mën (6700). onkomayo' (6255).*

Pera ferruginea (Schott) Muell. Arg.**. Tree of mountain forest, 400 m elev. *kahka'yo'; wanëyepi mën (6405):* wood used as a fuel. *kahka'yo' (tonkonam) (6615):* edible fruit. No name given *(6661).*

Phyllanthus sp.*. Tree of tall woody savanna, 90 m elev. *anchupayo' (6220).*

Sapium sp.**. Tree of mountain forest, 400 m elev. *mukonyo' (6673):* edible fruit. *kanwae (6650).*

Sebastiana corniculata (Vahl) Muell. Arg. Herb of tall woody savanna, 90 m elev. *o'mochahcho' (6168).*

Genus indet.*. Tree of tall woody savanna, 90 m elev. *anku' kayo' (6246). mahkukayo' (6214). mokokoyo' (6081).*

FABACEAE

Abrus precatorius L. Vine cultivated in village, 90 m elev. *o'mochachipeñ (6372):* seeds made into necklaces for sale to tourists.

Acosmium nitens (Sw.) Yakovl. Tree of granite outcrop margins, 370 m elev. *karamatëyo' (6381).*

Aeschynomene brasiliana (Poir.) DC. Shrub of tall woody savanna, 90 m elev. *tëpapa mën o'mochahcho' (6109).*

Aeschynomene sp. Shrub of granite outcrops, 370 m elev. *tëpapa yuchin mën (6379).*

Anadenanthera peregrina (L.) Speg.*. Dominant tree of tall woody savanna, 90 m elev. *mëhchë (6231).*

Andira inermis Benth. Climbing shrub or tree of tall woody savanna, 90 m elev. *wonka (winka) (6460).*

Arachis hypogaea L. Vine cultivated in agricultural fields, 90 m elev. *ko'nye* (no voucher): edible nuts.

Bauhinia ungulata L. Liana of mountain forest, 400 m elev. No name given *(6516):* wood used as firewood.

Brownea coccinea Jacq. vel aff.**. Tree of mountain forest, 400 m elev. *paapayo' (6708).*

Campsiandra laurifolia Benth. Tree of tall woody savanna, 90 m elev. *paariyo' (6553).*

Cassia moschata H.B.K.*.**. Tree of mountain forest, 400 m elev. and tall woody savanna and tree and shrub savanna, 90 m elev. *mañmañ (6566).*

Chamaecrista diphylla (L.) Greene. Herb of wet sand and gravel along road in tree and shrub savanna, 90 m elev. *ko'ne kañ mën (6434).*

Chamaecrista kunthiana (Schl. & Cham.) Irwin & Barneby. Herb of tree and shrub savanna, 90 m elev. No name given *(6200).*

Chamaecrista zygophylloides (Taub.) Irwin & Barneby var. *deamii* (Britton & Rose) Irwin & Barneby. Shrub of granite outcrops, 370 m elev. No name given *(6380).*

Copaifera sp. Tree of tree and shrub savanna, 90 m elev. *maarana pukun (6446):* resin used to treat cuts.

Crotalaria pilosa Mill. Herb of tree and shrub savanna, 90 m elev. *wareka marakaeyo' (6292):* dried fruits used as rattles.

Desmodium asperum Desv. Herb of mountain forest, 400 m elev. No name given *(6352).*

Desmodium barbatum (L.) Benth. Weak shrub of tree and shrub savanna, 90 m elev. *kuhpekwa mën (6192).*

Dipteryx punctata (Blake) Amsh.**. Tree of mountain forest, 400 m elev. *woñ kae (6674):* source of the tonka beans that are sold to *criollos.*

Eriosema violaceum (Aubl.) G. Don. Shrub of wet areas in tree and shrub savannas. *tiipase mën (6455).*

Hymenaea courbaril L.*. Tree of tall woody savanna, 90 m elev. *waayayo' (6256):* edible fruit.

Hymenolobium flavum Kleinh.*. Tree of tall woody savanna, 90 m elev. *tëpuru'kë mën (6239):* wood is durable and used in dwelling construction.

Hymenolobium petraeum Ducke*.**. Tree of mountain forest, 400 m elev. and tall woody savanna, 90 m elev. *tayeeche mën (tanyeeche mën) (6567). karamatëyo' (6161). wankayo' (6210). maarana kai' mën (6245).*

Indigofera lespedezoides H.B.K. Herb of tall woody savanna, 90 m elev. No name given *(6195).*

Indigofera suffruticosa Mill. Shrub of agricultural fields, 90 m elev. *pichañe mën (6325).*

Inga ingoides (Rich.) Willd.**. Tree of mountain forest, 400 m elev. *kota yëtëkën (6697):* edible fruit.

Inga sertulifera DC. Tree of mountain forest, 400 m elev. *poorarayo' (6340):* edible fruit.

Inga cf. *umbellifera* Steud.**. Tree of mountain forest, 400 m elev. *pororo (6715):* edible fruit. *moyoyo' (6631):* edible fruit.

Inga sp.**. Tree of mountain forest, 400 m elev. *e'nepa waripicheyo' (6486):* edible fruit. *pihkuankae (6621). echepeñyo' (6680):* edible fruit.

Machaerium sp.*. Tree of tall woody savanna, 90 m elev. *karamatiyo' (karamateyo') (6241):* wood used in dwelling construction. *kota pananyo' (6232).*

Macrolobium cf. *bifolium* (Aubl.) Pers.*.**. Tree of mountain forest, 400 m elev. and tall woody savanna, 90 m elev. *ayowankae (ayawankae) (6683). mayowankae (6152).*

Macrolobium multijugum (DC.) Benth. Tree of tree and shrub savanna, 90 m elev. *paareyo' (6581).*

Mimosa debilis H. & B. ex Willd. Shrub of tree and shrub savanna, 90 m elev. No name given *(6169).*

Table I

Continued

Mimosa microcephala H. & B. ex Willd. var. *lunaria* Barneby ined. Shrub of granite outcrops, 370 m elev. No name given (*6386*).

Mimosa orthocarpa Spruce ex Benth. Shrub of tree and shrub savanna, 90 m elev. No name given (*6276*).

Pachyrrhizus sp. Vine of mountain forest, 400 m elev. *o'mochahcho'* (*6361*): seeds made into necklaces.

Parkia pendula (Willd.) Benth.**. Tree of mountain forest, 400 m elev. *tanyeeche mën* (*6672*).

Pithecellobium polycephalum Benth.*. Tree of tall woody savanna, 90 m elev. *tayeeche mën* (*6228*).

Sclerolobium sp.**. Tree of mountain forest, 400 m elev. *che'ke* (*6678*): bark used as an unspecified medicinal.

Senna silvestris (Vell.) Irwin & Barneby var. *silvestris***. Tree of mountain forest, 400 m elev. *kuinkae* (*6608*). *che'ke* (*6588*): unspecified medicinal.

Swartzia laevicarpa Amsh.*.**. Tree of mountain forest, 400 m elev. and tall woody savanna, 90 m elev. *miriyon* (*6554*). *miriyunyo'* (*6211*): red exudate used as paint for decorative baskets. *miriyun* (*6462*): exudate used as a remedy for cuts.

Vigna lasiocarpa (Benth.) Verde. Vine of wet area on tree and shrub savanna, 90 m elev. *o'mochahcho'* (*6444*).

Zornia latifolia Sm. Herb of tree and shrub savanna, 90 m elev. No name given (*6280*).

FLACOURTIACEAE

Banara guianensis Aubl.**. Tree of mountain forest, 400 m elev. *ya'ra payayo'* (*6712*): edible fruit.

Casearia guianensis (Aubl.) Urb.*. Tree of tall woody savanna, 90 m elev. *kehkayo'* (*kehkoyo'*) (*6135*).

Casearia mollis H.B.K. Shrub of tree and shrub savanna, 90 m elev. *kehkoyo'* (*6568*).

Casearia sylvestris Sw. Shrub of tree and shrub savanna, 90 m elev. *karamateyo'* (*6570*): bark is burned to produce a black paint for decorative basketry.

Casearia ulmifolia Vahl ex Vent.*. Tree of tall woody savanna, 90 m elev. *tamu'ñe mën* (*6218*): wood used in framework of dwellings. *kuyiyo'* (*6100*): edible fruit.

Laetia suaveolens (Poepp.) Benth.**. Tree of mountain forest, 400 m elev. No name given (*6594*).

Ryania spruceana Monachino. Shrub of mountain forest, 400 m elev. No name given (*6494*).

GENTIANACEAE

Schultesia brachyptera Chamisso. Herb of wet tree and shrub savanna, 90 m elev. No name given (*6449*).

Schultesia guianensis (Aubl.) Malme. Herb of wet sand and gravel along road in tree and shrub savanna, 90 m elev. No name given (*6436*).

GESNERIACEAE

Chrysothemis dichroa Leeuwenberg. Herb on rocks along stream in mountain forest, 100–370 m elev.

takarapokë mën (*6305*). *topa' mën* (*6354*): collected for sale to *criollos* as an ornamental.

HELICONIACEAE

Heliconia psittacorum L.f. Herb of mountain forest, 100 m elev. *wëhtapu'kën* (*6456*).

HIPPOCRATEACEAE

Peritassa pruinosa (Seem.) A. C. Smith. Liana of mountain forest, 400 m elev. *kahka' kai mën* (*6506*): edible fruit.

Salacia impressifolia (Miers) A. C. Smith. Liana of mountain forest, 400 m elev. No name given (*6502*): edible fruit.

ICACINACEAE

Dendrobangia boliviana Rusby. Tree of tall woody savanna, 90 m elev. *mankayo' panako* (*6153*): edible fruit.

LACISTEMATACEAE

Lacistema aggregatum (Berg) Rusby. Tree of mountain forest, 400 m elev. *marahkwa kai' mën* (*6503*).

LAMIACEAE

Hyptis dilatata Benth. Herb of tree and shrub savanna, 90 m elev. *mara'po* (*6172*).

Hyptis mutabilis (Rich.) Briq. Herb of tree and shrub savanna, 90 m elev. *wana'pa mën tama'ñe mën tapokë mën* (*6272*).

Hyptis suaveolens (L.) Poit. Herb of tree and shrub savanna, 90 m elev. *tapokë mën* (*6270*).

LAURACEAE

Aiouea impressa (Meissner) Kostermans**. Tree of mountain forest, 400 m elev. *waikayo'* (*6491*).

Genus indet.**. Tree of mountain forest, 400 m elev. *mahkwayo'* (*6589*). *amahkoyo'* (*amahkwayo'*) (*6489*).

LECYTHIDACEAE

Lecythis corrugata Poiteau subsp. *rosea* (Spruce ex Berg) Mori**. Tree of mountain forest, 400 m elev. *chiwiri* (*6488*). *maipɨhpë* (*6647*): bark used to make tumpline for burden baskets.

LENTIBULARIACEAE

Utricularia subulata L. Herb of granite outcrops in mountains, 370 m elev. and wet sand and gravel

Table I

Continued

along road in tree and shrub savanna, 90 m elev. No name given (*6445*).

LOGANIACEAE

Strychnos* cf. *diaboli Sandw. Liana (or tree?) of mountain forest, 400 m elev. *mankowa kai' mën* (*6626*).
Strychnos fendleri Sprague & Sandw.* Tree of tall woody savanna, 90 m elev. *chirimiyo'* (*6112*).
Strychnos toxifera Schomb. Liana of mountain forest, 400 m elev. *mankowa* (*6505*): major ingredient in Panare curare.

LORANTHACEAE

Phthirusa retroflexa (R. & P.) Kuijt. Parasitic shrub on *Copaifera* sp. (*6446*) in tree and shrub savanna, 90 m elev. *iye ipun pë' mën* (*6447*): mashed leaves placed over aching body parts to relieve pain.

LYTHRACEAE

Ammannia auriculata Willd. Shrub of granite outcrops, 370 m elev. *sitipi mën* (*6390*).
Cuphea* cf. *o'donellii Lourt. Herb of tree and shrub savanna, 90 m elev. *mite' pichañe mën* (*6188*). No name given (*6438*).

MALPIGHIACEAE

Byrsonima coccolobifolia H.B.K. Tree of tree and shrub savanna, 90 m elev. *kehcho' kai' mën* (*6571*).
Byrsonima crassifolia (L.) H.B.K.*. Tree of tree and shrub savanna and tall woody savanna, 90 m elev. *ke'cho'* (*kehcho'*) (*6572*): edible fruit. *waayayo'* (*6261*): edible fruit; red coloring extracted from bark that is used to paint decorative baskets.
Byrsonima nitidissima H.B.K. Tree of mountain forest, 400 m elev. *pare'ka* (*6332*).
Byrsonima spicata (Cav.) DC.**. Tree of mountain forest, 400 m elev. *kehcho'* (*6707*): edible fruit.
Heteropsis sp. Liana of mountain forest, 400 m elev. *o'mochahcho'* (*6507*).
Hiraea bifurcata W. Anderson. Tree of granite outcrops, 400 m elev. *tëpopo mën* (*6548*).
Stigmaphyllon hypoleucum Miq. Liana of mountain forest, 400 m elev. No name given (*6487*).
Genus indet. Liana of mountain forest, 400 m elev. No name given (*6478*).

MALVACEAE

Gossypium barbadense L. Shrub cultivated in agricultural fields, 90 m elev. *tookëtë* (*6320*): source of cotton thread.

Hibiscus sororius L.f. Herb along streams of mountain forest, 400 m elev. *koi'cho' kai' mën* (*6501*): medicinal; the sap serves to cure a type of dermatitis called *koi'*.
Sida linifolia Cav. Herb of tree and shrub savanna, 90 m elev. No name given (*6193*).
Sida sp. Shrub of tree and shrub savanna, 90 m elev. *tato patan yemihtyo'* (*6451*): stems tied together serve as a broom.

MARANTACEAE

Calathea latifolia (Willd. ex Link) Kl. Herb of mountain forest, 100 m elev. *konono* (*6309*).
Calathea villosa Lindl. Herb of tall woody savanna, 90 m elev. *kerono* (*6087*).
Ischnosiphon arouma (Aubl.) Koern. Herb of mountain forest, 100 m elev. *manankye* (no voucher): material for weaving decorative baskets.
Maranta humilis Aubl. Herb of tall woody savanna, 90 m elev. *kerono* (*6084*).

MARCGRAVIACEAE

Marcgravia coriacea Vahl. Liana climbing over rocks in mountain forest, 400 m elev. *tëpopo mën* (*6540*).

MELASTOMATACEAE

Aciotis sp. Herb along streams in mountain forest, 400 m elev. *tu'ñen utu' panëto'*; *mechu* (*6356*): leaf decoction drunk as an unspecified medicinal.
Clidemia capitellata (Bonpl.) D. Don var. ***dependens*** (D. Don) Macbride. Shrub of mountain forest, 400 m elev. *tëpuru' kechañe mën* (*6302*).
Clidemia novemnervia (DC.) Triana. Shrub of tree and shrub savanna, 90 m elev. and mountain forest, 400 m elev. *o'mochahcho'* (*6299*): used for bathing oneself. *tëhpë puru'kë mën* (*6347*).
Clidemia sericea D. Don. Shrub of slopes of mountain forests, 100–200 m elev. *tëpuru' këchañe mën* (*6293*). *o'mochahcho'* (*6688*).
Comolia veronicaefolia Benth. s.l. Shrub of granite outcrops in mountain forest, 400 m elev. *to'pa mën warayun pi mën* (*6385*).
Leandra solenifera Cogn. Shrub of mountain forest, 400 m elev. *tëpuru' këchañe mën* (*6342*).
Miconia alata (Aubl.) DC. Shrub of mountain forest, 400 m elev. *mëhchahcho'* (*6648*).
Miconia prasina (Sw.) DC. Tree of mountain forest, 400 m elev. *kayookayo'* (*6702*).
Miconia stephanthera Ule. Shrub of tall woody savanna, 90 m elev. No name given (*6301*).
Miconia sp. Tree of mountain forest, 400 m elev. *anchapa kai' mën* (*6357*).

Table I
Continued

Pterolepis pumila (Bonpl.) Cogn. Herb of tree and shrub savanna, 90 m elev. *o'mochahcho'* (*6197*).
Rhynchanthera grandiflora (Aubl.) DC. Shrub of moist ground in tree and shrub savanna, 90 m elev. No name given (*6297*).
Tococa guianensis Aubl. Shrub along streams in mountain forest, 400 m elev. *o'mochahcho'* (*6499*): leaf decoction is drunk as an unspecified medicinal.

MELIACEAE

Trichilia lepidota Mart. subsp. *leucastra* (Sandw.) Penn.**. Tree of mountain forest, 400 m elev. *tu'kiipiyo'* (*6476*).

MENISPERMACEAE

Abuta velutina Gleason. Liana of mountain forest, 400 m elev. No name given (*6663*).

MONIMIACEAE

Siparuna cf. *guianensis* Aubl. Tree of tall woody savanna, 90 m elev. *takaropokë mën* (*6282*): used as a medicinal for bathing.

MORACEAE

Ficus trigona L.f. Tree of tall woody savanna, 90 m elev. *konwaeyo'* (*6278*).
Ficus gomelleira Kunth & Bouche. Tree along streams in tree and shrub savanna, 60–100 m elev. *kanwae* (*6575*).
Ficus cf. *ponnensis* Standl.**. Tree of mountain forest, 400 m elev. *kanwae* (*6682*).
Sorocea sprucei (Baill.) Macbr. subsp. *sprucei**.**. Tree of mountain forest, 400 m elev. and tall woody savanna, 90 m elev. *weyo'* (*weeyo'*) (*6696*). *iyepihpiyo'* (*6243*).

MUSACEAE

Musa ×*paradisiaca* L. Large herb cultivated in agricultural fields, 90 m elev. *kampure* (no voucher): edible fruit.

MYRISTICACEAE

Virola surinamensis (Rol.) Warb.**. Tree of mountain forest, 400 m elev. *kinake* (*6496*).

MYRTACEAE

Eugenia sp.*. Tree of tall woody savanna, 90 m elev. *ya'ra worokopinyo'* (*6253*): edible fruit; wood used in dwelling construction.

Myrcia inaequiloba (DC.) Legrand**. Tree of mountain forest, 400 m elev. *mahcho' kai' mën* (*6547*).
Plinia sp.**. Tree of mountain forest, 400 m elev. *tu'kinke mën* (*6671*). *mahcho' kai' mën* (*6367*): wood used as a fuel.

NYCTAGINACEAE

Guapira cuspidata (Heim.) Lundell*.**. Tree of mountain forest, 400 m elev. and tall woody savanna, 90 m elev. *tëpuru' kichañe mën* (*6238, 6234, 6149*). *to'pa mën tu'kinke* (*6605*). *tunkoyun, tëpopo mën* (*6549*).

OCHNACEAE

Ouratea guildingii (Pl.) Urban*. Tree or shrub of mountain granite outcrops, 370 m elev. and tall woody savanna, 90 m elev. *mankayo' kai mën* (*6552*). *tëpuru'ke mën* (*6552*).
Ouratea polyantha (Tr. & Pl.) Engl. Tree of tall woody savanna, 90 m elev. *tupuru'ke mën* (*6582*).
Sauvagesia ramosissima Spruce. Herb of granite outcrops, 370 m elev. *arepichani mën* (*6392*).

OLACACEAE

Chaunochiton angustifolium Sleumer. Shrub of granite outcrops, 370 m elev. *iye tu'nen uku' neto'* (*6355*): leaves used to prepare a medicinal drink and bath.
cf. *Heisteria* sp.**. Tree of mountain forest, 400 m elev. *mankoyo'* (*6656*): edible fruit.

ONAGRACEAE

Ludwigia nervosa (Poir.) Hara. Shrub of wet tree and shrub savanna, 90 m elev. *ku'pekwa mën wana'pa mën iye* (*6298*).
Ludwigia rigida (Miq.) Sandw. Shrub of wet tree and shrub savanna, 90 m elev. No name given (*6457*).

ORCHIDACEAE

Epidendrum sp. Terrestrial herb of *Selaginella* mats on granite outcrops in mountains, 370 m elev. *to'po mën* (*6395*).
Liparis nervosa (Thunb.) Lindl. Terrestrial herb of mountain forest, 400 m elev. *kaya* (*6349*).
Oncidium orthostates Ridl. Terrestrial herb of *Selaginella* mats on granite outcrops in mountains, 370 m elev. *to'po mën* (*6394*).
Pleurothallis sp. Epipetric herb along dry streambed in mountain forest, 400 m elev. *kaya to'pa mën sisikënë* (*6416*).

Table I

Continued

Polystachya estrellensis Rchb. f. Epiphytic herb of mountain forest, 400 m elev. *kaya (6348)*.

OXALIDACEAE

Biophytum casiquiarense R. Kunth. Shrub of mountain forest, 400 m elev. *poko mën (6602)*.

PASSIFLORACEAE

Passiflora cuneata Willd. Vine of mountain forest, 400 m elev. *kaso'kana (6653)*.

Passiflora cyanea Masters. Vine of mountain forest, 200 m elev. *kaso'kana (6420)*.

Passiflora quadriglandulosa Rodschied. Vine of mountain forest, 400 m elev. *kaso'kana (6537)*.

PIPERACEAE

Peperomia boomii Steyerm. Epipetric herb along dry streambed in mountain forest, 200 m elev. *kaya to'pa mën (6417)*.

Peperomia magnoliaefolia (Jacq.) A. Dietr. Epipetric herb along stream in mountain forest, 400 m elev. *paratayan kai' mën (6343)*.

Piper dilatatum L. C. Rich. Herb of wet areas on slopes of mountain forest, 200 m elev. *tëkarapokë mën yo' (6557)*.

Piper marginatum Jacq. var. **marginatum**. Herb of tall woody savanna, 90 m elev. *takoropokë mën (6117)*: leaf poultice used as a medicinal for the legs.

Piper piscatorum Trel. & Yunck. Herb of mountain forests, 400 m elev. *chawa (6544)*: used as a fish poison. *manchaawa (6337)*: leaves chewed as would be tobacco.

POACEAE

Andropogon angustatus (Presl) Steud. Herb of tree and shrub savanna, 90 m elev. No name given *(6448)*.

Andropogon bicornis L. Herb of wet tree and shrub savanna, 90 m elev. *wana'ipun kawë mën (6442)*.

Aristida longifolia Trin. Herb of tall woody savanna, 90 m elev. *wana'ipun (6313)*.

Axonopus aureus Beauv. s.l. Herb of tree and shrub savanna, 90 m elev. *wana'ipun si'si'kon (6374)*.

Axonopus fissifolius (Raddi) Kuhlm. Herb of tree and shrub savanna, 90 m elev. *wana'ipun (6173)*. *mïtë (6426)*.

Coix lacryma-jobi L. Cultivated along margin of agricultural fields, 90 m elev. *echepenyo' (6472)*: fruits used as beads in necklaces.

Echinolaena inflexa (Poir.) Chase. Herb of tree and savanna, 90 m elev. *arerechan (6287)*.

Eragrostis maypurensis (H.B.K.) Standl. Herb of tree and shrub savanna, 90 m elev. *wana' (6174)*.

Ichnanthus tenuis (Presl) Hitchc. & Chase. Herb of tree and shrub savanna and tall woody savanna, 90 m elev. *mïtë wana'ipun (6201)*. *soowo (6096)*.

Lasiacis anomala Hitchc. Herb of tall woody savanna, 90 m elev. *pe'rechan (6070)*. *are're' chipeñ (6114)*.

Lasiacis procerrima (Hack.) Hitchc. Herb of slopes of mountain forest, 300 m elev. *are're akëtënë (6418)*.

Metosetum cardonum Luces. Herb of granite outcrops, 370 m elev. *wana'ipun (6396)*.

Oplismenus burmanii (Retz) Beauv. Herb of tall woody savanna, 90 m elev. *mïtë (6277)*.

Panicum pyrularium Hitchc. & Chase. Herb of granite outcrops, 370 m elev. *mïtë (6400)*.

Panicum trichoides Sw. Herb of tall woody savanna, 90 m elev. *are're' mïtë (6106, 6115)*.

Paspalum contractum Pilger. Herb of tree and shrub savanna, 90 m elev. *wana'ipun (6288)*.

Pennisetum polystachion (L.) Schult. Herb of tree and shrub savannas, 90 m elev. *tson (6375)*.

Schizachyrium condensatum (H.B.K.) Nees. Herb of tree and shrub savanna, 90 m elev. *wana'ipun kawe mën (6452)*.

Setaria tenax (L. Rich.) Desv. Herb of mountain forest slopes, 100 m elev. *wana'ipun (6464)*.

Trachypogon plumosus (H. & B. ex Willd.) Nees. Herb of tree and shrub savanna, 90 m elev. *wana'ipun (6171, 6191)*.

Zea mays L. Herb cultivated in agricultural fields, 90 m elev. *ke'nya* (no voucher): edible grain.

POLYGALACEAE

Polygala sp. Herb of wet sand and gravel along road in tree and shrub savanna, 90 m elev. *ku'pëkwa mën (6440)*.

Polygala sp. Herb of granite outcrops, 370 m elev. No name given *(6409)*.

Polygala sp. Herb of granite outcrops, 370 m elev. No name given *(6414)*.

Polygala sp. Herb of granite outcrops, 370 m elev. No name given *(6411)*.

Polygala sp. Herb of wet sand and gravel along road in tree and shrub savanna, 90 m elev. *tëhpisaara kichka mën (6437)*.

Securidaca diversifolia (L.) Blake. Liana of mountain forest slopes, 200 m elev. No name given *(6553)*.

POLYGONACEAE

Coccoloba orinocana Howard. Shrub of granite outcrops, 370 m elev. *pankecho' (6391)*.

PROTEACEAE

Roupala montana Aubl.*. Tree of tall woody savanna, 90 m elev. *ya'ra pachehcho' (6240)*: medicinal for the head.

Table I
Continued

RUBIACEAE

Alibertia acuminata (Benth.) Sandw. var. *obtusiuscula* Steyerm.*. Tree or shrub of mountain forest, 400 m elev. and tall woody savanna, 90 m elev. *anoe* (*6363*): edible fruit. *puru'kechañe mën* (*6251*).

Amaioua corymbosa H.B.K.*.**. Tree of mountain forest, 400 m elev. and tall woody savanna, 90 m elev. *pare'ka* (*6364*): wood used as a fuel and dwelling support poles. *paarika* (*6603*).

Borreria capitata (R. & P.) DC. var. *tenella* (H.B.K.) Steyerm. Herb of tree and shrub savanna, 90 m elev. *pichañe mën* (*6431*).

Borreria densiflora DC. Herb of tree and shrub savanna, common around village margin, 90 m elev. *chooriyo'* (*6275*): serves as a medicinal for sores and wounds.

Borreria pygmaea Spruce ex K. Schum. Herb of granite outcrops, 370 m elev. *to'panke* (*6412*).

Diodia teres Walt. Herb of tree and shrub savanna, 90 m elev. *pichañe mën* (*6186*).

Duroia sprucei Rusby. Tree along streams in tree and shrub savanna, 90 m elev. *paareka* (*paarika*) (*6580*).

Genipa americana L. var. *caruto* (H.B.K.) K. Schum.*. Tree of tall woody savanna, 90 m elev. *anku'* (*6260, 6269*): edible fruit; black dye from mature fruit.

Geophila repens (L.) I. M. Johnston. Herb of tall woody savanna, 90 m elev. *kaya tukinke mën* (*6160*).

Guettarda divaricata (H. & B.) Standl.* Tree of tall woody savanna, 90 m elev. *kehkoyo'* (*kehkayo'*) (*6131*): edible fruit.

Guettarda spruceana Benth.**. Tree of mountain forest, 400 m elev. and tall woody savanna, 90 m elev. *tokëtë unkayo'* (*6652*): edible fruit. *tokëtë wankaeyo'* (*6466*): edible fruit.

Isertia parviflora Vahl**. Tree of mountain forest, 100–400 m elev. *pahpanyo'* (*6286*). *pokoyo'* (*6701*).

Ixora acuminatissima Muell. Arg. Shrub of granite outcrops in mountains, 370 m elev. *onkomiyo'* (*6403*).

Morinda tenuiflora (Benth.) Steyerm. var. *leiophylla* (Steyerm.) Steyerm. Shrub of tall woody savanna, 90 m elev. *tëpuru' ke mën* (*6065*): edible fruit.

Psychotria colorata (Willd. ex R. & S.) Muell. Arg. Shrub of mountain forest, 400 m elev. *tu'nen waara* (*6370*): used as a medicinal for coughing.

Psychotria hoffmanseggiana (Willd. ex R. & S.) Muell. Arg. Shrub of mountain forest, 400 m elev. *kehkeyon* (*6610*).

Randia hebecarpa Benth. Tree of tall woody savanna, 90 m elev. No name given (*6111*): edible fruit.

Richardia scabra L. Herb of tree and shrub savanna, 90 m elev. No name given (*6178*).

Rondeletia orinocensis Steyerm. Shrub of tall woody savanna, 90 m elev. *kehkoyo'* (*6289*).

Rudgea crassiloba (Benth.) Robinson*. Tree of tall woody savanna, 90 m elev. *onkomiyo'* (*6600, 6150, 6235*).

RUTACEAE

Galipea trifoliata Aubl.**. Tree of mountain forests, 400 m elev. *ehparayo'* (*6485*).

SAPINDACEAE

Allophylus occidentalis (Sw.) Radlk.*. Tree of tall woody savanna, 90 m elev. *tu'kipiyo'* (*6249*).

Cupania rubiginosa (Poir.) Radlk. Tree of mountain forest, 400 m elev. *tunkoyo'* (*6512*).

Serjania rhombea Radlk. Vine of mountain forest, 400 m elev. and tall woody savanna, 90 m elev. No name given (*6353*): used as a lashing material.

SAPOTACEAE

Elaeoluma glabrescens (M. & E.) Aubr.**. Tree of mountain forest, 400 m elev. *wënë iye* (*6640*). *punwoyo'* (*6604*).

Pouteria amygdalicarpa (Pittier) Penn.**. Tree of mountain forest, 400 m elev. *kankeyo'* (*6368, 6705*): edible fruit. *parunkayo'* (*6483*). *punwoyo'* (*6684*).

Pouteria aff. *engleri* Eyma**. Tree of mountain forests, 400 m elev. *punwoyo'* (*6666*).

Pouteria glomerata (Miq.) Radlk.*. Tree of tall woody savanna, 90 m elev. *pareka* (*6064*).

Pouteria reticulata (Engler) Eyma**. Tree of mountain forest, 400 m elev. *kañke* (*6710*). *maipïhpë* (*6703*).

Pouteria sp.**. Tree of mountain forest, 400 m elev. *chirinyo'* (*6668*). *mankoyo' kai mën* (*6664*). *tëpu'ke mën* (*6667*).

Pradosia surinamensis (Eyma) Penn.**. Tree of mountain forest, 400 m elev. *kahkayo'* (*6692*): edible fruit.

SCROPHULARIACEAE

Bacopa salzmannii (Benth.) Edwall. Herb of wet sand and gravel along road in tree and shrub savanna, 90 m elev. No name given (*6430*).

SIMAROUBACEAE

Picramnia spruceana Engler. Tree in mountain forest, 400 m elev. *kuruwaaru* (*6350*): a purple dye is expressed from the leaves by rubbing them; used for painting the body or decorative baskets.

Simarouba amara Aubl.**. Tree in mountain forest, 400 m elev. *tu'neyo'* (*6630*): used as a medicinal for snakebite.

SMILACACEAE

Smilax sp. Vine of tall woody savanna, 90 m elev. *këmëhkë tuku'yakë mën* (*6143*).

Table I

Continued

SOLANACEAE

Capsicum frutescens L. Shrub cultivated in agricultural fields, 90 m elev. *pimi* (no voucher): edible fruits.

Cestrum alternifolium L.f. Shrub of mountain forest, 100 m elev. *tëpuru'ke* (*6559*).

Solanum asperum L. C. Rich. Shrub of mountain forest, 100–370 m elev. No name given (*6358*).

Solanum bicolor R. & S. Shrub of mountain forest, 100 m elev. *tamu'ñe mën* (*6274*).

Solanum sacupanense Rusby. Shrub along streams in tree and shrub savanna, 60–100 m elev. No name given (*6579*).

Solanum subinerme Jacq. Shrub of mountain forest, 100–400 m elev. and tall woody savanna, 90 m elev. No name given (*6360*).

STERCULIACEAE

Guazuma tomentosa H.B.K. Tree of tree and shrub savanna, 90 m elev. *kuhpëkwa mën iye* (*6584*).

Helicteres guazumaefolia H.B.K. Shrub of tall woody savanna, 90 m elev. *ya'ra ipunyo'* (*6465*). No name given (*6259*).

Melochia villosa (Mill.) Fawc. & Rendle. Shrub of wet sand and gravel along road in tree and shrub savanna, 90 m elev. *wana'ipun* (*6435*).

THEACEAE

Caraipa llanorum Cuatr. subsp. *llanorum.* Tree of tall woody savanna, 90 m elev. *weyo'* (*weeyo'*) (*6291*).

THEOPHRASTACEAE

Clavija nobilis (Linden) Mez. Shrub of mountain forest, 400 m elev. *chami* (*6419*): edible fruit.

TILIACEAE

Apeiba schomburgkii Szyszyl.**. Tree of mountain forest, 400 m elev. *katunyo'* (*6477*).

Corchorus hirtus L. Shrub of tree and shrub savanna, margin of agricultural field, 90 m elev. *o'mochahcho'* (*6331*).

Luehea candida (DC.) Mart. Tree of tall woody savanna, margin of agricultural field, 90 m elev. *ya'ra ipunyo'* (*6330*): wood used as structural elements in dwellings.

Lueheopsis rosea (Ducke) Burret**. Tree of mountain forest, 400 m elev. *ya'ra ipunyo'* (*6695*).

Mollia stellaris Meijer. Tree of mountain forest slopes, 200 m elev. *ya'ra ipunyo'* (*6397*).

ULMACEAE

Trema micrantha (L.) Blume. Shrub along stream in agricultural field in tree and shrub savanna, 90 m elev. *anchapa* (*6316*).

VERBENACEAE

Aegiphila sp. Shrub of tall woody savanna, 90 m elev. *o'mochahcho'* (*6281*).

Lantana sp. Shrub of tall woody savanna, 90 m elev. *tapokë mën* (*6126*).

Lippia cf. *origanoides* H.B.K. Shrub of mountain forest, 400 m elev. *chawë* (*6338*).

Petrea arborea H.B.K. Liana of rocky granitic slopes of mountain forest, 100 m elev. *sërae sërae* (*6583*): indicated as having edible fruit, but this is probably incorrect as the fruits are dry.

Stachytarpheta sprucei Moldenke. Shrub of mountain forest, 400 m elev. No name given (*6264*).

Vitex capitata Vahl. Tree of tall woody savanna, 60–100 m elev. *sërae sërae* (*6578, 6565*): edible fruit.

VIOLACEAE

Rinorea sp. Tree along rivers in mountain forest, 400 m elev. *aruniyo'* (*6564*).

VITACEAE

Cissus sicyoides L. Vine of tall woody savanna, 90 m elev. *manman* (*mënmën*) (*6221*).

Cissus sulcicaulis (Baker) Planch. Vine of tall woody savanna, 90 m elev. *u'manman* (*6164*).

VOCHYSIACEAE

Erisma uncinatum Warming**. Tree of mountain forest, 400 m elev. *paarika* (*6659*). *yemuhcho'* (*6706*).

Qualea dinizii Ducke**. Tree of mountain forest, 400 m elev. *kitya* (*6515*). *mahcho' kai' mën* (*6649*).

Vochysia glaberrima Warm.**. Tree of mountain forest, 400 m elev. *yemuhcho'* (*6717*).

Vochysia venezuelana Stafleu. Tree of tree and shrub savanna, 100 m elev. *yemuhcho'* (*6562*).

XYRIDACEAE

Xyris lacerata Pohl ex Seub. Herb of tree and shrub savanna, 100 m elev. *wana'ipun* (*6290*).

ZINGIBERACEAE

Renealmia aromatica (Aubl.) Griseb. Herb of mountain forest, 100 m elev. *no'yayo'* (*6307*): mature fruits used to dye cloth purple.

Oxalis tuberosa Mol. (Oxalidaceae) in Mexico: An Andean Tuber Crop in Meso-America

Steven R. King and Hélio H. C. Bastien

Table of Contents

Abstract

KING, S. R. (Institute of Economic Botany, New York Botanical Garden, Bronx, New York 10458-5126, U.S.A.) and H. H. C. BASTIEN (77 Zacateros, San Miguel de Allende, GTO., Mexico). *Oxalis tuberosa* Mol. (Oxalidaceae) in Mexico: An Andean tuber crop in Meso-America. Advances in Economic Botany 8: 77–91. 1990. *Oxalis tuberosa* is a high altitude Andean tuber crop, cultivated from Venezuela to Argentina at elevations of 2500 to 4000 m. The results of recent fieldwork in the Transverse Neovolcanic Axis of Mexico document the origin, cultivation, distribution, diversity, utilization and nutritional value of *O. tuberosa* in Mexico. The probable period of introduction of *O. tuberosa* in Mexico appears to be post-Conquest. The adaptation of *O. tuberosa* to the Mexican highlands is discussed with reference to the introduction of this crop to other mountainous areas of the world.

Key words: *Oxalis tuberosa*; Mexico; tuber crop; nutrition; cultivation; utilization.

Resumen

Oxalis tuberosa es un tubérculo Andino bien conocido y esta cultivado desde Venezuela hasta la Argentina desde los 2500 hasta 4000 metros sobre el nivel del mar. Los resultados de investigaciones del campo en el Eje Neovolcánico Transversal de México estan presentados abajo anotando el cultivo, distribución, diversidad, utilización, valor nutritivo y origen de *O. tuberosa* en México. Una hipótesis esta ofrecida sobre la epoca probable de introducción de *O. tuberosa* a México. Las implicaciones de la adaptación de *O. tuberosa* a las alturas de México estan discutidas con referencia a la introducción de este cultivo a otras zonas montañosas del mundo.

Introduction

The Andes of South America are recognized as one of the major centers of origin of cultivated plants, of which the most well known is *Solanum tuberosum,* the potato. *Oxalis tuberosa* Mol. is one of several cultivated tuber crops of highland South America that also include *Tropaeolum tuberosum* R. & P. (Tropaeolaceae), *Ullucus tuberosus* Caldas (Basellaceae), *Lepidium meyenii* Walp. (Brassicaceae), *Arracacia xanthorriza* Bancroft (Apiaceae), *Polymnia sonchifolia* Poepp. & Endl. (Asteraceae), and *Mirabilis expansa* R. & P. (Nyctaginaceae). *Oxalis tuberosa* is cultivated from Venezuela to Argentina between the elevations of 2500 and 4000 m. People of highland South America consume roots and tubers on a daily basis as part of their unique agricultural adaptation to a mountainous environment.

The International Board of Plant Genetic Resources (IBPGR) has supported germplasm collection, maintenance and evaluation of several of these crops in Ecuador, Peru and Colombia (IBPGR 1983). *Oxalis tuberosa* and other Andean crops are also beginning to receive increased international agricultural research attention. This is because countries with mountainous environments in Africa, Asia and North America are seeking to expand their agricultural base (King & Vietmeyer, 1989; Vietmeyer, 1987). Basic data on *O. tuberosa* and other Andean roots and tubers has been published by Andean scientists (Blanco, 1977; Cárdenas, 1948; Cortes, 1977; León, 1964; Rea, 1985; Tapia, 1981). *Oxalis tuberosa* has been studied in Costa Rica, where it is reported to produce at elevations of 2400 m (Obregozo, 1956).

Origin and Introduction to Mexico

It is believed that *Oxalis tuberosa* is a post-Conquest introduction to Mexico. One Andean tuber specialist has stated that a red variety was introduced to Mexico during the colonial period but provided no supporting data or references (León, 1958). The Spanish did send orders to their ships requesting that they bring plants from the Old World to the New World and in the sixteenth century missionaries also sent plants and animals to the New World (Plucknett et al., 1987). It is probable that numerous plant species were transported between South American ports and the Mexican ports of Acapulco and Veracruz. Our hypothesis, based on the combined evidence presented below, is that *O. tuberosa* was introduced during the post-Conquest period.

One of the most important sources for evaluating the history and antiquity of Mexican food, medicinal, and useful plants is the beautifully illustrated Florentine Codex (Sahagun, 1979). This classic work was produced shortly after the arrival of the Europeans in Mexico. It is an important reference for students and scholars of Mexican natural history. We found no mention of any type of root crop similar to *Oxalis tuberosa*; the same lack of data was also noted by Dr. Hernandez-X. There is also no evidence of a plant that matches *O. tuberosa* in the De La Cruz–Badiano Aztec Herbal (Gates, 1939), a chronicle of Mexican plant resources, written in 1552.

A second major factor that points toward a post-Conquest introduction is the low diversity of cultivars encountered in Mexico. In the Andes more then 600 germplasm accessions have been collected in Peru and Bolivia (Tapia, 1981). There is also a large germplasm collection of 91 germplasm accessions of *Oxalis tuberosa* in Ecuador (Tola Cevallos et al., 1987). A number of these germplasm collections are likely to be duplicates but there is a tremendous documented diversity compared to the Mexican collections. The nutritional variability of *O. tuberosa* in Mexico is also limited. As was indicated above, Andean landraces exhibit a much greater nutritional variability than the Mexican samples.

A third factor suggesting the recent introduction of *Oxalis tuberosa* is the number of different local vernacular names associated with the same variety. The following names were used by farmers and vendors to identify the Mexican plants: "papa extranjera" (foreign potato), "papa colorado" (colored or red potato), "papa yuca" (manioc potato), "papa castillana" (Spanish potato), and "papa chirivia" (not translated potato). All of these names were applied to the red variety which we encountered and each of the names refers to *O. tuberosa* as a type of potato. Only in the village of Xometla, Vera Cruz, was a distinction made between a yellow and white type. By contrast, in Andean Peru, *O. tuberosa* is known as "oca" and cultivars bear distinct Quechua or Spanish names used to describe the characteristics of each cultivar (Rea & Morales, 1980). If *O. tuberosa* had been cultivated in Mexico for several centuries before the arrival of the Spanish it is likely that indigenous and/or regional names, not based on the Spanish word "papa" (potato), would now still be used. We encountered no such names, only those listed above.

One final, important part of this question involves the literature on the origin, cultivation and diversity of potato cultivars in Mexico. Ugent (1968) cited the same Transverse Neovolcanic Axis as being the dominant region of potato cultivation in the late 1960's. In this area he documented 22 distinct potato cultivars and examined the question of the history of the potato in Mexico.

Based on taxonomic, linguistic and historical data he concluded that there is little or no evidence to support potato cultivation prior to the Spanish conquest. Given that a fairly large diversity of potato varieties is present today and that they were introduced after the Spanish entered Mexico, it seems unlikely that *Oxalis tuberosa* could have preceded the potato and still exhibit such a low degree of cultivar variation.

Martinez (1936) does not include *Oxalis tuberosa* in his book on useful plants of Mexico but he does mention it as being an edible tuber from Peru, in his catalog of common and scientific names of Mexican plants (1937). Bukasov (1930) was first to note its presence in Mexico and he pointed out the morphological distinctions between the Andean and Mexican *O. tuberosa*. Subsequent research on *O. tuberosa* from Mexico and six Andean countries has revealed that the differences noted by Bukasov reflected more the limited variability he observed rather than regional morphological distinctions (Cárdenas, 1958).

Bukasov provided no data on the importance of *Oxalis tuberosa* in Mexican agriculture. León (1958) reported it as having been introduced to Mexico. Williams (1978), however, incorrectly identified *Oxalis tuberosa* in Mexico as *Tropaeolum tuberosum,* because of tuber similarity. This mistaken identification was based on market tuber specimens because these two species are in different families and are otherwise quite distinct. No further research on the cultural and agricultural importance of *O. tuberosa* in Mexico has been published.

Methods

A survey of the Mexican Transverse Neovolcanic Axis was undertaken in collaboration with Hélio Bastien, a Mexican agronomist. Numerous farm sites were visited and farmers interviewed. The data in Table I records the localities where voucher specimens and tubers for nutritional analysis were collected. Fresh tuber material was also collected and sent to the Hazelton Biomedical Laboratories in Madison, Wisconsin, for nutritional analysis. Areas to be surveyed were selected on the basis of information provided by regional field agronomists, market vendors and the agricultural knowledge of our Mexican team. A total of 16 farms were surveyed in three states of Central Mexico: Veracruz, Michoacán and México. *Oxalis tuberosa* was found in cultivation in fourteen of these farms. To discover the extent of tuber diversity in these areas a color photograph (Fig. 1) of the variation of Andean *O. tuberosa* was used as a comparative tool for dis-

Table I

Herbarium voucher and tuber specimen collection data

Collection #	Eleva-tion (m)	Date (D/M)	Locality	Material collected	Institution deposited
K-682	2760	8/9/86	Approx. 19°32'N and 97°08'W; Tembladeras, Veracruz.	HBV-tubers for lab.	NY, MEXU & CHAPA.
K-683	2500	10/9/86	Approx. 19°58'N and 97°20'W; Xometla, Veracruz.	HBV-tubers for lab.	NY, MEXU & CHAPA.
K-684	2500	10/9/86	Approx. 19°58'N and 97°20'W; Xometla, Veracruz.	HBV-tubers for lab.	NY, MEXU & CHAPA.
K&B-687	2400	18/9/86	Approx. 19°40'N and 101°40'W; San Gregorio, Michoacán.	HBV-tubers for lab.	NY, MEXU & CHAPA.
K&B-689	2850	21/9/86	Approx. 19°38'N and 100°41'W; La Garnica, Michoacán.	HBV-tubers for lab.	NY, MEXU & CHAPA.
K-690	3300	25/9/86	Approx. 19°35'N and 99°58'W; Ojo de Aqua, Mexico.	HBV-tubers for lab.	NY, MEXU & CHAPA.

cussion. This technique proved very useful as many informants were highly interested in the diversity present in the Andes.

Mexican Distribution of *Oxalis Tuberosa*

We encountered *Oxalis tuberosa* (Fig. 2) throughout the Mexican Transverse Neovolcanic Axis between the elevations of 2400 to 3300 m. The highest density of cultivation was encountered in the vicinity of Toluca at an elevation of 2500 to 2800 m, where fields of up to one-half hectare were not uncommon.

Our field research indicates (Fig. 3) that this cultivated species is a common minor crop throughout the Transverse Neovolcanic Axis. We found no specimen of this cultigen in the collections of the UNAM herbarium (MEXU). This is not unusual as most major herbaria contain very few cultivated plants (Prance, 1986). The herbarium at the Colegio de Postgraduados in Chapingo (CHAPA) did include one specimen, collected by Efraim Hernández-X. near Chapingo.

Botanists working in the northern and southern highland regions of Mexico have not reported seeing *Oxalis tuberosa* in cultivation or in markets (M. Nee & D. Stevens, pers. comm.). It should be possible, however, to cultivate it in areas above 2000 m which have enough moisture to support its development. Further investigations in other highland regions need to be carried out.

Cultivar Diversity

The cultivar diversity of *Oxalis tuberosa* in Mexico appears to be low, especially when compared to the variation present in the Andes. In the present study we encountered two cultivars, based on tuber yield, taste, texture and subtle color differences in the phloem.

In the village of Xometla (Veracruz) at 2500 m, farmers recognized two types of tubers: yellow and white. The color difference was only present in the phloem of the tuber and not in the epidermis, which in both varieties was red. When these two varieties are viewed in cross section side by side it is possible to note a slight color difference (Fig. 4).

The distinction between the cultivars was explained as follows by a farmer. The yellow variety produces better yields than the white variety and the tubers are larger, in both length and width. The white variety was also said to be more "ágria" (sour) than the yellow and was therefore not as desirable as the latter. These two varieties reportedly could be distinguished by the farmers on the basis of the aerial portion of the plant. Herbarium vouchers (*King 683, 684*) of the two varieties are not easily distinguishable.

There were reports from other informants of tubers in the above mentioned yellow and white colors. Two informants in the cities of Orizaba and Toluca reported that, infrequently, plants were harvested with yellow and white tubers, intermixed with red, all as part of one individual

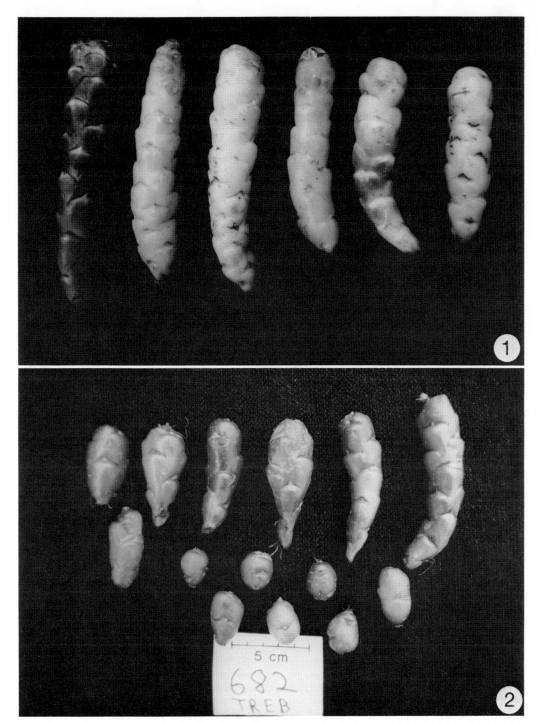

FIG. 1. Diversity of Andean *Oxalis tuberosa* from one market in Pasto, Colombia. The tubers in this photograph identified as "papa extranjera" by Mexican informants.

FIG. 2. Tubers of Mexican *Oxalis tuberosa* collected from one field. All tubers collected or observed were red to pink in color. Tembladeras, Veracruz, 2760 m (*King & Bastien 682*).

FIG. 3   Map of the localities where *O. tuberosa* was encountered during this research, August 1987. ★ = Mexico City.

plant. As this research was conducted two months prior to the harvest season these accounts will require verification.

A third informant also mentioned that the red tubers are at times placed in the sun for five to seven days and that they then turn white. This is, according to our informant, to sweeten the tubers. This practice also has been observed in the Andean region, where it is also reported to sweeten the taste of the tubers (Cortes, 1977).

Data from our analysis, presented below, in-dicate that the Mexican varieties are not, in general, high in oxalic acid. The color change of tubers left out in the sun has been observed in the Andean region not only with *Oxalis tuberosa* varieties but also in the case of other Andean tubers such as *Ullucus tuberosus* and *Tropaeolum tuberosum*.

In the other regions of Mexico, where *Oxalis tuberosa* is cultivated, only the one red variety was mentioned. While further fieldwork is needed during the harvest season, it does appear that

FIG. 4.   Two varieties of *O. tuberosa* recognized by farmers of Xometla, Veracruz. Cross-sectional view of yellow variety, top tuber (*King 683*) and white variety, bottom four tubers (*King 684*). The color distinction refers to the phloem tissue; the epidermis of both varieties was red. 2500 m.

FIG. 5.   Stolons of *O. tuberosa* forming tubers, six months after planting. San Gregorio, Michoacán, 2670 m (*King & Bastien 687*).

there is limited diversity of its cultivars in Mexico, especially when compared to the Andean zone.

Soils and Vegetation

The soils of the Transverse Neovolcanic Axis have been termed andosol and volcanic (Ugent, 1968). The soils are generally dark brown to black, slightly acidic, and are considered fertile (Ugent, 1968). It has been reported that the pH in potato field soils in the region of the Nevado de Toluca range from 5.0–5.7 (Ugent, 1968). Ugent also reported that the pH in the *Pinus* and *Abies* forest soils were less acidic, ranging from 5.8–6.0. The natural vegetation of the highland regions where *Oxalis tuberosa* is cultivated consists of forest stands of *Pinus* and *Abies* combined with grassland dominated by *Festuca, Muhlenbergia, Stipa* and *Agrostis.*

Ecogeographic Limits of the Crop

We encountered *Oxalis tuberosa* under cultivation between 19°20′ and 20°30′ North and 96°50′ and 102°18′ West. Our research suggests, however, that it is likely to be encountered throughout the Transverse Neovolcanic Axis at elevations of 2000 m and above. The lowest limit for successful cultivation found in this study was 2400 m in the village of San Gregorio, Michoacan. Researchers at the Instituto Nacional de Investigaciones sobre Recursos Bióticos (INIREB) in Jalapa, Veracruz have tried unsuccessfully to cultivate *O. tuberosa* as part of their collection of useful plants at their Francisco Javier Clavijero botanical garden at an elevation of 1300 m. Plants grown from tubers outdoors attained a height of 30 cm, then wilted and died, primarily due to the heat.

The upper limit of cultivation for *Oxalis tuberosa* that we observed was 3300 m. In the villages of Conejo, Veracruz and Raices, Mexico, potatoes were the main crop, but three to nine plants of *O. tuberosa* were planted at the edge of several potato fields. Local farmers told us that *O. tuberosa* did not grow or yield well at this elevation and that in villages at lower elevations we would encounter many people growing it. The temperature at these elevations was more severe, with frosts and snow occurring occasionally. In the Andes, *O. tuberosa* is cultivated up to elevations of 4000 m and is resistant to the extreme damage caused by freezing temperatures (Tapia, 1981). It is likely that the clones of *O. tuberosa* cultivated in Mexico represent a variety that is not well adapted to freezing temperatures and that the northern latitude lengthens the cultivation cycle, increasing the exposure to seasonal low temperatures.

Agricultural Cycle of *Oxalis Tuberosa*

Oxalis tubers are planted from February to March, and harvested from October to December. The maturation time ranges from seven to nine months; plantings above elevations of 3000 m take up to eleven months to mature. In August, six months after planting, we encountered plants with numerous stolons that were in the process of forming tubers (Fig. 5). In the Peruvian Andes the amount of time between planting and harvesting is similar but planting occurs in September and October and the harvest is in April and May (King, 1987).

In Mexico, *Oxalis tuberosa* is propagated vegetatively, as it is in the Andes, from tubers selected from the previous year's harvest. In the region of Toluca several small tubers 2–4 cm in diameter are placed in each hole 5–10 cm deep. These small tubers are known as "el ripio," which means the waste or unusable portion. Seed tubers are selected not for their large size, but because they are leftovers, unsuitable for consumption, but just as useful for propagation. Farmers reported that *O. tuberosa* does not produce fruit. In the Andes, production of viable botanical seed has been recorded, but there are no reports of farmers propagating by seed (Cárdenas, 1958).

Cultivation Systems

In this investigation we observed ten distinct variations of interplanting *Oxalis* tuberosa, each utilizing a specific mixture or configuration of crops. The full list presented in Table II describes these variations observed at fourteen farm or house garden sites. One of the most unusual methods involved planting *Oxalis* tubers with maize in the same hole. The two plants grow together, maize rapidly overtopping the *Oxalis* (Figs. 6, 7). Farmers utilizing this system explained that the maize plant protects the *Oxalis*

Table II

Interplanting systems of *Oxalis tuberosa* in Mexico

Location	Eleva-tion (m)	Mode	Other plants
Los Raices, Mexico	3300	5–10 *O. tuberosa* plants at the end of each row of potatoes.	*Solanum tuberosum*
Oje de Aqua, Mexico	3300	Mostly *O. tuberosa* with other food plants at low density.	*Psium sativum, Triticum aestivum, Avena sativa*
La Garnica, Michoacán	2850	*O. tuberosa* in rows with cucurbit vines trailing around.	*Cucurbita ficifolia*
San Gregorio, Michoacán	2400	*O. tuberosa* between rows of maize.	*Zea mays*
San Gregorio, Michoacán	2580	*O. tuberosa* in same row as maize, spaced between maize.	*Zea mays*
El Conejo, North slopes of Cofre de Perote, Veracruz	2800	*O. tuberosa* planted in dooryard garden with herbs.	Medicinal and culinary herbs
Tembladeras, Veracruz	2755	*O. tuberosa* mixed in low density dooryard garden.	*Vicia faba*
Tembladeras, Veracruz	2760	High density monocrop.	none
Xometla, Veracruz	2500	*O. tuberosa* planted in same hole as maize.	*Zea mays*
Xometla, Veracruz	2580	*O. tuberosa* mixed with ornamentals, maize.	ornamentals *Zea mays*

from damage due to periodic hail storms. In this system the *Oxalis* tubers are harvested before the maize is fully mature. To our knowledge Andean tubers are never planted in this manner.

Crop spacing varies according to the method employed, each farmer applying his own experience and expertise to planting. One 7 by 15 m plot was densely planted, with almost indistinguishable rows. This dense planting of *Oxalis* tubers contrasted sharply with the well spaced rows of potatoes growing in the area. This densely planted site was, however, the exception among the ten methods observed. The average spacing between plants was 15–40 cm and the average distance between rows was 90–100 cm. The largest plot we observed was one-half hectare. Yields reported by a Mexican farmer for a one quarter hectare plot varied from 500 to 1000 kg.

Inputs, Insects, and Diseases

None of the farmers interviewed used fertilizer on their plots of *Oxalis tuberosa*. They did report using manure fertilizer on their other crops, most of which were grown for their cash value. The low market price of *O. tuberosa* was one of the reasons cited for not applying fertilizer.

Insect predation was not reported to be a problem for *Oxalis*. None of the plants we examined showed indications of insect attack and no farmers reported using insecticide on their plants. *Oxalis tuberosa* in the Andes has been reported to be affected by *Heterodera rostochensis*, the golden nematode (Wellman, 1972) and by chrysomelid beetle larvae (Kay, 1973). In addition, four fungi, *Colletotrichum* sp., *Phyllosticta* sp., *Puccinia oxalidis* and *Urocystis oxalidis* have been documented as plant pests (Kay, 1973).

Recent research on Andean tubers has shown that these crops are affected by mycoplasma-like bodies and a variety of viruses. Atkey and Brunt (1982) have conducted experiments with severely diseased *Oxalis tuberosa* from Bolivia and identified mycoplasma-like bodies, which caused premature death of the plant. A number of viruses have also been identified that affect *Ullucus tuberosus*, another Andean tuber crop (Brunt et al., 1982a, 1982b). No data is yet available on viral diseases in Mexican *O. tuberosa*. Research on the viral status of Mexican *O. tuberosa* is important in view of experiments by Brunt and co-workers that indicate that virus-free plants may show dramatic yield increases.

Fasciation of *Oxalis tuberosa* stems was common in six of the fourteen sites investigated (Fig. 8) (*King 684*). We were unable to observe tubers of these plants. Fasciated plants of *O. tuberosa* and *Tropaeolum tuberosum* have been observed

FIG. 6. *Oxalis tuberosa* and maize sown in the same hole; the maize is protecting *O. tuberosa* from hail damage. Xometla, Veracruz, 2550 m.

FIG. 7. Field of *O. tuberosa* interplanted with maize as in Figure 6. Xometla, Veracruz, 2550 m.

FIG. 8. Stem of fasciated *O. tuberosa* plant, six months after planting. Numerous plants in this condition were observed. Tembladeras, Veracruz, 2760 m.

in the Andean region by White (1975), who has suggested that several factors interact to produce this morphological condition. These factors include nutrient and hormone imbalance, meristematic disturbances from polyploidy, and pathogens, particularly insect galls (Gorter, 1965). The Mexican farmers reported that fasciated *Oxalis* plants produced as well as the non-fasciated plants and that this anomaly was not considered detrimental.

Utilization, Market Value and Cultural Acceptance

In many areas of its cultivation, in Mexico, *Oxalis tuberosa* was regarded to be more like a "fruit" than a "vegetable." In the markets it is often sold in the fruit section, not in the area where tubers such as potatoes are sold. It is commonly eaten raw with a small amount of salt, lemon and powdered chili pepper (*Capsicum*). *Oxalis tuberosa* is also used to make preserves, and in some instances is combined with roots of *Pachyrhizus tuberosa* (jícama, or yam-bean, Leguminosae) and fruits such as apples, peaches or pears. A second type of preserve is also prepared using vinegar, cucumbers and onions. It was also reported to be used in a pork stew and fried in a manner similar to fried potatoes.

In the Andes *Oxalis tuberosa* is added to stews and soups or it is steamed and served as a sweet. The tubers are also sun dried or freeze-dried for a variety of dishes. A type of fermented beverage generically known as chicha is also prepared from *O. tuberosa* in the Andes. A few cultivars are also eaten raw, in both the Andes and Mexico. In New Zealand *Oxalis* tubers are boiled, baked, or fried and served like potatoes (Vietmeyer, 1987).

In Mexico the tubers are sold in the markets from November to February. The price is always less than potatoes by twenty to fifty percent. The price ranged from seventy to one hundred pesos per kg ($US 0.10 to 0.30) in August 1986. Several farmers reported that they planted only a small amount to eat and sell because there was not enough demand for it at the markets. One farmer, near the village of Tancitaro, Michoacan reported that twenty years ago he planted a hectare but that its popularity and cash value has declined since that time. In the markets of Orizaba and Pátzcuaro, *Oxalis* tubers are sold both inside the market by middle men and outside on the street by the producers themselves.

In the market in Toluca, several vendors carried *Oxalis tuberosa* when it was in season. They claimed there was a large demand for it during the Christmas season. These vendors and some farmers reported the tubers to be used in piñatas at festivals, suggesting a possible cultural role of *O. tuberosa.*

Comparative Nutritional Value

Fresh tuber material was submitted for nutritional analysis to an independent laboratory. Results are presented in Table III and compared with pooled data from Andean cultivars (King & Gershoff, 1987). Table III presents the proximate nutritional composition of four samples of *Oxalis tuberosa* from three distinct localities in Mexico. The percent protein value shows a range from 4–4.7%, which is 50% lower than the maximum value of 8.4% for Andean cultivars. The Mexican tubers also show a much lower degree of nutritional variation in protein content compared to the protein variation present in the Andean cultivars.

Other apparent distinctions between the Mexican and Andean proximate analyses are a higher ash content and a higher percentage of crude fiber in Mexico. The crude fiber in Mexican samples had a maximum of 21.5% compared to the Andean maximum of 5.1%. The carbohydrate value of the Mexican samples is also slightly lower than the Andean.

Amino acid analyses have not been conducted on Mexican *Oxalis tuberosa,* but the amino acid values for two Andean varieties have been determined (King and Gershoff, 1987). These samples contained a good balance of essential amino acids, with valine and tryptophan, respectively, being the limiting amino acids. In order to assess the general nutritional value of the Mexican tubers, the essential amino acids must be determined; these will be the subject of future reports.

A number of varieties of Andean *Oxalis tuberosa* are processed in order to make them edible (Hodge, 1951). It is the high level of calcium oxalate that is reputed to cause the bitter taste (Hodge, 1951), which must be removed through leaching and freeze-drying. *Oxalis* tubers do contain oxalic acid (King, unpubl. results) but no

Table III

Nutritional variability of Andean and Mexican *Oxalis tuberosa*

Component (%) (100/g)	Andes		Mexico			
	Min.	Max.	1	2	3	4
Protein	3.0	8.4	4.0	4.4	4.7	4.7
Moisture	80.2	84.6	87.8	95.5	95.8	87.4
Fat	0.5	0.6	1.6	<0.1	<0.1	1.5
Ash	1.9	3.5	4.9	11.1	9.5	5.5
Crude fiber	4.0	5.1	11.4	17.7	21.4	9.5
Carbohydrates	80.2	84.6	75.3	66.6	64.0	79.3
Calories/100 g	368.7	364.0	326.0	284.0	276.0	343.0

All values are presented on a dry weight basis, except moisture, which is presented on a fresh weight basis. Andean data from King & Gershoff (1987).

reports on the levels have been encountered. Material collected for this research was tested for levels of oxalic acid. One sample contained 79 ppm (7.9 mg/100 g), similar to levels in *Solanum tuberosum* (Hodgkinson, 1977). While this sample is only an indicator, it is the only quantitative figure published to date on the level of oxalic acid in Mexican *Oxalis* tubers.

Germplasm Status of *Oxalis Tuberosa*

There is no collection of Mexican *Oxalis tuberosa* germplasm in Mexico. Germplasm material from our field research has been sent to an in vitro tissue culture germplasm collection of Andean tubers at the University of San Marcos, in Lima, Peru. Mexican farmers who cultivate *Oxalis* are interested in receiving germplasm of the Andean varieties for experimentation, as are researchers at the Colegio de Postgraduados at Chapingo. The request for Andean germplasm for experimentation in Mexico has been forwarded to the germplasm banks in Peru.

Extensive collection of Mexican *Oxalis tuberosa* during the harvest season would perhaps discover as yet unrecognized varieties and/or increase the range in which they are presently known to occur. The maintenance of these collections in Mexico would be the next logical step towards systematically evaluating the diversity of *O. tuberosa* from distinct geographic areas.

Implications for International Agriculture

Before we examine the significance of *Oxalis tuberosa* in world agriculture, it is useful to men-

tion the current role of a better known tuber crop of Andean origin. The potato is now a major food crop, under cultivation in 130 countries, with a world production of 290 million tons annually (International Potato Center, 1984). This production figure ranks potatoes fourth in volume after wheat, maize and rice. To understand why, up to now, only the potato has become a major crop of world importance while other nutritious tuber crops from the Andean region have not is outside the scope of this research. What is important to take into account is that the potato was unknown outside of the Andes only 400 years ago.

The fact that *Oxalis tuberosa* is a widespread minor crop in the central highlands of Mexico demonstrates that it is adaptable to mountainous environments outside of the Andes. As was mentioned earlier, it has been cultivated as a minor crop for the past 20 years in New Zealand. Small scale growers are also beginning to experiment with it on the northwestern coast of the United States (Vietmeyer, pers. comm.).

More recently, in 1987, during an international workshop on mountain agriculture and crop genetic resources in Kathmandu, Nepal, agronomists from several Himalayan countries requested germplasm material of *Oxalis tuberosa* and other Andean tubers for experimentation (ICIMOD, 1987). We sent sample material from the Andean region to a tissue culture laboratory in Nepal. It is clear that there is a growing world interest in *O. tuberosa* and other Andean crops.

Currently, the agricultural research and development priorities of the International Potato Center in Lima, Peru do not include projects focusing on *Oxalis tuberosa* or other Andean tuber crops. The development of this crop will continue for the present only through the interest and efforts of individuals and agronomic research programs interested in new crops.

Conclusion

Oxalis tuberosa is a widespread minor crop in the Central Volcanic Cordillera of Mexico. There appear to be very few cultivars present in the region. *Oxalis tuberosa* has been integrated into a number of distinct cultivation systems and is especially well suited to elevations between 2400 and 3000 m. It is utilized in a variety of dishes and preparations and is of fair to good nutritional value although amino acid analyses still need to be performed. It is probable that *O. tuberosa* was introduced through Spanish trade networks from South America during the past 200 to 300 years. Finally, researchers are encouraging the cultivation and utilization of *O. tuberosa* in other areas of the world. If given research and development priority, this crop could serve ultimately to expand the food base that supports the global community.

Acknowledgments

The fieldwork for this study was made possible in part by a grant from the Lawrence/Conoco Fund, a grant from the Mellon Foundation to the Institute of Economic Botany, and a grant from the Armand G. Erpf Fund to Dr. Gershoff, Dean, Tufts University School of Nutrition. The authors wish to thank Luis Amaya Acosta, Charles Peters and the Instituto Nacional de Investigaciones sobre Recursos Bióticos (INIREB) for their warmth and logistical support and Dr. Stanley Gershoff for his participation in the nutritional analysis. Special thanks is also due to Ms. A. Parra-King for logistical support in Michoacan. Thanks are also due to Michael Balick, David Williams, Ellen Dean and the late Tim Plowman for their assistance and suggestions on the manuscript. This is publication number 79 in the series of the Institute of Economic Botany of The New York Botanical Garden.

Literature Cited

Atkey, P. T. & A. A. Brunt. 1982. The occurrence of mycoplasma-like bodies in severely diseased Oca (*Oxalis tuberosa*) plants from Bolivia. Phytopathol. Z. **103**: 294–300.

Blanco, O. 1977. Investigación en el mejoramiento de tuberculos menores. En curso de cultivos Andinos. Serie informe de Cursos, Conferencias, y Reuniones, No. 117. IICA, La Paz, Bolivia.

Brunt, A. A., R. J. Barton, R. J. Philips & R. A. C. Jones. 1982a. *Ullucus* virus C, a newly recognized comovirus infecting *Ullucus tuberosus* (Basellaceae). Ann. of Appl. Biol. **101**: 73–78.

———, S. Philips, R. A. C. Jones, & R. H. Kenten. 1982b. Viruses detected in *Ullurcus tuberosus* (Basellaceae) from Peru and Bolivia. Ann. of Appl. Biol. **101**: 65–71.

Bukasov, S. M. 1930. The cultivated plants of Mexico, Guatemala and Colombia. Bull. Appl. Bot. Genet. and Pl. Breed. Suppl. 47. Transl. from Russian.

Cárdenas, M. 1948. Plantas alimenticias nativas de los Andes de Bolivia. Folia Universitaria (Universidad de Cochabamba, Bolivia) **2(2)**: 36–51.

———. 1958. Informe sobre trabajos hechos en Bolivia sobre oca, ulluco y mashua. Estudio sobre tuberculos alimentícias de los Andes. Commun. Turrialba **63**: 5–21. IICA.

Cortés, H. 1977. Avances en la investigación de la Oca. En I Congresso International de cultivos Andinos, Ayacucho. Serie de Cursos, Conferencias y Reuniones No. 178, IICA, Lima, Peru.

Gates, W. 1939. The De La Cruz–Badiano Aztec Herbal of 1552. Maya Soc. Publ. 22.

Gorter, C. J. 1965. Origin of fasciation. Encyc. Pl. Physiol. **15(2)**: 330–351.

Hodge, W. H. 1951. Three native tuber foods of the Andes. Econ. Bot. **5**: 185–201.

Hodgkinson, A. 1977. Oxalic acid in biology and medicine. Academic Press, New York.

IBPGR. 1982. Plant genetic resources in the Andean region. Proc. of meeting of IBPGR, IICA and JUNAC. Lima, Peru.

International Center for Integrated Mountain Development (ICIMOD). 1987. Mountain agriculture and crop genetic resources. Report of the International Workshop on Mountain Agriculture and Crop Genetic Resources, 16–19 February 1987, Kathmandu, Nepal. Kefford Press, Singapore. 47 pp.

International Potato Center. 1984. Potatoes for the developing world. Lima, Peru, 150 pp.

Kay, D. E. 1973. Root crops. Tropical Products Institute, London.

King, S. R. 1987. Four endemic Andean tuber crops: Promising food resources for agricultural diversification. Mountain Res. Develop. **7(1)**: 43–52.

——— & S. N. Gershoff. 1987. Nutritional evaluation of three underexploited Andean tubers: *Oxalis tuberosa*, *Ullucus tuberosus* and *Tropaeolum tuberosum*. Econ. Bot. **41**: 503–511.

——— & N. D. Vietmeyer. 1989. An evaluation of Andean root and tuber crops: Genetic resources for mountainous environments. Proc. Workshop on Mountain Agriculture and Crop Genetic Resources, 16–19 February 1987, Kathmandu, Nepal. India Book House, New Delhi, India.

León, J. 1958. Estudios sobre tubérculos alimentícios nativos de la region andina. Commun. Turrialba **63.**

———. 1964. Plantas alimentícias andinas. IICA, Zona Andina, Bol. Tecn. No. 6. Lima, Peru.

Martínez, M. 1936. Plantas utiles de Mexico. Mexico, Botas. 400 pp.

———. 1937. Catálogo de nombres vulgares y científicas de plantas Mexicanas. Mexico, Botas. 551 pp.

Obregozo, G. 1956. La estructura y variabilidad de las ocas Peruanas. Tesis de magister, IICA. Turrialba, Costa Rica.

Plucknett, D. L., N. J. Smith, J. T. Williams & N. M. Ashishetty. 1987. Gene banks and the world's food. Princeton University Press.

Prance, G. T. 1986. La taxonomia y su relacion a las ciencias agrícolas. Proc. of Meeting of the director and curators of Colombian herbariums. Medellín, June 1986.

Rea, J. 1985. Recursos fitogenéticos agrícolas de Bolivia. Base para establecer el sistema. CIRF, La Paz, Bolivia.

——— **& D. Morales.** 1980. Catalogo de tubérculos andinos. Publicación IBTA, La Paz, Bolivia.

Sahagun, Fr. B. De. 1979. Codice Florentino. El Manuscrito 218–20 de la colección palatina de la biblioteca medicea Laurrenziana. (Edition facsimilar.) Secretaria de gobernación, Mexico. 3 vols.

Tapia, M. E. 1981. Los tubérculos andinos. *En:* M. E. Tapia (ed.), Avances en las tubérculos alimenticios de los Andes. IICA-CIID, Lima, Peru.

Tola Cevallos, J., C. Nieto, E. Parolta & R. Castillo. (In press). Status of valuable native Andean crops in high mountain agriculture of Ecuador. Proc. of International Workshop on Mountain Agriculture and Crop Genetic Resources, 16–19 February 1987. Kathmandu, Nepal. India Book House, New Delhi.

Ugent, D. 1968. The potato in Mexico: Geography and primitive culture. Econ. Bot. 22(2): 108–123.

Vietmeyer, N. 1986. Lesser-known plants of potential use in agriculture and forestry. Science 232: 1379–1384.

———. 1987. Lost crops of the Incas. Encyclopedia Britannica Yearbook of Science and the Future, Chicago, Illinois.

Wellman, F. L. 1972. Tropical American plant diseases. Scarecrow Press, Metuchen, New Jersey.

White, J. W. 1975. Notes on the biology of *Oxalis tuberosa* and *Tropaeolum tuberosum.* Unpubl. Honors Thesis. Harvard Economic Botany Library.

Williams, L. O. 1978. Brief botanical notes. The Añu in Mexico. Econ. Bot. **32(1):** 104.

Local Product Markets for Babassu (*Orbignya phalerata* Mart.; Palmae) and Agro-Industrial Change in Maranhão, Brazil

Peter H. May

Table of Contents

Abstract

MAY, P. H. (The Ford Foundation, Praia do Flamengo, 100, 12º andar, 22210 - Rio de Janeiro - RJ, Brazil). Local product markets for Babassu (*Orbignya phalerata* Mart.; Palmae) and agro-industrial change in Maranhão, Brazil. Advances in Economic Botany 8: 92–102. 1990. This study describes market conditions as they affect the volume and flow of babassu palm (*Orbignya phalerata*) products to their end uses (chiefly soap and chemical production and siderurgical applications). The terms under which such goods are traded in the market are crucial in dictating whether extractive plant resources such as babassu can be managed for sustained production. The study traces the evolution and current trends in local babassu product markets in the state of Maranhão, Brazil, which are being transformed by agro-industrial change. The conclusions suggest that analysis of development potential for tropical plants having promising economic value should incorporate detailed study of price formation at the producer level, as well as the property rights and production relations which determine who benefits from technical change.

Key words: *Orbignya phalerata*; babassu; Maranhão; extractive plant products; oil; kernels.

Introduction

Research on extractive plant products has seldom probed into the intricacies of market conditions as they affect the volume and flow of such products to their end users. Market characteristics include such features as the terms of trade between producers, intermediaries and industries, and how these relations have responded historically to changes in national and global

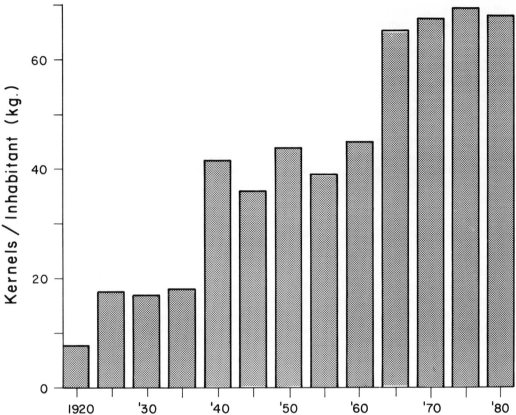

FIG. 1. Babassu kernels marketed per rural inhabitant, Maranhão: 1920–1980. Data are 5-year moving averages from IBGE, 1984 and previous volumes.

economies as well as in local land uses. Wild plant products are mostly extracted by peasants in remote regions having but limited interaction with national or foreign markets. When market channels do develop, as they did, for example, in the case of natural rubber, producers suffer from the monopsony power of middlemen who often also control access to land (Alves Pinto, 1984). This is unfortunate, for the terms under which extractive products are traded in the market and the tenure rights over the resources from which they are derived determine, in large measure whether such resources may be managed for sustained production.

The literature on babassu palm (*Orbignya phalerata* Mart.) is no exception in this regard. Babassu and other wild species of the genus *Orbignya* occur widely in the tropical Americas (Balick et al., 1987), providing a broad range of products important in market and subsistence economies (May et al., 1985a). It is estimated that babassu kernels, used in local industries in Maranhão and other Northeast Brazilian states to produce oil and feedcake, generated $135 million in revenues for the regional economy in 1979 (May, 1986). Over 400,000 rural households depend on babassu for some share of their incomes (Mattar, 1979). The multi-layered babassu fruit also is highly touted as a source for charcoal, tar and starch (MIC/STI, 1979).

Despite the present and potential importance to domestic markets of babassu kernels and charcoal for industrial purposes (chiefly soap and chemical production and siderurgical applications), studies of babassu have been limited to assessment of the technical characteristics of products and their prospective final markets. The structure of existing marketing channels is analyzed purely descriptively and only anecdotal information regarding price formation has been available in the literature. In like fashion, the miserable working conditions of babassu fruit gatherers are noted (Braga & Dias, 1968; Valverde, 1957), yet there is scanty recognition that the form of control over both babassu markets and the stands themselves may contribute to the gatherers' poverty or the potential for managing babassu stands in the future. The present study traces the evolution and current trends in local babassu product markets and producers' access to land in the state of Maranhão, Brazil, to the end of suggesting avenues for incorporating such information in economic botany research and development policy.

Supply Response to Market Expansion

Babassu fruits were initially harvested by indigenous groups as a hedge against famine (Steward, 1963), but since the early 1920's have increasingly been gathered by settlers to extract kernels to be sold as raw material for oil expression. Production was enhanced historically by an increase in the rural population and in demand for raw materials in the babassu oil industry. These factors resulted in more land being colonized by babassu palms and more fruit being collected for kernel extraction.

Babassu forests were formed in relatively fertile areas as the palm dominated areas cleared for shifting cultivation (Anderson, 1983). These palm forests literally blanket much of the landscape in zones occupied by agriculturalists in the transition zone between the semi-arid Northeast and the humid tropics of Northern Brazil. Intensified fruit harvesting in areas of fairly long-term human occupation explains much of the historical growth in kernel production. The average volume of kernels marketed per rural inhabitant in Maranhão more than quadrupled from an average of 15.3 kg per year in the period between 1920 and 1935 to 66.8 kg between 1965 and 1980 (Fig. 1).

Babassu kernel output has passed through three discrete phases. The first, between 1920 and 1935, corresponds with the period when kernels were predominantly traded as an export commodity. The second, from 1940 to 1960, is the period when babassu oil was mostly pressed from kernels transported to southern Brazil. The most recent period corresponds with the expansion of the oil industry in the babassu zone itself. Expansion in oil product markets as a result of economic development and population growth in Brazil as a whole increased the local demand for kernels, making babassu resale a more attractive activity for landowners, who came to rely on the kernels for an increasing share of their farm revenues during the 1960's (May, 1986).

Since the local development of the babassu industry, however, peasants have not significantly increased their production intensity of babassu kernel (1960–1980 in Fig. 1). Local demand for kernel had a significant effect on production behavior at the outset, but later growth in kernel output appears to have been due chiefly to dramatic population growth through migration of colonists to frontier lands rather than greater per-capita production. Transport costs from more distant sources, along with the rudimentary nature of kernel extraction combine to keep prices high.

Conditions in final product markets have also contributed to the stabilization of per-capita kernel production rates. Although the price of babassu oil has been sustained by restrictive import tariffs for substitutes such as coconut oil, the price-elasticity of babassu oil consumption in Brazil is low. This implies that, even were it possible to substantially reduce the cost of delivering babassu kernel to the oil industry, there would be little increase in consumption for traditional uses.[1] The final markets for such traditional products as oil-based soap and cooking oil have begun to shrink as synthetic detergents and less fatty edible oils have gained market share. To compensate for shrinking markets, the price industrialists and merchants pay for kernels has declined in real terms. This may explain why per-capita kernel production levels have not increased.

It is to the industry's advantage to make mar-

---

[1] This contention is based on an econometric analysis of babassu oil consumption response to price and to change in national product per capita for the period 1960–1984. The elasticity coefficient for oil price was positive but low at 0.15, indicating that a 10% decrease in oil price would only result in a 1.5% growth in consumption. Babassu oil consumption responded negatively to growth in per-capita national product in Brazil, which indicates that, as the economy grows, there is a tendency to substitute other products for babassu oil.

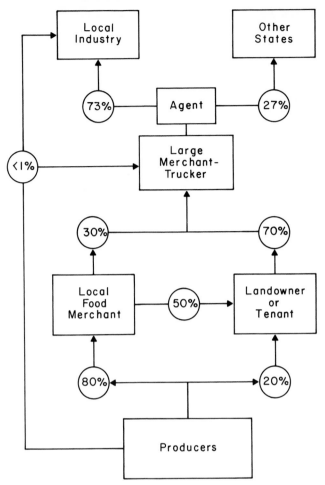

FIG. 2.   Intermediary structure of babassu kernel marketing in Maranhão (adapted from Mendes, 1979).

keting channels more efficient, so as to boost the value of babassu kernel to the producer, thus stimulating greater production intensity. Unfortunately, this is no simple process, as the other actors in the marketing chain: landowners, merchants and truckers, also have their own interests at stake. Landowners' decisions, in proscribing peasants' freedom to gain access to and market babassu products, critically affect supply of kernel to the industry. At root, the future of the oil industry thus appears to lie in the outcome of the interaction between landowner decisions and peasant labor allocation. The distribution of returns from kernel marketing at different levels in the chain is an important factor affecting this interaction.

Price Formation Along the Kernel Marketing Chain

Babassu kernel production and peasants' use of other products of the palm represent an integral part of the rural economy in Maranhão. Babassu kernel sale is one of the few available cash-generating options for the peasantry of the babassu zone during the period between annual crop harvests. Supply thus appears to respond fairly consistently with the degree to which kernel extractivism complements the peasant's lifestyle and economic opportunities. When the production relations around which these strategies are built begin to break down, the ability of kernel sales to fulfill their customary place in the house-

Table I

Historical marketing margins for babassu kernels

Source	Producer[a]	Land-owner–store-keeper[b]	Merchant[b]
Valverde (1957)	n.a.	20%	n.a.
Nicholls & Paiva (1966)	n.a.	36–50%	n.a.
Braga & Dias (1969)	45%	42%	22%
FAO (1974)	50–70%	7–33%	10–38%
Mendes (1978)	42–55%	21–40%	31%[c]
FTI (1982)	43–63%	26–50%[d]	14–25%
May (1986) average[e]	61%	25%	17.5%
Range	46–73%	9–44%	2–35%

Sources: Noted in table.

[a] Percent of final industry purchase price (equivalent to merchant's sale price).

[b] Gross margin = (sale price − purchase price)/sale price.

[c] Based on range of prices offered by oil firms in São Luís in February 1987.

[d] Latter margin (50%) was divided equally between a stand renter and a landowner.

[e] Weighted by the share of kernels marketed each month by respondents.

hold economy is threatened. Subsequently, labor will be allocated in different ways.

Babassu producers formerly lived on remote, isolated farms or state lands. Their atomism as production-consumption units has been progressively modified with the onset of agrarian change. In the past, landowners and merchants penetrating the subsistence economy were able to set kernel prices fairly arbitrarily, paying a smaller share of the industry's purchase price to the producer than they could if access to markets was not impeded. This situation was mitigated somewhat by the opening of transport corridors and lines of communication between urban and rural areas.

There are several alternate paths through which kernels are marketed. In rare cases, fruit collectors sell kernels directly to an industry or merchant/trucker. In the majority of cases, however, there are at least two intermediary levels in the marketing chain: the kernels are purchased by a storekeeper, resold to a merchant/trucker, and finally to the industry. In most cases, an additional link is added by the landowner, who pays a kernel-purchasing agent or storekeeper a per-

centage of the proceeds and/or allows him to plant crops or keep livestock on the property. The landowner then sells the kernels to the merchant/trucker. Another level may be added between the merchant and the industry in the form of an agent who lines up the sale and arranges for kernels to arrive at the plant, receiving a commission (Fig. 2).

An increase in the number of intermediary links effects a reduction in the share of the industry's purchase price received by the kernel producer and is widely considered to be the principal factor contributing to the costliness of the raw material. The degree of market integration achieved during the past decades has apparently improved the bargaining position of the peasantry with respect to their share of the proceeds, partially compensating the decline in real price of kernels. A comparison of data from the farm-level surveys conducted for this study with that from earlier research illustrates this point.

Braga and Dias (1968) found that producers received only 45 percent of the industry buying price in the mid-1960's, resulting in higher margins for intermediaries than those I found in my surveys. Surveys in the 1970's by FAO (1974) and Mendes (1981) report results similar to those I obtained (Table I). The price paid to producers surveyed in the *município* of Chapadinha between October 1983 and May 1984 averaged 61 percent of that paid by the industry, when weighted by the proportion of kernels sold monthly. The weighted gross margins on resale by landowners and storekeepers to large kernel merchants averaged 25 percent, while gross margins at the final leg in the kernel marketing chain were only 17.5 percent.[2]

Considerable variations are to be found in price policies at the level of the individual landholding. I discovered an average monthly range of 53 percent from the low to high prices paid for kernel between properties within the Chapadinha survey area. Prices in Chapadinha averaged somewhat higher than those in a second survey *município*, Lima Campos, but varied over twice as widely. Variation in quality within survey areas

---

[2] Gross margin is calculated as the difference between purchase and sales price over the sales price. This calculation describes the share of resale prices that make up the intermediary's profit before deducting transport and storage costs.

is minimal; consequently, this price variability is due principally to the greater isolation of some estates in Chapadinha from market information or access.

Despite the general prohibition against kernel marketing off the premises, respondents in Lima Campos agreed that collectors are motivated to carry kernels they extract off the property to sell elsewhere, even at only slightly better prices. Threat of expulsion is the principal sanction against resident laborers selling to other kernel buyers. Such coercion seems much more effective in Chapadinha. There, distances to paved roads and between rural stores tend to be greater than in other areas. Traditional landowners in Chapadinha engage in collusion to prohibit "leakage" of extractive products which constitute a greater share of their farms' revenues than is true in Lima Campos. Most landowners interviewed in Chapadinha said they would "call attention" to a producer's transgression from the residence contract, with a repeated offense being grounds for expulsion.

Harvest seasonality also influences kernel supply and pricing and hence individual returns. The seasonality of kernel production, corresponding to fruiting and harvest cycles, confers severe discontinuity on industrial operations (Fig. 3). Babassu fruits begin to mature and fall from the bunch starting in July and August. Gathering and kernel extraction are concentrated in the six-month period between October and March. It is clear from the graph (Fig. 3C), however, that a few firms are able to maintain some level of supply nearly year-long. This is explained by the fact that, in particularly moist areas, babassu produces year-round, and by storage by many households of babassu fruit for kernel extraction during rainy periods. Kernels, once extracted, are subject to increasing rancidity and weight loss so that producers and merchants rarely hold stocks for lengthy periods. Recent difficulties in securing working capital have made it financially untenable for industries to hold stocks.

Seasonal price relationships reflect the pattern of supply. Although some firms close down at the end of the peak harvest period, competition for the reduced raw material supply between the firms that remain in operation tends to push prices upward. A comparison of industry purchase prices per kg between the beginning of the harvest in October and its tail end in May from 1973 to 1983 reveals that there were real kernel price increases over the eight-month period in most years (Table II). As the harvest cycle begins anew in August and September and idle plants resume operations, there may be a brief spurt of even higher prices, but these tend to flatten out or actually decline—as they did in 1983—as the harvest begins to generate new supplies.

Prices for kernel increased rapidly in the survey year after this initial stabilization due to scarcities in foreign markets which allowed oil extractors to export a fair share of their overall production. The shortfall in oil supply to serve domestic demand placed a strong upward pressure on internal oil prices, as occurred ten years previously during a similar export market opening (Table II). Kernel prices at the producer level also increased dramatically, more than doubling in real terms between October and May.

The effect on producer prices of these oil price increases was considerably delayed, however. While the industry's buying price for kernel increased substantially in real terms as early as December, a matching response in local producer prices only began to take hold in February. By this time, producers had already sold over 70 percent of the total volume of kernels they reported extracting from October to May. By March, peasant families were already devoting more of their total time to weeding their rice fields and preparing for the coming harvest than to gathering babassu fruit.

The price increase failed to stimulate a marked increase in output among producers in the survey. Although there was a slight increase in February over January kernel sales, this appears to have been more a result of the freeing up of labor after the first weeding in rice fields than a response to price increases for kernel. A similar rise in kernel output has been observed for years in which price increases were less notable (Anderson & Anderson, 1983; Leal, 1972).

Output in April through June was negligible despite the continuing increase in real prices. It appears that the lion's share of the increased returns from the price hike were shared among merchants and landowners rather than filtering down to the producer level. Peasants would have benefited if the prices they received reflected the industry's increased purchase price early enough in the harvest season. By the time local prices increased, most babassu had already been har-

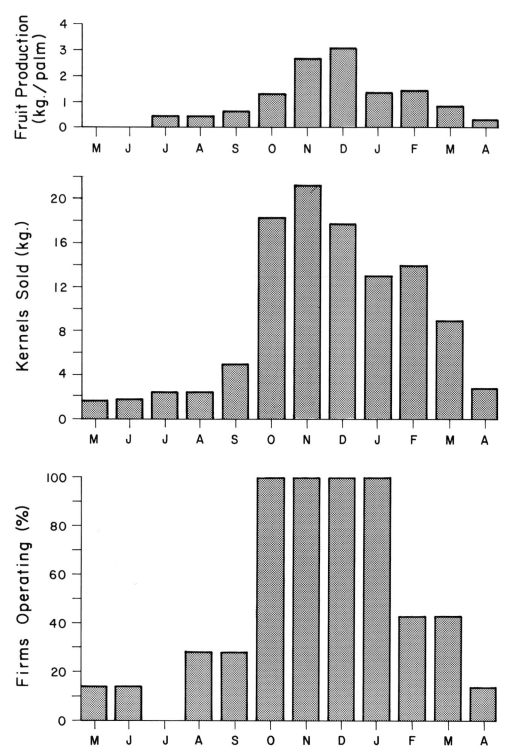

FIG. 3A–C. Seasonality in the babassu industry. **A,** Monthly fruit production per palm. Figure derived from Anderson (1983). **B,** Kernels sold by producers. July–Sept. kernel sales are producers' estimates; other months obtained through weekly survey and record-keeping by producers. **C,** Proportion of firms reporting activity at or near capacity.

Table II

Seasonal fluctuation in price per kg of babassu kernels, placed at the plant in São Luís

Year	Oct. price Cr$	May price Cr$	Change (%)	Inflation rate (%)	Real chg. (%)
1973–1974	1.75	3.45	97	15	82
1974–1975	3.10	2.50	(−)19	12	(−)7
1975–1976	2.40	3.20	33	30	3
1976–1977	4.20	5.70	36	25	11
1977–1978	5.65	7.45	32	21	11
1978–1979	8.25	12.00	45	28	17
1979–1980	16.75	19.75	18	56	(−)38
1980–1981	21.50	26.00	21	52	(−)31
1981–1982	36.50	92.50	153	44	109
1982–1983	132.00	150.00	13	104	(−)91
1983–1984	310.00	1050.00	239	91	148
Average			49	34	19

Sources: 1973–1977 from Mendes (1979); 1977–1981 from FTI (1982); 1982–1984 from CACEX, unpubl. data.

vested, and peasants were far more concerned with the rice harvest, which would feed them until babassu fruits began to fall again in August and September.

Since babassu kernel extraction is closely meshed with the overall peasant economy, a short-term reallocation of labor in response to price surges is not to be expected. Year-to-year variation in the productivity of babassu palms—associated with rainfall conditions as long as 18 months prior to fruit maturation—probably has a greater impact on the amount of kernels extracted than do momentary price shifts. When rainfall conditions are good, rice productivity is greater, and farmers will tend to devote less time to babassu gathering than to agricultural activities. However, when rainfall is low, babassu production will not be affected for some time, and peasants are able to absorb some of their crop losses by devoting more labor to babassu. Obviously, when drought conditions persist for some time, the crop and babassu harvests are both poor, but such climatic events are rare in the babassu zone.

Some analysts have suggested that there has been a progressive deterioration in the terms of trade between babassu kernels and goods that peasants obtain in exchange for them. Such deterioration, they suggest, might have motivated more permanent shifts in peasant production strategies. For example, Kono (*in* FIPES, 1982) blames the apparent stabilization in kernel supplies since 1980 on a decline in the kernel-to-

rice price ratio. This theory suggests that former kernel producers had been induced to allocate labor to other activities than babassu extraction by the relative price shift. To check this hypothesis, I compared the price ratio between rice and babassu kernels in 1973 and 1983 in Maranhão. There had indeed been some erosion in the relative prices of the two commodities: whereas one kg of babassu kernel was able to purchase 3.8 kg of rice in 1973, babassu kernel buying power fell to 2.3 kg of rice in 1983, declining 39 percent.

The findings of my research (May, 1986) suggest that babassu production intrinsically complements other sources of rural income by making use of otherwise idle household labor during the period between annual crop harvests, a period when few income generating alternatives exist. Relative price changes ultimately may effect alteration in the patterns through which rural households allocate labor. To bring about such a change, however, babassu prices would have to fall below what must be an extremely low labor opportunity cost before such a relative shift would bring about a substantial reduction in kernel extraction. Greater emphasis in research and policy considerations should be given to changes in land use and in peasant access to resources as determinants of supply of extractive plant products.

In the early 1970's, expansion in the road network and generous rural credit terms to large-scale investors stimulated land-use changes throughout the region. Agro-industries began to produce sugar, alcohol and cellulose. Raw material production for these industries occupied the lands previously covered with babassu stands. The trunks of palms removed from these areas were used to produce inferior quality cardboard. Other projects aimed at fattening beef cattle on planted pastures. In the period from 1967 to 1984, government subsidies and land concessions were provided for ranches occupying 1.35 million ha in the babassu zone (May, 1986). This rapid expansion of cattle pastures was fueled by government-built infrastructure, abundant credit, attractive fiscal incentives, cheap land, and a general desire to consolidate landholdings with minimum investment.

Pasture expansion in the babassu zone is often also accompanied by clearcutting of the palm stands. The need for regular suppression of juvenile palms to reduce competition with pasture grasses leads ranchers to view babassu as a weed that should be eradicated. Besides invasion by juvenile palms, people who trespass to gather babassu fruits in ranches are perceived as interfering with pasture management. These people are blamed for starting wildfires, cutting fences, and leaving behind fragments of fruit husks that can cause injury to the hooves of cattle. To rid themselves of such incursions, ranchers are increasingly clearcutting the palm forests.

Changes in land use and rural enterprises have restricted access by peasants to areas for babassu gathering and crop production. The complementarity between these sources of income has been undermined by such change. Babassu kernel resale becomes less attractive to landowners as their enterprises shift toward more intensive uses of land. The oil industry has in turn suffered from reduced raw material supply, further restricting production of babassu oil. This process has stimulated the final product market to opt for other alternatives. Industrialists within the babassu zone have therefore intensified their historical search for improvements in babassu fruit processing technology, hoping to reduce kernel extraction costs and obtain additional sources of revenue from by-products.

Implications of Whole Babassu Fruit Markets

Recent industrial innovations promise a more complete utilization of babassu fruit for strategic products such as charcoal, ethanol and tar as well as the traditional oil and feedcake. Modernizing landowners appear to find whole fruit marketing more attractive than the traditional system of kernel resale (Anderson & Anderson, 1983). This shift has motivated a permanent alteration in methods of babassu production and resource access in recent years. Such changes, described in detail below, are compatible with broader changes in rural enterprise organization and land use underway in the babassu zone, but incompatible with peasant agriculture and extractivism.

None of the whole fruit processing firms were operating at full capacity during the period my field research was in progress. Nevertheless, their four years of wide-ranging efforts to secure supplies of whole fruit through 1980 and their continuing purchase of charcoal and husks had brought about lasting changes in production relations for babassu collection as well as in the distribution of returns from product marketing.

Landowners who have converted their operations primarily or exclusively to pasture have found it convenient to alter arrangements for collection and marketing of babassu. Among the more prominent production schemes being adopted is that of consignment of the fruit collection enterprise to one or more of the rancher's herdsmen. These in turn guard the perimeters of the properties from incursions by collectors seeking free entry. The herdsmen then contract labor on a daily basis to gather fruit. In this system, the returns from babassu constitute a substantial portion of the herdsmen's salaries. They thus have an incentive to pay the minimum possible rate to collectors.

In the absence of demand for whole fruit, the herdsmen or landowners consign the fruit to collection centers, whether on their property or in a nearby community. Individual entrepreneurs, sometimes financed by a kernel or charcoal buyer, set up collection centers as a mini-factory. There, women are engaged to break the fruit manually. In payment, they are entitled to half of the kernels they extract, but must sell these to the collection center operator at his price, usually 10 percent or more below the going local kernel price. In addition to being required to split the value of the kernels they extract, the women are often also required to convert the husks to charcoal, which remains the property of the center operator.

In other areas, the growing market for charcoal

and fruit husks has provided an incentive for landowners to insist that these by-products now become theirs as rent for the right of access to fruit collection. A more recent phenomenon is that of large-scale purchases of whole fruit to be used for fuel in other states without removal of the kernel. Such buyers are having increasing success in contracting for supplies. Landowners of substantial babassu groves in pastures are reported to find it more convenient to harvest fruit off the ground with front-end loaders rather than paying workers to collect at piece-rate. Even where they continue to be employed, gatherers of whole fruit are rarely allowed to retain fruit for subsistence purposes.

Studies of the impacts of whole fruit markets point to increasing dependence on babassu for incomes that are being eroded by expulsion from agriculture (Anderson & Anderson, 1983; Cunha, 1979). Yet this income is being earned by men, rather than women. The technical innovation in the babassu industry demands that fruit be gathered in large volume (up to two tons daily per worker) and transported in continuous flows to a central processing unit rather than broken manually at home and sold when convenient. Granted that women would be just as capable as men of accomplishing the task, the reduction in employment needs for whole fruit processing over manual kernel extraction will likely result in less babassu income being controlled by women.

However, it is not the innovation in the industry alone that is causing problems of resource access and distribution in the babassu zone. Conversion of land use to pasture and expulsion of peasants from rural properties, occurring in tandem with the innovative industry, imply drastic alterations in rural employment and income distribution. The distributive implications of the transformation in babassu processing technology are eloquently summarized by rural workers union members:

Everybody suggests that babassu is not profitable, that babassu impoverishes everyone. Babassu put everyone out on the street, but . . . it was the livestock projects that were implanted in areas where babassu stands were cut down. . . . How many people have gone hungry because of these palms that have been cut down? How can we say that we're going to modernize whole fruit processing, and that this will make the landowner, the great proprietor, preserve babassu stands? How can we say that this is going to mean better conditions for the caboclo

(peasant)? . . . It's going to worsen the situation; it means more unemployment, on account of a product that, like it or not, is natural. . . . We have to make better progress, industrialize, change the systems that have been there for so long, but we have to study also some means to employ people, with the goal that . . . you don't take this all at once from the caboclo, *because it means another form of mortality, another system of hunger. Let's pay attention to this because I will never concede that all of a sudden with this whole fruit, this is going to give income to the* caboclo. *This income is going to enter into the capitalist's pocket.*

This is the problem that I see. Now, your proposition I accept—industrialization is an easier means to do work. I want to see this done and also that we work to help the suffering people that are going to end up without jobs.[3]

To reverse the current process of clearcutting of babassu stands and denial of access to resources by peasants reliant on babassu requires fundamental shifts in development policy. Rather than plan large-scale industrialization of babassu fruit processing, efforts currently underway by agricultural researchers in the region to develop technology adaptable to the farm and community level should be supported. Agroforestry practices involving integration of babassu with pastures and other crops have been identified, and deserve further research to enhance productivity (May et al., 1985b). Most important and even more difficult, the peasants' property rights to the palm forest must be secured. Contractual arrangements for access to babassu stands that have until now been left up to a structurally imperfect market require policy intervention to assure that the problems described here may be resolved.

Conclusion

The data presented above offer an historical perspective on adjustments in local markets for products of a major native plant resource that have occurred in response to changes in regional development patterns and in industrial technology. Such research is valuable in pinpointing the probable impacts of proposals to make greater use of native plants as resources for local economic development. Avenues for ameliorating

---

[3] Representatives of the Federation of Agricultural Workers of Maranhão, quoted from round-table discussion in FIPES (1982). Author's translation.

negative social effects can be identified where they might arise. It is therefore suggested that future studies of tropical plants having promising economic value incorporate detailed study of price formation at the producer level, as well as the property rights and production relations which determine who benefits from technical change.

Acknowledgments

This is publication number 135 in the series of the Institute of Economic Botany of the New York Botanical Garden. Research for this article was conducted with partial support of the New York Botanical Garden under USAID Grant #DAN-SS42-G-SS-1089-00.

Literature Cited

Alves Pinto, N. P. 1984. Política da borracha no Brasil; a falência da borracha vegetal. Hucitec, São Paulo.

Anderson, A. 1983. The biology of *Orbignya martiana* (Palmae) a tropical dry forest dominant in Brazil. Doctoral Dissertation. Botany Dept., University of Florida, Gainesville.

——— **& E. S. Anderson.** 1983. People and the palm forest: Biology and utilization of Babassu palms in Maranhão, Brazil. Botany Dept., University of Florida, Gainesville.

Balick, M. J., C. B. Pinheiro & A. B. Anderson. 1987. Hybridization in the babassu palm complex: I. *Orbignya phalerata* × *O. eichleri*. Amer. J. Bot. **74(7):** 1013–1032.

Braga, H. C. & C. Dias. 1968. Aspectos socio-econômicos do Babaçu. Instituto de Oleos, Rio de Janeiro.

Cunha, B. 1979. Industrialização integral do coco babaçu: efeitos sobre a renda e o emprego dos pequenos produtores no estado do Maranhão. Masters Thesis. Universidade Federal de Viçosa, Minas Gerais.

FAO. 1974. Draft report of the Brazil babassu project. 3 vols. FAO Investment Center, Rome.

FIPES. 1982. Realidade e perspectivas do babaçu: mesa redonda. Fundação Instituto de Pesquisas Econômicas e Sociais, São Luis.

FTI. 1982. Babaçu: avaliação técnico-econômica como fonte de energia. Fundação de Tecnologia Industrial, Núcleo de Estudos e Engenharia Econômica, Rio de Janeiro.

IBGE. 1984. Anuário estatístico do Brasil: 1983. Fundação Instituto Brasileira de Geografia e Estatística, Rio de Janeiro.

Leal, M. 1972. Indústrias de óleos vegetais do Maranhão e Piauí. Governo do Estado do Maranhão, Secretaria de Planejamento, São Luís.

Mattar, H. 1979. Industrialization of the babaçu palm nut: the need for an ecodevelopment approach. Conference on Ecodevelopment and Ecofarming. Seminar Center of the German Foundation for International Development, Berlin.

May, P. 1987. A modern tragedy of the non-commons: Agro-industrial change and equity in Brazil's babassu palm zone. Latin American Program Dissertation Series, Cornell University, Ithaca, New York.

———, **A. Anderson, M. Balick & J. M. Frazão.** 1985a. Subsistence benefits from the babassu palm (*Orbignya martiana*). Econ. Bot. **39:** 113–129.

———, ———, **J. M. Frazão & M. Balick.** 1985b. Babassu in the agroforestry systems in Brazil's Mid-North region. Agroforestry Syst. **3:** 275–295.

Mendes, A. M. 1979. Monografia do babaçu. Comissão Estadual de Planejamento Agrícola, Estado do Maranhão, São Luís.

MIC/STI. 1979. Coco do babaçu: matéria prima para carvão e energia. Ministério da Indústria e do Comércio, Brasília.

———. 1982. Mapeamento e levantamento do potencial das ocorrências de babaçuais, Estados do Maranhão, Piauí, Mato Grosso e Goiás. Ministério da Indústria e do Comércio, Brasília.

Nicholls, W. H. & R. M. Paiva. 1966. The Itapecuru valley of Maranhão: Caxias. *In:* Ninety-nine fazendas: The structure and productivity of Brazilian agriculture. Chapter 2. Center for Latin American Studies, Vanderbilt University, Nashville.

Steward, J. H. 1963. Handbook of South American Indians. U.S. Government Printing Office, Washington, D.C.

Valverde, O. 1957. Geografia econômica e social do babaçu no Meio-Norte. Rev. Bras. de Geog. **19:** 381–418.

The Poisonous Anacardiaceae Genera of the World

John D. Mitchell

Table of Contents

Abstract

MITCHELL, J. D. (Herbarium and Institute of Economic Botany, The New York Botanical Garden, Bronx, New York 10458-5126). The poisonous Anacardiaceae genera of the world. Advances in Economic Botany **8**: 103–129. 1990. Twenty-five of the approximately seventy-six genera of Anacardiaceae are reported to be poisonous. Eight of these genera require further study in order to confirm their toxicity to humans. Dermatitis induced by contact with Anacardiaceae oleoresins is a cell-mediated, delayed hypersensitivity reaction. A brief discussion of contact dermatitis, other medical problems, and the chemistry of Anacardiaceous oleoresins is summarized. A few of these genera have important medical applications. A major emphasis of this paper is the identification of poisonous genera. Regional keys to specimens with flowers or fruits, generic descriptions, and distribution maps for all of the poisonous genera are provided.

Key words: Anacardiaceae; catechols; contact dermatitis; mango; oleoresins; poison ivy; resin canals; rengas trees.

Introduction

The Anacardiaceae are a moderately large family of approximately 76 genera and 600 species. Particularly well known are those members which produce poisonous sap that induces contact dermatitis in susceptible individuals. The temperate genus *Toxicodendron* is perhaps the

Advances in Economic Botany 8: 103–129, 1990
© 1990 The New York Botanical Garden

most notorious and has been the subject of numerous studies by botanists, chemists and the medical profession (e.g., Baer, 1979; Dunn et al., 1986; Evans & Schmidt, 1980; Gillis, 1971; Lejman et al., 1984; Mitchell & Rook, 1979; Mitchell et al., 1985; Muenscher, 1972). However, the majority of poisonous Anacardiaceae are tropical (Maps I–III) and poorly known.

Twenty-five (thirty-three percent) of the approximately seventy-six anacardiaceous genera are convincingly documented as being poisonous. A few additional genera may also be poisonous since they are related to known poisonous genera (e.g., *Cardenasiodendron,* a segregate of *Loxopterygium*; *Bonetiella,* formerly placed in *Pseudosmodingium*; *Drimycarpus,* a segregate of *Holigarna*).

Poisonous Anacardiaceae belong to four of the five tribes as defined by Engler (1882, 1892). The poisonous genera *Anacardium, Gluta, Mangifera* and *Swintonia* are members of the tribe Anacardieae. *Holigarna, Melanochyla* and *Semecarpus* are in the small tribe Semecarpeae, *Comocladia, Metopium, Toxicodendron,* etc., are in the largest tribe, Rhoeae, and *Spondias* is in the tribe Spondiadeae. Apparently, none of the genera in the small tribe Dobineae are poisonous.

The primary purpose of this paper is to summarize what is known about the poisonous genera of Anacardiaceae with special emphasis placed on their identification, without, however, discussing individual species in detail. Likewise, a detailed review of the medical and chemical literature on poisonous Anacardiaceae is beyond the scope of this paper.

Induced Dermatitis

The cause of many cases of contact dermatitis, particularly in the tropics, is the handling or ingestion of anacardiaceous products. Among the most economically important species that can produce dermatitis are the cashew-nut tree (*Anacardium occidentale* L.), the mango tree (*Mangifera indica* L.), the Brazilian pepper tree (*Schinus terebinthifolius* Raddi), the varnish or lacquer tree (*Toxicodendron vernicifluum* (Stokes) Barkley), the Japanese wax tree (*Toxicodendron succedaneum* (L.) Kuntze), the dhobi or marking-nut tree (*Semecarpus anacardium* L.f.) and the red quebracho (*Schinopsis quebracho-colorado* (Schlecht.) Barkley & Meyer.) (Figs. 4, 5).

The principle function of anacardiaceous oleoresins is probably to serve as a defense system against herbivory by insects and vertebrates (Joel, 1980; Levy & Primack, 1984).

The offending oleoresins are present in virtually all plant parts with resin canals. These secretory or resin canals are usually located in the roots, bark, petiole, leaf blade, occasionally in the stem cortex and pith pedicels, calyx, corolla, and exocarp and mesocarp of the drupe (Fig. 1) (Barkley & Reed, 1940; Copeland, 1961; Ding Hou, 1978; Gillis, 1971; Joel & Fahn, 1980a, 1980b; Möbius, 1899; Venning, 1948).

Oleoresins are not present in the stamens and pollen grains (Copeland, 1961; Copeland & Doyel, 1940). Very large resin or oil cavities are present in the mesocarp of *Anacardium occidentale* and *Semecarpus anacardium* (Roth, 1977). The canals originate schizogenously, lysigenously or schizolysigenously (Copeland, 1961; Copeland & Doyel, 1940; Venning, 1948).

Dermatitis due to contact with poisonous Anacardiaceae is a source of misery for millions and an economic burden as well. For example, Western Poison Oak (*Toxicodendron diversilobum* (Torrey & Gray) Greene) is a major culprit in workman's compensation cases in California (Lampe & McCann, 1985).

The first time an individual comes into contact with these oleoresins (allergens) a case of dermatitis usually does not develop. However, higher concentrations of oleoresins such as occur in cashew nutshell liquid act as primary irritants (Baer, 1983; Evans & Schmidt, 1980). The second exposure to the allergens will often produce an inflammatory response (Dunn et al., 1986; Lampe & Fagerstrom, 1968; Lewis & Elvin-Lewis, 1977). Worse yet, an individual sensitized to one species of poisonous Anacardiaceae will probably develop dermatitis after an initial contact with other species, even of a different genus (Behl & Captain, 1979; Lampe & McCann, 1985). Conversely, some individuals are very fortunate in that they do not respond to multiple exposures to anacardiaceous oleoresins (Baer, 1979).

The Immune Response

The immune response to the oleoresins is a cell-mediated, delayed, hypersensitivity reaction (Baer, 1983; Dunn et al., 1986; Lampe & McCann, 1985; Lewis & Elvin-Lewis, 1977). The

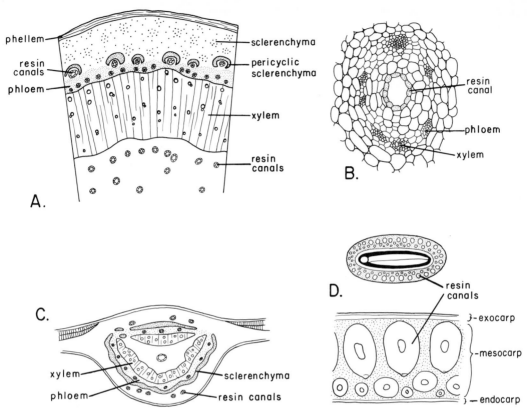

FIG. 1. Resin canals and their locations in Anacardiaceae tissues. **A.** Transverse section of *Gluta renghas* L. petiole (×30) (based on fig. XIX, p. 18 *in* Goris (1910)). **B.** Transverse section of mesocarp of *Toxicodendron vernicifluum* (Stokes) F. A. Barkley (=*Rhus vernicifera* DC.). Magnification ca. ×200 (based on fig. 28. II, p. 236 *in* Möbius (1899)). **C.** Transverse section of *Anacardium humile* St. Hil. leaf blade (×30) (adapted from fig. X, p. 13 *in* Goris (1910)). **D.** Schematic diagram of *Toxicodendron vernicifluum* drupe (×5) and resin canals in the mesocarp (ca. ×30) (adapted from fig. 27. IV, p. 234 *in* Möbius (1899)).

oleoresins or their derivatives act as haptens that covalently bind to skin-proteins (Dunn et al., 1986; Lampe & McCann, 1985). The resulting skin-protein-oleoresin complex becomes the antigen that is recognized by the body's T-cells (Thymus-dependent cells). The immune response is also directed by T-cells (Lampe & McCann, 1985). The hypersensitivity reaction can vary from mild to severe (viz., requiring hospitalization), depending on the type of oleoresin, the location of contact, amount of skin or other tissues exposed and the individual's immune system (Hardin & Arena, 1974; Lampe & McCann, 1985; Morton, 1971).

The latency period or reaction time between contact and immune response usually varies from twelve hours to five days (Baer, 1979; Behl &

Captain, 1979; Gillis, 1971; Hardin & Arena, 1974; Lampe & Fagerstrom, 1968; Lampe & McCann, 1985; Lewis & Elvin-Lewis, 1977). In man and susceptible animals the latency period increases with age (Lejman et al., 1984). In *Toxicodendron* the latency period also increases with length of time (in years) since the previous episode of contact dermatitis (Johnson et al., 1972).

Dermatitis

The immune response is characterized by an initial erythematous edema followed by the eruption of papules, vesicles and/or bullae. The arrangement of vesicles and bullae is frequently linear (Behl & Captain, 1979; Levy & Primack, 1984). The skin eruptions often itch, burn, sting and exude serous fluid (Lampe & McCann, 1985). Contact with this serous fluid will not spread the dermatitis to other parts of the body (Lampe & Fagerstrom, 1968; Lampe & McCann, 1985). Occasionally, an indelible black stain remains on the skin after contact with poisonous sap (Morton, 1971).

The nature of the dermatitis varies depending on what part of the body is exposed to oleoresin. If contact is made with the eyelids or scrotum, prominent erythematous edema is the result. On the other hand, the soles of the feet, the palms of the hand and the scalp covered with hair are relatively insensitive to oleoresins (Behl & Captain, 1979; Lampe & Fagerstrom, 1968; Lampe & McCann, 1985).

Other Symptoms

The ingestion of leaves and drupes (especially unripe ones) may lead to violent gastroenteritis, hemorrhoids, headaches, respiratory distress and sometimes death (Allen, 1975; Campbell, 1983; Lampe & Fagerstrom, 1968; Levey & Primack, 1984; Morton, 1971, 1976; Watts & Breyer-Brandwijk, 1962; Westbrooks & Preacher, 1986).

The oleoresins may be spread by contaminated clothing, dogs, cats, other animals, lacquered furniture, sawdust, and smoke (Ding Hou, 1978; Hardin & Arena, 1974; Lampe & McCann, 1985; Morton, 1971; Rendle, 1929). The smoke from burning poisonous Anacardiaceae may cause respiratory tract inflammation, severe dermatitis and temporary blindness (Ding Hou, 1978; Levy & Primack, 1984; Morton, 1971; Westbrooks & Preacher, 1986). Some people believe that it is unwise to shelter beneath rengas trees (collective term for *Gluta, Melanochyla, Semecarpus,* and *Swintonia*) when it is raining (Corner, 1952).

Some additional medical problems have been associated with Anacardiaceae. Unidentified volatile compounds emanating from the inflorescences of *Schinus terebinthifolius* and *Mangifera indica* apparently produce asthma-like reactions in some individuals (Campbell, 1983; Morton, 1971).

One species, *Schinopsis quebracho-colorado* contains chemical compounds that induce photodermatitis (Lampe & Fagerstrom, 1968). Photodermatitis is caused by chemical compounds, mostly furocoumarins, which become primary irritants or allergens when exposed to certain wavelengths of visible or ultra-violet light (Evans & Schmidt, 1980).

Medical Role of the Poisonous Anacardiaceae

The poisonous genera can be divided into three groups: 1) a primary group, well documented as causing severe contact dermatitis and other serious medical problems; 2) a secondary group of genera that usually cause less severe dermatitis and other symptoms (Table I); and 3) a third group of genera which may or may not be inducers of contact dermatitis. The evidence for considering them to be poisonous is not very well substantiated (Table II).

The roles that the poisonous Anacardiaceae play in medicine are not entirely negative. They serve as experimental models in the study of the immune system (Baer, 1979; Baer et al., 1967; Dunn et al., 1986; Johnson et al., 1972). Two of the most dreaded genera, *Toxicodendron* (Duke & Ayensu, 1985; Kim & Ju, 1982) and *Semecarpus* (Goudgaon et al., 1985; Indap et al., 1983; Prasad et al., 1985), have potential application in the war against cancer. In addition, cashew nut shell liquid (derived from *Anacardium occidentale*) has several uses, such as the control of mosquito larvae and schistosomiasis vectors, and as an antimicrobial agent (see refs. in Mitchell & Mori, 1987).

The toxic oleoresins have important applications in industry. *Semecarpus anacardium* shell liquid (oil) is used as an indelible ink to mark laundry (Behl & Captain, 1979), *Toxicodendron* spp. and *Gluta* spp. are the sources of lacquer for fine furniture in Asia (Behl & Captain, 1979; Burkill, 1935), and cashew nut shell liquid, a byproduct of cashew nut processing, is used in the manufacture of brake linings, various plastics, etc. (Mitchell & Mori, 1987).

Chemistry of the Dermatitis-inducing Compounds

Virtually all of the anacardiaceous oleoresins that induce contact dermatitis are mixtures of phenolic compounds. They vary primarily in the length, branching, number and position of double bonds, in the hydrocarbon side chain and in the number and position of hydroxy groups on

Table I

Anacardiaceae genera of primary and secondary importance as inducers of contact dermatitis

Primary	Secondary
Anacardium	*Mangifera*
Comocladia	*Schinopsis*
Gluta (incl. *Melanorrhoea*)	*Schinus*
Holigarna	
Lithrea	
Loxopterygium	
Mauria	
Melanochyla	
Metopium	
Pseudosmodingium	
Semecarpus	
Smodingium	
Swintonia	
Toxicodendron	

Table 2

Anacardiaceae genera that may induce contact dermatitis, but require further substantiation

Astronium	*Parishia*
Blepharocarya	*Pentaspadon*
Campnosperma	*Spondias*
Cotinus	*Trichoscypha*

the benzene ring (Baer, 1979; Baer et al., 1980; Baylis et al., 1981; Behl & Captain, 1979; fig. 2 in Evans & Schmidt, 1980; Gross et al., 1975; Lloyd et al., 1980; Murthy et al., 1968; Robinson, 1980; Tyman & Morris, 1967). The unrelated families Ginkgoaceae and Proteaceae also contain dermatitis-inducing phenolic compounds of very similar structure (Baer, 1983; Evans & Schmidt, 1980; Woods & Calnan, 1976).

The oleoresins of only twelve of the tweny-five known or suspected poisonous genera are documented. In most of these genera only one to a few species per genus have been studied. The genera which have been analyzed are 1) *Anacardium,* 2) *Campnosperma,* 3) *Gluta* (incl. *Melanorrhoea*), 4) *Holigarna,* 5) *Lithrea,* 6) *Mangifera,* 7) *Metopium,* 8) *Pentaspadon,* 9) *Schinus,* 10) *Semecarpus,* 11) *Smodingium,* and 12) *Toxicodendron.* The references that document their chemistry are contained in the generic accounts section of this paper.

CATECHOL DERIVATIVES AND SIMILAR STRUCTURES

Catechols

Pentadecyl-catechols (urushiols) have been isolated from several genera, such as *Gluta, Holigarna* (fig. 2, table II in Evans & Schmidt, 1980), *Lithrea* (Gambaro et al., 1985), *Metopium* (fig. 2, table II in Evans & Schmidt, 1980; Gross et al., 1975), *Semecarpus* (fig. 2, table II in Evans & Schmidt, 1980; fig. 2 in Woods & Calnan,

1976), *Smodingium* (fig. 2, table II in Evans & Schmidt, 1980), and *Toxicodendron* (fig. 2 in Baer, 1983; fig. 2, table II in Evans & Schmidt, 1980; fig. 2 in Woods & Calnan, 1976).

Heptadecyl catechols are found in several genera, such as *Gluta* (fig. 2, table II in Evans & Schmidt, 1980; fig. 2 in Woods & Calnan, 1976), *Lithrea* (Gambaro et al., 1985), *Metopium* (fig. 2, table II in Evans & Schmidt, 1980; Gross et al., 1975), *Semecarpus* (fig. 2, table II in Evans & Schmidt, 1980), and *Toxicodendron* (fig. 7.3 in Baer, 1979; fig. 2, table II in Evans & Schmidt, 1980).

Resorcinols

Pentadecyl resorcinols have been isolated from *Anacardium occidentale* (fig. 7.3 in Baer, 1979; fig. 2, table II in Evans & Schmidt, 1980).

Heptadecyl resorcinols are characteristic of *Mangifera indica* (Cojocaru et al., 1986).

Salicylic acid derivatives (anacardic acid, etc.) have been isolated from the resins of *Anacardium occidentale* and *Pentaspadon* spp. (fig. 2, table II in Evans & Schmidt, 1980).

Other Types of Phenols

Pentadecyl phenols have been extracted from *Anacardium occidentale* (fig. 7.3 in Baer, 1979; fig. 2, table II in Evans & Schmidt, 1980).

Quinols have been isolated from *Campnosperma auriculatum* (B1.) Hook. (Thomson, 1971).

Identification of Poisonous Genera of Anacardiaceae

Poisonous Anacardiaceae are trees, shrubs, suffrutices or vines with usually alternate (opposite in *Blepharocarya*), simple or compound, exstipulate leaves. The flowers are small, usually with actinomorphic flowers and usually have either intra- or extrastaminal disks. The ovule is apotropous, and there is only one ovule per locule. The fruit is variable, but fundamentally a drupe.

Resin canals are found in most of the tissues of these genera. The resin often hardens and turns black when exposed to air, and Anacardiaceae can often be recognized by the presence of black resin splotches on their leaves and inflorescences (Ding Hou, 1978).

Keys to Poisonous Genera of Anacardiaceae

The keys are regional and artificial in order to facilitate the identification of genera.

KEY TO POISONOUS GENERA OF ANACARDIACEAE OF THE UNITED STATES, CANADA, WEST INDIES, MEXICO AND CENTRAL AMERICA (BASED ON FLOWERING MATERIAL)

1. Leaves always simple.
  2. Style one.
    3. Leaves not lepidote, without stellate hairs. Stigma unlobed, not discoid.
      4. Leaves broadly to narrowly obovate. Petals lanceolate to linear. Disk absent. Stamens 9–12.
      .... 1. *Anacardium.*
      4. Leaves usually lanceolate. Petals elliptic or obovate. Extrastaminal disk prominent. Stamens
      5. .... 11. *Mangifera* (introd.).
    3. Leaves lepidote on both surfaces, stellate pubescent abaxially. Stigma lobate, discoid. ....
    .... 4. *Campnosperma.*
  2. Styles three. .... 6. *Cotinus.*
1. Leaves compound.
  5. Leaflets with intramarginal vein. Styles (3–)4–5. Ovary (3–)4–5 locular. .... 22. *Spondias.*
  5. Leaflets without intramarginal vein. Styles one or three. Ovary unilocular.
    6. Calyx of three sepals; corolla of three petals. .... 5. *Comocladia*
    6. Calyx of five sepals; corolla of five petals.
      7. Leaves trifoliolate. Scandent shrubs or climbing vines. ....
      .... 24. *Toxicodendron* section *Toxicodendron*
      7. Leaves with more than three leaflets. Trees and shrubs.
        8. Stamens 10.
          9. Leaflets usually sessile or subsessile. Calyx lobes deeply incised. ...19. *Schinus* (introd.).
          9. Leaflets petiolate. Calyx with shallow lobes. .... 12. *Mauria.*
        8. Stamens 5.
          10. Sepals accrescent in pistillate flowers. Ovule attached at apex of locule. .. 2. *Astronium.*
          10. Sepals not accrescent. Ovule attached at base on locule.
            11. Leaflets 3–7(–9). Inflorescences as long or longer than leaves. ....
            .... 14. *Metopium* (Fig. 4).
            11. Leaflets 9–27. Inflorescences shorter than leaves.
              12. Inflorescence erect. Habitat arid tropical scrub or tropical dry forest. .....
              .... 17. *Pseudosmodingium.*
              12. Inflorescence pendant. Habitat tropical moist forest. ....
              .... 24. *Toxicodendron* section *Venenata.*

KEY TO POISONOUS GENERA OF ANACARDIACEAE OF THE UNITED STATES, CANADA, WEST INDIES AND CENTRAL AMERICA (BASED ON FRUITING MATERIAL)

1. Leaves always simple.
  2. Pedicels plumose, often not bearing fruits. .... 6. *Cotinus.*
  2. Pedicels not plumose, usually bearing fruits.
    3. Leaves lepidote on both surfaces and stellate pubescent abaxially. Drupe often two-locular. ...
    .... 4. *Campnosperma.*
    3. Leaves not lepidote, without stellate hairs. Drupe always one locular.
      4. Hypocarp sigmoid or pyriform. Leaves broadly to narrowly obovate. Drupe reniform, mesocarp woody. .... 1. *Anacardium.*

   4. Hypocarp absent. Leaves usually lanceolate. Drupe oblong, mesocarp fleshy. . . . .
. . . . 11. *Mangifera* (introd.).
1. Leaves compound.
  5. Leaflets with intramarginal vein. Drupe (3–)4–5 locular. . . . . 22. *Spondias*.
  5. Leaflets without intramarginal vein. Drupe one locular.
   6. Drupe reniform, with two broad lateral wings. . . . . 17. *Pseudosmodingium*.
   6. Drupes globose or obovoid, lacking lateral wings.
    7. Calyx enlarged and persistent in fruit. . . . . 2. *Astronium*.
    7. Calyx not enlarged nor persistent in fruit.
     8. Exocarp, separating from mesocarp when drupe is ripe. Endocarp bony.
      9. Exocarp usually pink or red; mesocarp not white, resinous, without black striations.
       Seed suspended from apex or locule. . . . . 19. *Schinus* (introd.).
      9. Exocarp white, cream-colored or grey; mesocarp white, waxy with black striations.
       Seed basally attached to locule. . . . . 24. *Toxicodendron*.
     8. Exocarp not separating from mesocarp. Endocarp chartaceous or cartilaginous.
      11. Seed suspended from apex of locule. El Salvador to Panama. . . . . 12. *Mauria*.
      11. Seed attached at base of locule. Mexico to Guatemala and Belize and the West Indies.
       12. Leaflets 3–7, margins always entire. Leaves not aggregated toward ends of branch-
       es. Infructescence as long as or longer than leaves. . . . . 14. *Metopium* (Fig. 4).
       12. Leaflets 7–29, margins entire, sinuate, dentate or spinose. Leaves densely aggre-
       gated toward ends of branches. Infructescence usually shorter than leaves. . . . .
. . . . 5. *Comocladia*.

KEY TO POISONOUS GENERA OF ANACARDIACEAE OF SOUTH AMERICA
(BASED ON FLOWERING MATERIAL)

1. Leaves simple.
  2. Leaves lepidote and stellate pubescent. . . . . 4. *Campnosperma*.
  2. Leaves not lepidote or stellate pubescent.
   3. Androecium zygomorphic. Style one; stigma one.
    4. Leaves broadly to narrowly obovate. Petals lanceolate to linear. Disk absent. Stamens 6–12.
    . . . . 1. *Anacardium*.
    4. Leaves usually lanceolate. Petals elliptic or obovate. Extrastaminal disk prominent. Stamens
    5. . . . . 11. *Mangifera* (introd.).
   3. Androecium actinomorphic. Style one; stigma 3 or styles three.
    5. Stamens 5. Styles lateral. . . . . 18. *Schinopsis* pro parte.
    5. Stamens 10. Style(s) central.
     6. Ovule suspended from apex of locule.
      7. Calyx deeply incised. Flowers unisexual (rarely bisexual). . . 19. *Schinus* subgenus *Duvaua*.
      7. Calyx with shallow lobes. Flowers bisexual. . . . . 12. *Mauria simplicifolia* Kunth.
     6. Ovule attached at base of locule. . . . . 9. *Lithrea* post parte.
1. Leaves compound.
  8. Leaflets with intramarginal vein. Styles (3–)4–5. Ovary (3–)4–5 locular. . . . . 22. *Spondias*.
  8. Leaflets without intramarginal vein. Styles one or three. Ovary unilocular.
   9. Stamens 10.
    10. Ovule suspended from apex of locule.
     11. Calyx deeply incised. Flowers unisexual (rarely bisexual). . . . .
     . . . . 19. *Schinus* subgenus *Euschinus*.
     11. Calyx with shallow lobes. Flowers bisexual. . . . . 12. *Mauria*.
    10. Ovule attached at base of locule. . . . . 9. *Lithrea* pro parte.
   9. Stamens 5.
    12. Style(s) central.
     13. Ovule attached at base of locule. . . . . 24. *Toxicodendron*.
     13. Ovule suspended from apex of locule. . . . . 2. *Astronium*.
    12. Styles lateral.
     14. Leaflets 10–31, sessile, margins entire. . . . . 18. *Schinopsis* post parte (Fig. 3).
     14. Leaflets 7–13, petiolulate or sessile, margins crenulate, serrulate or entire. (Note: where
     Schinopsis and *Loxopterygium* overlap in distribution the leaflet margins of *Loxopter-*
     ygium are always serrulate or crenulate.) . . . . 10. *Loxopterygium*.

KEY TO POISONOUS GENERA OF ANACARDIACEAE OF SOUTH AMERICA
(BASED ON FRUITING MATERIAL)

1. Hypocarp sigmoid or pyriform. Drupe reniform, mesocarp woody. (Hypocarp lacking in *Anacardium microsepalum* Loesener.) ....1. *Anacardium.*
1. Hypocarp absent. Drupe not reniform, mesocarp not woody.
    2. Drupe usually more than 6 cm long. Leaves always simple, with hair-like fibers visible when broken.
    .... 11. *Mangifera* (introd.).
    2. Drupe always less than 6 cm long (except for *Spondias dulcis* Parkinson which has drupes up to 10 cm in length). Leaves compound or simple, without hair-like fibers.
        3. Drupe with a single lateral wing.
            4. Endocarp bony, very thick. Lateral wing without prominent veins. ..... 18. *Schinopsis* (Fig. 3).
            4. Endocarp bony, but thin. Lateral wing with prominent veins. (In *Loxopterygium gardneri* Engler, the lateral wing is lacking.) ....10. *Loxopterygium.*
        3. Drupe lacking a lateral wing.
            5. Calyx enlarged and persistent in fruit. ....2. *Astronium.*
            5. Calyx not enlarged and persistent in fruit.
                6. Leaves always simple, lepidote and stellate pubescent. .... 4. *Campnosperma.*
                6. Leaves simple or compound, not lepidote or stellate pubescent.
                    7. Drupe with (3–)4–5 locules. ....22. *Spondias.*
                    7. Drupe unilocular.
                        8. Exocarp separating from mesocarp when drupe is ripe. Endocarp bony.
                            9. Exocarp white, cream-colored or light grey.
                                10. Leaves simple or compound, sometimes simple and compound leaves on same plant. Compound leaves 3–5 foliolate. Mesocarp resinous. .. 9. *Lithrea.*
                                10. Leaves always compound, 7–25 foliolate. Mesocarp waxy. ....
                                .... 24. *Toxicodendron.*
                            9. Exocarp pink, red, dark blue or black. ....19. *Schinus.*
                        8. Exocarp not separating from mesocarp. Endocarp chartaceous. ....12. *Mauria.*

KEY TO POISONOUS GENERA OF ANACARDIACEAE OF
AFRICA, MADAGASCAR, THE MASCARENE ISLANDS AND THE SEYCHELLES
(BASED ON FLOWERING MATERIAL)

1. Leaves simple.
    2. Calyx not calyptriform (not cap-like, nor dehiscent).
        3. Leaves lepidote. Androecium actinomorphic. Stigma lobate, discoid. ....
        .... 4. *Campnosperma* (Madagascar and the Seychelles only).
        3. Leaves not lepidote. Androecium zygomorphic. Stigma unlobed, not discoid.
            4. Leaves broadly to narrowly obovate. Petals lanceolate to linear. Disk absent. Stamens 9–11.
            .... 1. *Anacardium* (introd.).
            4. Leaves usually lanceolate. Petals elliptic or obovate. Extrastaminal disk present. Stamens 5.
            .... 11. *Mangifera* (introd.).
    2. Calyx calyptriform (cap-like and dehiscent). ....7. *Gluta* (Madagascar only).
1. Leaves compound.
    5. Leaves trifoliolate. ....21. *Smodingium.*
    5. Leaves with more than three leaflets.
        6. Leaflets with intramarginal veins. Ovary with three to five locules. .... 22. *Spondias* (introd.).
        6. Leaflets without intramarginal veins. Ovary unilocular.
            7. Stamens 4–5. Calyx shallowly 4–5 lobed. Petals 4–5. ....25. *Trichoscypha.*
            7. Stamens 10. Calyx deeply incised, with 5 lobes. Petals 5. ....19. *Schinus* (introd.).

KEY TO POISONOUS GENERA OF ANACARDIACEAE OF
AFRICA, MADAGASCAR, THE MASCARENE ISLAND AND THE SEYCHELLES
(BASED ON FRUITING MATERIAL)

1. Leaves simple.
    2. Pear-shaped hypocarp present. ....1. *Anacardium* (introd.).
    2. Hypocarp absent.
        3. Leaves lepidote. Drupe often two locular. ... 4. *Campnosperma* (Madagascar and Seychelles only).

  3. Leaves not lepidote. Drupe always unilocular.
    4. Mesocarp thick, fleshy. Drupe often more than 8 cm long. .... 11. *Mangifera* (introd.).
    4. Mesocarp thin, resinous. Drupe less than 6 cm long. ....7. *Gluta* (Madagascar only).
1. Leaves compound.
  5. Leaves trifoliolate. Drupe samaroid. .... 21. *Smodingium*.
  5. Leaves with more than three leaflets. Drupe not samaroid.
    6. Leaflets with intramarginal veins. Drupe with (3–)4–5 locules. .... 22. *Spondias* (introd.).
    6. Leaflets without intramarginal veins. Drupe unilocular.
      7. Drupe 1 cm long or longer. Endocarp cartilaginous or crustaceous. Leaflets petiolulate. ....
      ....25. *Trichoscypha.*
      7. Drupe less than 1 cm long. Endocarp bony. Leaflets usually sessile. ......19. *Schinus* (introd.).

KEY TO POISONOUS GENERA OF ANACARDIACEAE OF TEMPERATE EURASIA
(BASED ON FLOWERING MATERIAL)

1. Leaves simple.
  2. Leaves evergreen. Androecium zygomorphic. Style one; stigma one. .... 11. *Mangifera.*
  2. Leaves deciduous. Androecium actinomorphic. Styles three; stigmas three. .... 6. *Cotinus.*
1. Leaves compound.
  3. Stamens 10. ....19. *Schinus* (introd.).
  3. Stamens 5. .... 24. *Toxicodendron.*

KEY TO POISONOUS GENERA OF ANACARDIACEAE OF TEMPERATE EURASIA
(BASED ON FRUITING MATERIAL)

1. Leaves simple.
  2. Leaves evergreen. Drupe often more than 8 cm long. .... 11. *Mangifera.*
  2. Leaves deciduous. Drupe less than 1 cm long. .... 6. *Cotinus.*
1. Leaves compound.
  3. Exocarp pink or red when ripe. Mesocarp resinous without black striations. ......19. *Schinus* (introd.).
  3. Exocarp white, cream-colored, grey or brown. Mesocarp waxy with black striations. ....
  .... 24. *Toxicodendron.*

KEY TO POISONOUS GENERA OF ANACARDIACEAE OF
TROPICAL ASIA, MALESIA, AUSTRALIA AND PACIFIC ISLANDS
(BASED ON FLOWERING MATERIAL)

1. Leaves simple.
  2. Ovary inferior. .... 8. *Holigarna* (Fig. 4).
  2. Ovary superior.
    3. Calyx calyptriform. Stamens 4–infinity. ....7. *Gluta.*
    3. Calyx not calyptriform. Stamens 4–11.
      4. Hypanthium present. .... 13. *Melanochyla.*
      4. Hypanthium absent.
        5. Leaves lepidote. Ovary often two locular. .... 4. *Campnosperma.*
        5. Leaves not lepidote. Ovary always unilocular.
          6. Style one; stigma one. Disk extrastaminal or lacking.
            7. Leaves with slightly thickened marginal vein. Androecium actinomorphic. ......
            .... 23. *Swintonia.*
            7. Leaves without marginal vein. Androecium usually zygomorphic.
              8. Extrastaminal disk present. Petals elliptic or obovate, with glandular ridges
              adaxially. .... 11. *Mangifera.*
              8. Extrastaminal disk absent. Petals lanceolate to linear, without glandular ridges.
              .... 1. *Anacardium* (introd.).
          6. Styles three; stigmas three. Disk intrastaminal.
            9. Ovary usually hairy. Leaves often papillose abaxially, vein axils without domatia.
            ....20. *Semecarpus* (Fig. 2).
            9. Ovary glabrous. Leaves not papillose abaxially, vein axils with domatia abaxially.
            .... 24. *Toxicodendron* section *Simplicifolia.*
1. Leaves compound or occasionally unifoliolate.
  10. Leaves opposite. .... 3. *Blepharocarya* (endemic to Australia).

10. Leaves alternate.
    11. Leaflets with intramarginal veins. Ovary (3–)4–5 locular. . . . . 22. *Spondias* (sensu stricto).
    11. Leaflets without intramarginal veins. Ovary unilocular.
        12. Stamens 10. . . . . .19. *Schinus* (introd.).
        12. Stamens 4 or 5.
            13. Flowers tetramerous. . . . . 15. *Parishia.*
            13. Flowers pentamerous.
                14. Plants with bisexual flowers only. Leaflets often hairy in abaxial vein axils. . . .
                . . . . 16. *Pentaspadon.*
                14. Plants dioecious. Leaflets glabrous or papillate (not hairy) in abaxial vein axils.
                . . . . 24. *Toxicodendron.*

Key to Poisonous Genera of Anacardiaceae of Tropical Asia, Malesia, Australia and the Pacific Islands (Based on Fruiting Material)

1. Leaves simple.
    2. Drupe inferior. . . . . 8. *Holigarna* (Fig. 5).
    2. Drupe superior.
        3. Hypocarp present.
            4. Drupe reniform. Leaf surface not papillose abaxially. . . . . .1. *Anacardium* (introd.).
            4. Drupe not reniform. Leaf surface usually papillose abaxially. . . . . 20. *Semecarpus* (Fig. 2).
        3. Hypocarp absent.
            5. Leaves lepidote. Drupe often two locular. . . . . 4. *Campnosperma.*
            5. Leaves not lepidote. Drupe unilocular.
                6. Calyx deciduous, calyptriform. . . . . .7. *Gluta.*
                6. Calyx persistent, not calyptriform.
                      7. Petals accrescent and reflexed in fruit. . . . . 23. *Swintonia.*
                      7. Petals deciduous.
                          8. Drupe less than 8 mm long; exocarp separating from mesocarp when fruit is ripe; mesocarp waxy. . . . . 24. *Toxicodendron* section *Simplicifolia.*
                          8. Drupe more than 1 cm long; exocarp not separating mesocarp when fruit is ripe. Mesocarp fleshy or resinous.
                              9. Drupe glabrous; mesocarp thick, juicy, without black resin. Leaves glabrous abaxially. . . . . 11. *Mangifera.*
                              9. Drupe pubescent; mesocarp thin, not juicy, with abundant black resin. Leaves papillose or pubescent abaxially. . . . . 13. *Melanochyla.*
1. Leaves compound or occasionally unifoliolate.
    10. Leaves opposite. Infructescence bearing involucriform cupules. . . . .
    . . . . 3. *Blepharocarya* (endemic to Australia).
    10. Leaves alternate. Infructescence not bearing involucriform cupules.
        11. Leaflets with intramarginal veins. Drupe with (3–)4–5 locules. . . . . . 22. *Spondias* (sensu stricto).
        11. Leaflets without intramarginal veins. Drupe unilocular.
            12. Sepals enlarging to several times the length of the drupe (when in fruit). . . . . . . 15. *Parishia.*
            12. Sepals not enlarging when in fruit.
                13. Exocarp separating from mesocarp when fruit is ripe, white, gray, tan, brown or red.
                    14. Drupe pink or red. Domatia absent. Seed suspended from apex of locule. . . . . . .
                  . . . . .19. *Schinus* (introd.).
                    14. Drupe white, gray, tan or brown. Domatia present in abaxial vein axils. Seed attached at the base of the locule. . . . . 24. *Toxicodendron.*
                13. Exocarp not separating from mesocarp when fruit is ripe, purple or black. . . . .
    . . . . 16. *Pentaspadon.*

Generic Accounts

1. *Anacardium* L. (Cashew Nut Trees); Map IB

Botanical ref.: Mitchell and Mori (1987).
Medical ref.: Bedello et al. (1985); Behl and Captain (1979); Chopra (1965); Levy and Primack (1984); Mitchell and Rook (1979); Perry and Metzger (1980); Watt and Breyer-Brandwijk (1962); Woodruff (1979); Woods and Calnan (1976).
Chemical ref.: Baylis et al. (1981); Evans and Schmidt (1980); Hegnauer (1964); Lloyd et al. (1980); Murthy et al. (1968); Sood et al.

(1986); Tyman and Morris (1967); Watt and Breyer-Brandwijk (1962).

Suffrutices (*subshrubs*) to very large *trees*. *Leaves* simple, alternate, aggregated toward branch tips. *Inflorescences* thyrsoid, terminal and/ or axillary. *Flowers:* male and bisexual flowers on same plant (andromonoecious); perianths pentamerous; disk or nectary lacking; stamens (or staminodes) 6–12 with one or four much larger than the rest; ovary unilocular, the style one, the stigma unlobed, one. *Drupe* reniform usually subtended by a fleshy hypocarp (lacking in *Anacardium microsepalum*).

For a more detailed description and a discussion of *Anacardium* taxonomy consult Mitchell and Mori (1987).

Distribution: A neotropical genus of ten species distributed from Honduras to southeastern Brazil and Paraguay (Map IB). *Anacardium occidentale* (cashew) is cultivated or adventive throughout the humid, lowland tropics.

Chemical constituents: cardols, cardanols, anacardols, anacardic acid, pentadecyl phenols, etc.

2. *Astronium* Jacquin; Map IA

Botanical ref.: Barkley (1968); Mattick (1934).
Medical ref.: Mitchell and Rook (1979); Woods and Calnan (1976).

Habit medium-sized to large *trees. Leaves,* alternate, imparipinnate. *Inflorescences* terminal or axillary. *Flowers:* unisexual or bisexual flowers on same plant or on different individuals (polygamous). Male flowers: perianths pentamerous; disk intrastaminal 5-lobed; stamens 5; pistillode lacking. Female flowers: perianths pentamerous; disk intrastaminal 5-lobed; staminodes 5; ovary unilocular, the ovule subapical, the styles 3, the stigmas discoid. *Drupe* with five persistent, enlarged sepals.

Astronium was most recently revised by Barkley (1968).

Distribution: A neotropical genus of approximately ten species distributed from southern Mexico to northern Argentina (May IA).

Chemical constituents: The author is not aware of any identifications of dermatitis-producing compounds.

3. *Blepharocarya* Mueller, Map IIA

Botanical ref.: Airy Shaw (1965); Ding Hou (1978); Wannan et al. (1987).

Medical ref.: Womersley (1983); Woods and Calnan (1976).

Habit *trees. Leaves* opposite, paripinnate. *Inflorescences* thyrsoid, terminal and axillary, bearing oppositely branched glomerules. *Flowers:* male and female flowers on different plants (dioecious). Male flowers: perianths tetramerous; disk cupular; stamens 8. Female flowers: perianths tetra- to pentamerous; disk annular; ovary unilocular, laterally compressed, with hairy margins.

Blepharocarya is sometimes segregated as a separate family Blepharocaryaceae (Airy Shaw, 1965). However, Ding Hou (1978) maintains that the opposite leaves and unusual inflorescence are insufficient characters to warrant removing it from the Anacardiaceae. Wannan et al. (1987) provide additional evidence from the study of resin (secretory) canals, inflorescences of male and female plants and pollen to support its placement in the Anacardiaceae.

Distribution: A genus of two species endemic to Queensland and Northern Territory, Australia (Map IIA).

Chemical constituents: The author is not aware of any identifications of toxic compounds.

4. *Campnosperma* Thwaites; Map IIIB

Botanical ref.: Ding Hou (1978).
Medical ref.: Corner (1952); Mitchell and Rook (1979).
Chemical ref.: Hegnauer (1964); Mitchell and Rook (1979); Thomson (1971).

Medium-sized to large *trees. Leaves* simple, alternate, aggregated toward branch tips, lepidote, sometimes stellate pubescent abaxially. *Inflorescences* axillary. *Flowers:* unisexual or bisexual flowers on same plant or on different individuals (polygamous); perianths tri- to pentamerous; disk intrastaminal, annular or lobed; stamens 6–10; ovary unilocular, the ovule apical, the style one, the stigma one, lobate-discoid. *Drupe* often two locular, but only one locule bearing seed.

Distribution: A pantropical genus of approximately ten species distributed as follows: Costa Rica to Ecuador (one species), Amazonia (one species), Madagascar (one species), the Seychelles (one species), Sri Lanka (one species), and the rest of the genus ranges from Thailand through Malesia to Micronesia and Melanesia (Map IIIB).

5. *Comocladia* Browne; Map IIIB

Medical ref.: Adams (1972); Marcano F. (1977); Mitchell and Rook (1979); Morton (1981); Standley (1923).

Shrubs to small *trees*. *Leaves* alternate, imparipinnate, clustered at the ends of the branches; leaflets entire, sinuate, dentate or spinose. *Inflorescences* axillary. *Flowers:* perianths tri- or tetramerous; disk intrastaminal, trilobed; stamens three; ovary unilocular, the ovule basal, the styles three. *Drupe* oblong, with a fleshy mesocarp.

Comocladia is currently being revised by David Young.

Distribution: A neotropical genus of approximately fifteen species in Mexico, Guatemala and the West Indies (Map IIIB).

Medical problems: Contact with the sap of *Comocladia* often produces a severe case of dermatitis.

Chemical constituents: The author is not aware of any chemical studies of toxic compounds of *Comocladia*.

6. *Cotinus* Miller; Map IIA

Medical ref.: Lampe and McCann (1985); Mitchell and Rook (1979).

Shrubs or small *trees*. *Leaves* simple, alternate. *Inflorescences* terminal. *Flowers:* male and female flowers usually on different individuals (dioecious); perianths pentamerous; disk intrastaminal; stamens 5 (often absent in female flower); ovary unilocular, the ovule basal, the styles three, lateral. *Drupe* laterally compressed, subreniform.

Distribution: Three to four species, one species in the central United States and the rest in temperate Eurasia (Map IIA).

Medical problems: Dermatitis due to contact with *Cotinus* is rarely mentioned in the literature.

Chemical constituents: The author is not aware of any chemical studies of dermatitis-producing compounds of *Cotinus*.

7. *Gluta* L. (incl. *Melanorrhoea* Wallich) Burma Lac Tree, Rengas Trees; Map IIA

Botanical ref.: Ding Hou (1978).
Medical ref.: Behl and Captain (1979); Burkill (1935); Corner (1952); Ding Hou (1966,

1978); Evans and Schmidt (1980); Mitchell and Rook (1979).

Chemical ref.: Behl and Captain (1979); Du and Oshima (1985); Evans and Schmidt (1980); Hegnauer (1964); Jefferson and Wangchareontrakul (1985); Mitchell and Rook (1979); Woods and Calnan (1976).

Small to large *trees*. *Leaves* simple, alternate. *Inflorescences* axillary. *Flowers:* bisexual; calyx calyptriform; petals vary in number from four to eight; disk lacking; stamens from four to more than one hundred, often inserted on a conical torus; ovary unilocular, sessile or supported by a gynophore, the style one, the stigma one. *Drupe* sometimes associated with large wing-like petals.

Ding Hou (1978) has lumped *Melanorrhoea* with *Gluta* on the basis that there are several species with intermediate characters. For example, the presence or absence of persistent petals associated with the fruit varies within individuals of a few species.

Distribution: A Malesian-centered genus with one species in Madagascar, and the rest being distributed from tropical Asia to New Guinea (with one species) (Map IIA).

Medical problems: Species of *Gluta* (Rengas Trees) are avoided by most people. The dermatitis produced by contact with sap from *Gluta* spp. is frequently debilitating.

Chemical constituents: Glutarenghol, Laccol, Moreakol, Renghol, Thitsiol, etc.

8. *Holigarna* (Buchanan) Hamilton ex Roxburgh, Fig. 2; Map IB

Medical ref.: Behl and Captain (1979); Chopra et al. (1949); Evans and Schmidt (1980); Gamble (1918); Kirtikar et al. (1935); Mitchell and Rook (1979).

Chemical ref.: Evans and Schmidt (1980); Hegnauer (1964); Majima (1922).

Habit *trees*. *Leaves* simple, alternate, the petioles with one or two pairs of deciduous appendages. *Inflorescences* terminal or axillary. *Flowers:* perianths pentamerous; disk intrastaminal, obscure in bisexual flowers; stamens five; pistillode lacking in male flowers; ovary inferior, unilocular; the ovule apical, the styles three. *Drupe* sometimes subtended by a hypocarp, mesocarp contains an abundance of black resin.

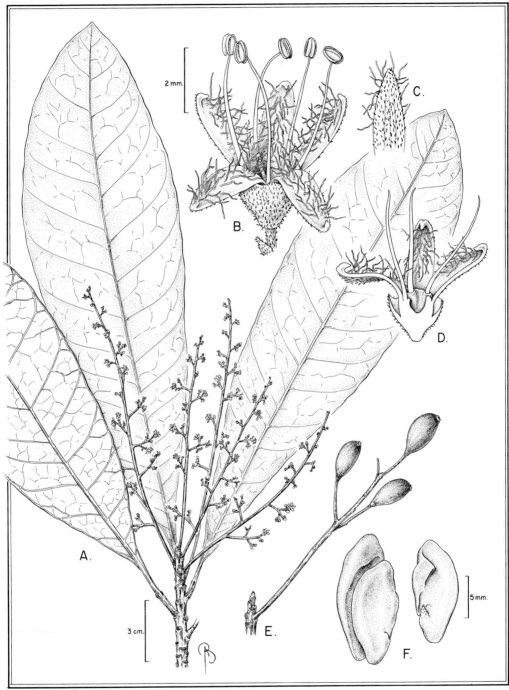

FIG. 2. *Holigarna ferruginea* March. **A.** Habit, male plant (*Saldanha 13076*). **B.** Male flower at anthesis (*Saldanha 13076*). **C.** Abaxial surface of petal (*Saldanha 13076*). **D.** Male flower with anthers missing, sectioned, revealing intrastaminal disk and absence of pistillode (*Saldanha 13076*). **E.** Infructescence (*Saldanha 13713*). **F.** Embryo with large cotyledons (*Saldanha 13713*).

Holigarna is a genus in need of critical revision.

Distribution: About eight species distributed from India to Indochina (Map IB).

Chemical constituents: Heptadecyl catechols.

9. *Lithrea* Miers ex H. & A.; Map IIIA

Botanical ref.: Barkley (1962b).
Medical ref.: Gambaro et al. (1986); Mitchell and Rook (1979); Woods and Calnan (1976).
Chemical ref.: Gambaro et al. (1986).

Shrubs to small *trees. Leaves* simple or imparipinnate, alternate. *Inflorescences* axillary. *Flowers:* male and female usually on different individuals (dioecious); perianths pentamerous; disk intrastaminal, ten lobed; stamens ten; ovary unilocular, the ovule basal, the style one, the stigmas three. *Drupe* exocarp white or pale grey.

Lithrea was revised by Barkley (1962b).

Distribution: Central Chile (one species), southern Brazil and Bolivia to northern Argentina (two to three species) (Map IIIA).

Chemical constituents: Pentadecyl catechols and heptadecyl catechols.

10. *Loxopterygium* Hook.; Map IIB

Botanical ref.: Barkley (1962a).
Medical ref.: Barkley (1962a); MacBride (1951); Mitchell and Rook (1979); Record (1939).

Medium-sized *trees. Leaves* imparipinnate, alternate with entire, crenulate or serrate leaflets. *Inflorescences* thyrsoid, terminal or axillary. *Flowers:* male and female flowers on different individuals (dioecious); perianths pentamerous; disk intrastaminal, prominently five-lobed in male flowers; stamens five (rudimentary in female flowers); ovary unilocular, the ovule lateral, the styles three, lateral. *Drupe* with a prominently veined lateral wing (lacking in *Loxopterygium gardneri* Engler).

Loxopterygium was revised by Barkley (1962a).

Distribution: A neotropical genus of approximately five species distributed from northern South America to eastern Brazil and Argentina (Map IIB).

Chemical constituents: The author is not aware of any chemical analyses of dermatitis-producing compounds of *Loxopterygium*.

11. *Mangifera* L., Mango Trees; Map IB

Botanical ref.: Ding Hou (1978); Mukherjee (1985).
Medical ref.: Behl and Captain (1979); Goldstein (1968); Keil et al. (1946); Levy and Primack (1984); Mitchell and Rook (1979); Morton (1971); Watt and Breyer-Brandwijk (1962).
Chemical ref.: Cojocaru et al. (1986); Goldstein (1968); Keil et al. (1946); Vasistha and Siddiqui (1938).

Habit large *trees. Leaves* simple, altenate. *Inflorescences* thyrsoid, terminal or axillary. *Flowers:* male and bisexual flowers on same plant (andromonoecious); perianths tetra- to pentamerous; disk extrastaminal; stamens (or staminodes) five to twelve with one (rarley three to five or all fertile) larger than the others; ovary unilocular, the style one, the stigma one. *Drupe* usually very large with a thick, fleshy mesocarp.

The most up-to-date treatment of *Mangifera* is in Ding Hou (1978).

Distribution: A tropical Asian, Malesian-centered genus of thirty-five species with the following distribution: Sri Lanka, India, Nepal, S. China through Malesia to the Solomon Islands (Map IB). One species, *Mangifera indica* is cultivated throughout the tropics and subtropics of the world.

Chemical constituents: Cardols and other phenolic compounds.

12. *Mauria* Kunth; Map IIIA

Medical ref.: Hoyos F. (1985); MacBride (1951); Mitchell and Rook (1979); Record (1939); Steyermark and Huber (1978).

Shrubs or *trees. Leaves* simple or imparipinnate. *Inflorescences* thyrsoid, terminal or axillary. *Flowers:* bisexual; perianths pentamerous; disk intrastaminal, ten-lobed; stamens ten; ovary unilocular, the ovule apical or subapical, the style one, the stigmas three. *Drupe* somewhat oblong, laterally compressed, the mesocarp thin, fleshy.

Mauria is in need of a critical monographic treatment.

Distribution: A neotropical Andean-centered genus of about eight species distributed from El Salvador south to Venezuela and Bolivia (Map IIIA).

Medical problems: The sap of *Mauria* spp. is recorded as being poisonous by several authors.

Chemical constituents: The author is not aware of any chemical analyses of the dermatitis-producing compounds of *Mauria*.

13. *Melanochyla* Hook. (Rengas Trees); Map IA

Botanical ref.: Ding Hou (1978).

Medical ref.: Burkill (1935); Corner (1952); Ding Hou (1966, 1978); Mitchell and Rook (1979).

Trees, frequently with stilt roots. *Leaves* simple, alternate, often papillose abaxially. *Inflorescences* terminal and/or axillary. *Flowers:* male and female flowers on separate individuals (dioecious); male flowers: hypanthium present; perianths tetra- to pentamerous; disk intrastaminal four- or five-lobed; female flowers: hypanthium sometimes partly or completely concealing ovary; perianths tetra- to pentamerous; ovary unilocular, the ovule sub-apical, the style one, the stigmas three. *Drupe:* mesocarp and endocarp with large resin cavities filled with black varnish.

Distribution: A Malesian genus of seventeen species distributed from the Malay Peninsula to Sumatra, Java and Borneo (Map IA).

Chemical constituents: The author is not aware of any chemical analyses of the dermatitis-producing compounds of *Melanochyla*.

14. *Metopium* Browne (Poisonwood); Fig. 3; Map IA

Botanical ref.: Barkley (1937).

Medical ref.: Campbell (1983); Hardin and Arena (1974); Lampe and Fagerstrom (1968); Levy and Primack (1984); Mitchell and Rook (1979); Morton (1971); Tomlinson (1986); Westbrooks and Preacher (1986); Woods and Calnan (1976).

Chemical ref.: Gross et al. (1975); Evans and Schmidt (1980).

Small *trees. Leaves* imparipinnate, alternate. *Inflorescence* axillary. *Flowers:* male and female flowers on separate individuals (dioecious); perianths pentamerous; disk intrastaminal, annular; stamens five; ovary unilocular, the ovule basal, the style one, the stigmas three. *Drupe* oblong, the mesocarp thin, fleshy.

Metopium was revised by Barkley (1937).

Distribution: Three species distributed as following: Southern Florida, the Bahamas, the Greater Antilles, Aruba, Bonaire, Curacao and southern Mexico to Guatemala (Map IA).

Medical problems: Contact dermatitis produced by *Metopium* can be very severe.

Chemical constituents: Penta- and heptadecyl catechols.

15. *Parishia* Hook.; Map IB

Botanical ref.: Ding Hou (1978).

Medical ref.: Mitchell and Rook (1979); Senear (1933).

Habit *trees. Leaves* imparipinnate, alternate. *Inflorescences* terminal and/or axillary. *Flowers:* male and female flowers on separate individuals (dioecious); perianths tetramerous; disk intrastaminal, usually shape variable; stamens four; ovary unilocular, hairy, the ovule subapical, the style 3(–4)-lobed, the stigmas three. *Drupe* unilocular; associated with four enlarged wing-like sepals.

Distribution: A tropical Asian, Malesian genus of five species distributed as follows: the Andaman Islands, Burma, Thailand, Malay Peninsula, Sumatra, Borneo and the Philippines (Map IIB).

Chemical constituents: The author is not aware of any chemical analyses of dermatitis-producing compounds of *Parishia*.

16. *Pentaspadon* Hook. f. (incl. *Microstemon* Engler); Map IA

Botanical ref.: Ding Hou (1978).

Medical ref.: Burkill (1935); Evans and Schmidt (1980); Mitchell and Rook (1979).

Chemical ref.: Evans and Schmidt (1980); Hegnauer (1964).

Large *trees. Leaves* imparipinnate, alternate. *Inflorescences* axillary. *Flowers:* bisexual; perianths pentamerous; disk intrastaminal, ten-grooved or crenulate; stamens five; ovary unilocular, the ovule lateral, the style one, the stigma bi-lobed. *Drupe:* mesocarp contains an abundance of resin canals.

Three species of *Pentaspadon* are treated by Ding Hou (1978).

Distribution: Six species distributed from

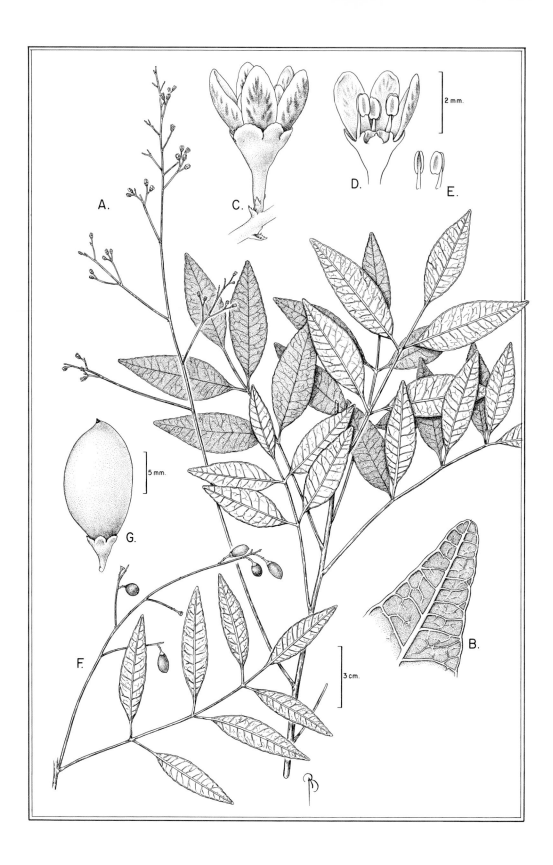

A.

C.

D.

E.

2 mm.

5 mm.

G.

F.

B.

3 cm.

Thailand and Vietnam south through Malesia to the Solomon Islands (Map IA).

17. *Pseudosmodingium* Engler; Map IIB

Botanical ref.: Barkley and Reed (1940).
Medical ref.: Barkley and Reed (1940); Mason and Mason (1987); Mitchell and Rook (1979); Record (1939); Standley (1923).

Small *trees. Leaves* imparipinnate, alternate. *Inflorescences* erect, axillary, paniculate. *Flowers:* unisexual or bisexual flowers on same plant or on different individuals (polygamous); perianths pentamerous; disk intrastaminal, cupuliform; stamens five; ovary unilocular, the ovule basal, the styles three. *Drupe* reniform with two lateral, chartaceous wings.

Pseudosmodingium is currently being revised by Hector Narave at XAL.

Distribution: Seven species endemic to southern Mexico (Map IIB).

Chemical constituents: The author is not aware of any chemical analyses of *Pseudosmodingium.*

18. *Schinopsis* Engler; Fig. 4; Map IIIB

Botanical ref.: Meyer and Barkley (1973).
Medical ref.: Lampe and Fagerstrom (1968); Mitchell and Rook (1979); Rendle (1929); Woods and Calnan (1976).

Small to large *trees. Leaves* simple, pari- or imparipinnate. *Inflorescences* terminal or axillary. *Flowers:* male and female flowers on different individuals (dioecious); perianths pentamerous; disk intrastaminal, five-lobed; stamens five; ovary unilocular, laterally compressed, the ovule subapical, the styles three, lateral. *Drupe* with a shiny, lateral wing.

Distribution: Seven species distributed from western Peru and eastern Brazil south to central Argentina (Map IIIB).

Chemical constituents: The author is not aware of any chemical analyses of *Schinopsis* oleoresins.

19. *Schinus* L. (Brazilian Pepper Tree, California Pepper Tree); Map IB

Botanical ref.: Barkley (1944, 1957); Paviani (1965).
Medical ref.: Campbell (1983); Lampe and Fagerstrom (1968); Levy and Primack (1984); Mitchell and Rook (1979); Morton (1971, 1976); Stahl et al. (1983); Watt and Breyer-Brandwijk (1962); Westbrooks and Preacher (1986); Woods and Calnan (1976).
Chemical ref.: Skopp and Schwenker (1984); Stahl et al. (1983).

Shrubs or *trees. Leaves* alternate, simple, pari- or imparipinnate. *Inflorescences* thyrsoid, terminal or axillary. *Flowers:* unisexual or bisexual flowers on same plant or on different individuals (polygamous), or male and female flowers always on different individuals (dioecious); perianths pentamerous; disk intrastaminal, ten-lobed; stamens ten; ovary unilocular, the ovule apical or subapical, the styles one or three, the stigmas three. *Drupe* with exocarp separating from mesocarp when fruit is ripe.

Distribution: A neotropical genus of about twenty-four species. The hypothetical natural distribution of *Schinus* is from Ecuador and eastern Brazil south to southern Argentina and Chile (Map IB). *Schinus molle* and *Schinus terebinthifolius* have been introduced throughout the warm temperate zone, the subtropics and the tropics.

Medical problems: Contact dermatitis, gastrointestinal disturbances and respiratory tract inflammation are associated with *S. terebinthifolius* and, to a lesser extent, *S. molle* L.

Chemical constituents: Cardanols.

20. *Semecarpus* L.f. (Dhobi Nut or Marking Nut Trees, Rengas Trees) Fig. 5; Map IIIA

Medical ref.: Behl and Captain (1979); Burkill (1935); Chopra et al. (1949); Corner (1952); Ding Hou (1966, 1978); Mitchell and Rook (1979); Woods and Calnan (1976).
Chemical ref.: Backer and Haack (1938); Ev-

←

FIG. 3. *Metopium venosum* (Griseb.) Engler. **A.** Habit, male plant (*Ekman 3513*). **B.** Leaflet apex (*Ekman 3513*). **C.** Male flower at anthesis (*Ekman 3513*). **D.** Male flower at anthesis, sectioned, revealing intrastaminal disk and pistillode (*Ekman 3513*). **E.** Dorsal and lateral views of stamen (*Ekman 3513*). **F.** Infructescence (*Leon 11802*). **G.** Fruit (*Leon 11802*).

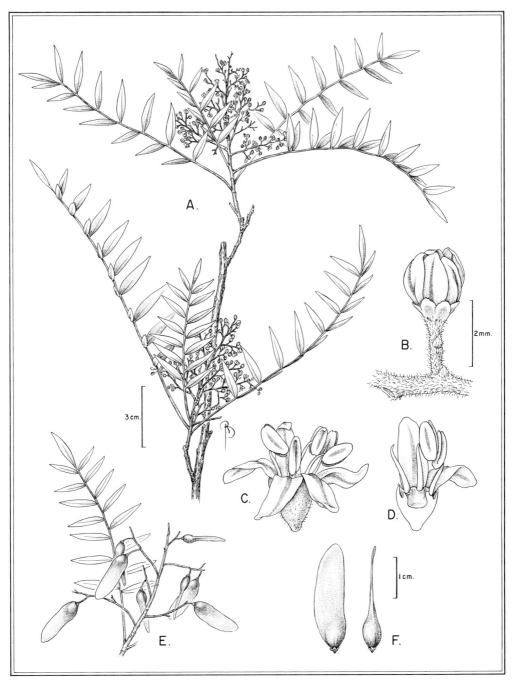

FIG. 4. *Schinopsis quebracho-colorado* (Schlecht.) Barkley & Meyer. **A.** Habit, male plant (*Arenas 1771*). **B.** Immature male flower (*Arenas 1771*). **C.** Male flower at anthesis (*Arenas 1771*). **D.** Male flower at anthesis, sectioned, revealing intrastaminal disk and absence of pistillode (*Arenas 1771*). **E.** Infructescence (*Meyer 8828*). **F.** Fruit (*Meyer 8828*). Note prominent lateral wing.

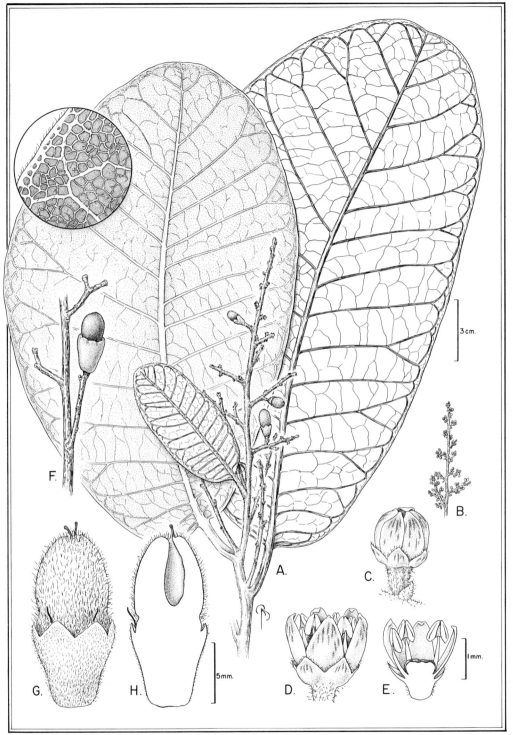

FIG. 5. *Semecarpus anacardium* L.f. **A.** Habit, female plant (*Ramamoorthy 1852*). **B.** Inflorescence with male flowers (*Sannasena 42*). **C.** Immature male flower (*Sannasena 42*). **D.** Male flower at anthesis (*Sannasena 42*). **E.** Male flower at anthesis, sectioned, revealing intrastaminal disk and absence of pistillode (*Sannasena 42*). **F.** Infructescence (*Erlanson 5324*). Note drupe subtended by fleshy hypocarp. **G.** Immature fruit (*Ramamoorthy 1842*). **H.** Immature fruit sectioned (*Ramamoorthy 1842*).

ans and Schmidt (1980); Goudgaon et al. (1984); Hegnauer (1964); Indap et al. (1983); Prasad et al. (1985).

Trees, shrubs or *treelets* (rarely monocaulous). *Leaves* simple, alternate, frequently papillose abaxially. *Inflorescences* terminal and/or axillary, rarely cauliflorous. *Flowers:* male and female flowers usually on separate individuals (dioecious); perianths tetra- to pentamerous; disk intrastaminal, cupuliform, flat or slightly convex, frequently four or five-notched; stamen four or five; pistillode often lacking in male flowers; ovary usually hairy, unilocular, the ovule basal, the styles three. Hypocarp fleshy. *Drupe* with a mesocarp containing large resin cavities.

Semecarpus is in need of taxonomic revision.

Distribution: A large Indopacific genus of about sixty species distributed from India, Sri Lanka, S. China south through Malesia to Australia, New Caledonia and the Fiji Islands (Map IIIA).

Chemical constituents: Bhilawanols, Renghol, various penta- and hepta-decyl catechols.

21. *Smodingium* Meyer ex Sonder; Map IIA

Botanical ref.: Palgrave (1981).
Medical ref.: Dyer (1975); Mitchell and Rook (1979); Palgrave (1981); Watt and Breyer-Brandwijk (1962); Woods and Calnan (1976).
Chemical ref.: Evans and Schmidt (1980).

Shrubs to small *trees. Leaves* alternate, trifoliolate. *Inflorescences* (large, terminal and axillary panicles. *Flowers:* male and female flowers on separate individuals (dioecious); perianths pentamerous; disk intrastaminal, annular; stamens five; ovary unilocular, laterally compressed, the ovule apical, the styles lateral, three. *Drupe* encircled with a chartaceous wing.

Distribution: A monotypic (*Smodingium argutum* Meyer ex Sonder) genus endemic to South Africa (Map IIA).

22. *Spondias* L. (sensu stricto excl. *Solenocarpus* W. & A.); Map IIIB

Medical ref.: Woods and Calnan (1976).

Trees. Leaves alternate, imparipinnate, rarely bipinnate; leaflets with a slightly thickened intramarginal vein. *Inflorescences* paniculate terminal and/or axillary. *Flowers:* bisexual or male and female flowers on separate individuals (dioecious); perianths tetra- to pentamerous; disk intrastaminal, crenulate, flat, or papillose; stamens eight to ten; ovary four to five locular, the styles four to five. *Drupe:* the mesocarp fleshy, the endocarp woody or bony with a fibrous matrix.

Spondias is in need of a worldwide taxonomic treatment.

Distribution: A pantropical genus of about ten species in tropical America, tropical Asia, Malesia and Oceania (Map IIIB). Three species, *Spondias dulcis* Parkinson, *Spondias mombin* L. and *Spondias purpurea* L. are cultivated virtually throughout the lowland tropics of the world for their edible fruits.

The case for *Spondias*-caused dermatitis is weak.

Chemical constituents: The author is not aware of any analyses of dermatitis-producing compounds.

23. *Swintonia* Griff. (Rengas Trees); Map IIIA

Botanical ref.: Ding Hou (1978).
Medical ref.: Burkill (1935); Ding Hou (1966, 1978).

Trees. Leaves alternate, simple with a thickened margin. *Inflorescences* terminal and axillary. *Flowers:* all bisexual, or male and bisexual flowers on same plant (andromonoecious); perianths pentamerous; disk extrastaminal, five-lobed; stamens five; ovary unilocular, the style one, lateral. *Drupe* subtended by five recurved, enlarged petals.

Distribution: Twelve species distributed from the Andaman Islands and Burma east to Indochina and south to Malesia (Map IIIA).

Chemical constituents: The author is not aware of any analyses of dermatitis-producing compounds.

24. *Toxicodendron* Miller (Poison Ivy, Poison Oak, Poison Sumac, Japanese Varnish Tree, Oriental Wax Tree, etc.); Map IIB

Botanical ref.: Gillis (1971); Möbius (1899).
Medical ref.: Baer (1979); Behl and Captain (1979); Evans and Schmidt (1980); Fuller and McClintock (1986); Gillis (1971); Kligman (1958); Lampe and Fagerstrom (1968);

Mitchell and Rook (1979); Muenscher (1972); Nakamura (1985).

Chemical ref.: Baer (1979); Du et al. (1984); Evans and Schmidt (1980); Gross et al. (1975); Hegnauer (1964); Jefferson and Wangchareontrakul (1985); Loev (1952).

Trees, shrubs and *vines. Leaves* alternate, trifoliolate, multifoliolate (rarely unifoliolate). *Inflorescences* pendent, axillary panicles. *Flowers:* male and female flowers on separate individuals (dioecious); perianths pentamerous; disk intrastaminal, five-lobed; stamens five; ovary unilocular, the ovule basal, the style one, the stigma one. *Drupe* exocarp separates from mesocarp when ripe.

Toxicodendron is divided into three sections: 1) *Toxicodendron,* 2) *Venenata* and 3) *Simplicifolia* (Gillis, 1971). Gillis (1971) and others have argued strongly for the segregation of *Toxicodendron* from *Rhus* L.

Toxicodendron sections *Toxicodendron* and *Simplicifolia* were monographed by Gillis (1971). The larger section *Venenata* needs to be revised.

Distribution: A predominantly temperate genus of about thirty species in temperate North America south to Bolivia and temperate East Asia south to New Guinea (Map IIB). The center of diversity is East Asia.

Chemical constituents: Penta- and heptadecyl catechols, primarily.

25. *Trichoscypha* Hook.; Map IB

Medical ref.: Palgrave (1981).

Trees, shrubs, treelets (rarely un-branched treelets) and *lianas. Leaves* alternate, imparipinnate. *Inflorescences* terminal and/or axillary (rarely cauliflorous). *Flowers:* male and female flowers on separate individuals (dioecious); perianths tetra- to pentamerous; disk intrastaminal, cupuliform, frequently pubescent; stamens four to five; ovary unilocular, the ovule subapical, the styles three or four, reflexed, erect or compressed. *Drupe* with resin that turns black and hardens. *Trichoscypha* is in need of critical revision.

Distribution: A large genus of about seventy-five species restricted to Tropical Africa (Map IB).

Chemical constituents: The author is not aware of any chemical analyses of dermatitis-producing compounds.

Acknowledgments

I thank Bobbi Angell for the preparation of the figures and maps. I am grateful to Scott Mori and Anders Barfod for their careful reviews of the manuscript and to Melodie Greer and Lauren Weaver for typing drafts of the manuscript. This is publication number 52 in the series of the Institute of Economic Botany of the New York Botanical Garden.

Literature Cited

Adams, C. D. 1972. Flowering plants of Jamaica. University of the West Indies, Mona, Jamaica.

Airy Shaw, H. K. 1965. Diagnoses of new families, new names, etc., for seventh edition of Willis's 'Dictionary'. Kew Bull. **18(2):** 249–273.

Allen, B. M. 1975. Common malaysian fruits. Longman Malaysia SDN, Berhad, Kuala Lumpur, Malaysia and Singapore.

Backer, H. J. & N. H. Haack. 1938. Le principe toxique des fruits de renghas *Semecarpus heterophylla* Bl. Recl. Trav. Chim. Pays-Bas **57:** 225.

Baer, H. 1979. The poisonous Anacardiaceae. *In* A. D. Kinghorn (ed.), Toxic plants. Columbia University Press, New York.

———. 1983. Allergic contact dermatitis from plants. *In* R. F. Keeler & A. T. Tu (eds.), Handbook of natural toxins. Vol. I. Plant and fungal toxins. Marcel Dekker, New York and Basel.

———, **M. Hooton, H. Fales, A. Wu & F. Schaub.** 1980. Catecholic and other constituents of the leaves of *Toxicodendron radicans* and variation of urushiol concentrations within one plant. Phytochemistry **19:** 799–802.

———, **R. Watkins, A. Kurtz, J. Bycks & C. Dawson.** 1967. Direct contact sensitivity to catechols and cutaneous toxicity of catechols related to the active principles of poison ivy. J. Immunol. **99:** 365–369.

Barkley, F. A. 1937. A monograph of the genus *Rhus* and its allies in North and Central America including the West Indies. Ann. Missouri Bot. Gard. **24:** 265–498.

———. 1944. *Schinus* L. Brittonia **5(2):** 160–198.

———. 1957. A study of *Schinus* L. Lilloa **28:** 5–110.

———. 1962a. Anacardiaceae: Rhoideae: *Loxopterygium.* Lloydia **25(2):** 109–122.

———. 1962b. Anacardiaceae: Rhoideae: *Lithraea.* Phytologia **8(7):** 329–365.

———. 1968. Anacardiaceae: Rhoideae: *Astronium.* Phytologia Phytologia 16(2): 107–152.

——— **& M. J. Reed.** 1940. *Pseudosmodingium* and *Mosquitoxylum.* Amer. Midl. Nat. **24(3):** 666–679.

Baylis, C. J., S. W. D. Odle & J. H. P. Tyman. 1981. Long chain phenols. Part 17. The synthesis of 5-((z)-pentadec-8-enyl) resorcinol, 'cardol monoene', and of 5-((zz)-pentadec-8,11-dienyl) resorcinol dimethyl ether, 'cardol diene' dimethyl ether. J. Chem. Soc., Perkin Trans. **1**: 132–141.

Bedello, P. G., M. Goitre, D. Cane, G. Roncarolo & V. Alovisi. 1985. Allergic contact dermatitis to cashew nut. Contact Dermatitis **12(4)**: 235.

Behl, P. N. & R. M. Captain. 1979. Skin-irritant and sensitizing plants found in India. S. Chand & Co., Ram Nagar, New Delhi.

Burkill, I. H. 1935. Dictionary of the economic products of the Malay Peninsula. 2 vols. London.

Campbell, G. R. 1983. An illustrated guide to some poisonous plants and animals of Florida. Pineapple Press, Englewood, Florida.

Chopra, R. N. 1965. Poisonous plants of India. Revised and enlarged. Indian Council of Agricultural Research, New Delhi.

———, **O. L. Badwhar & D. Ghosh.** 1949. Poisonous plants of India. Vol. I. Government of India Press, Calcutta.

Cojocaru, M., S. Droby, E. Glotter, A. Goldman, H. E. Gottlieb, B. Jacoby & D. Prusky. 1986. 5-12 Heptadecenyl-resorcinol the major component of the antifungal activity in the peel of Mango (*Mangifera indica*) fruit. Phytochemistry **25(5)**: 1093–1096.

Copeland, H. F. 1961. Observations on the reproductive structures of *Anacardium occidentale*. Phytomorphology **11**: 315–325.

——— **& B. E. Doyel.** 1948. Some features of the structure of *Toxicodendron diversiloba*. Amer. J. Bot. **27(10)**: 932–938.

Corner, E. J. H. 1952. Wayside trees of Malaysia. Government Printing Singapore Office.

Ding Hou. 1966. Report of a study-trip on Anacardiaceae to Malaysia and Singapore in 1966. Rijksherbarium, Leiden.

———. 1978. Anacardiaceae. *In* Flora Malesiana ser. I(8): 395–577.

Du, Y. & R. Oshima. 1985. Analysis of long-chain phenols in the sap of the Burmese Lac tree (*Melanorrhoea usitata*) by capillary gas-liquid chromatography. J. Chromatogr. **218(2)**: 378–383.

———, ———, **H. Iwatsuki & J. Kumanotani.** 1984. High resolution of gas liquid chromatographic analysis of urushiols of Lac tree (*Rhus vernicifera*) without derivitization. J. Chromatogr. **295(1)**: 179–186.

Duke, J. A. & E. S. Ayensu. 1985. Medicinal plants of China. Vol. I. Reference Publications, Inc., Algonac, Michigan.

Dunn, I. S., D. J. Liberato, N. Castagnoli, Jr. & V. S. Byers. 1986. Influence of chemical reactivity of urushiol-type haptens on sensitization and the induction of tolerance. Cell Immunol. **97**: 189–196.

Dyer, R. A. 1975. The genera of southern African flowering plants. Vol. I. Dicotyledons. Department of Agricultural Technical Services, Pretoria.

Engler, A. 1882. Anacardiaceae. *In* A. et. C. DeCandolle, Monog. phan. **4**: 171–500.

———. 1892. Anacardiaceae. *In* A. Engler & K. A. E. Prantl (eds.), Nat. Pflanzenfam. ed. 1, **3(5)**: 183–278.

Evans, F. J. & R. J. Schmidt. 1980. Plants and plant products that induce contact dermatitis. Planta Med. **38**: 289–316.

Fuller, T. C. & E. McClintock. 1986. Poisonous plants of California. Univ. of California Press, Berkeley.

Gambaro, V., M. C. Chamy, E. von Brand & J. A. Garbarino. 1986. 3-(Pentadec-10-enyl)-catechol, a new allergenic compound from *Lithraea caustica*, (Anacardiaceae). Planta Med. **44(1)**: 20–22.

Gamble, J. S. 1918. Flora of the presidency of Madras. Part 2 (Celastraceae to Leguminosae-Papilionatae). Published under the authority of the Secretary of State for India in Council, London.

Gillis, W. T. 1971. Systematics and ecology of poison-ivy and the poison oaks (*Toxicodendron*, Anacardiaceae). *Rhodora* **73**: 72–159; 161–237; 370–443; 465–540.

Goldstein, N. 1968. The ubiquitous urushiols—Contact dermatitis from mango, poison ivy and other 'poison' plants. Cutis **4**: 679–685.

Goris, M. A. 1910. Contribution a l'étude des Anacardiacées de la tribu des Mangiferées. Ann. Sci. Nat. Bot. **11**: 1–29.

Goudgaon, N. M., J. B. Lamture & U. R. Nayak. 1985. *Semecarpus anacardium* as an anticancer agent, epoxy derivatives of the monoene and diene bhilawanols. Indian Drugs **22(11)**: 556–560.

Gross, M., H. Baer & H. Fales. 1975. Urushiols of poisonous Anacardiaceae. Phytochemistry **14**: 2263–2266.

Hardin, J. W. & J. W. Arena. 1974. Human poisoning from native and cultivated plants. Duke University Press, Durham, North Carolina.

Hegnauer, R. 1964. Anacardiaceae. *In* R. Hegnauer, Chemotaxonomie der Pflanzen. Band 3. Birkhauser Verlag, Basel and Stuttgart.

Hoyos, F. J. 1985. Flora de la Isla Margarita, Venezuela. Sociedad y Fundación La Salle de Ciencias Naturales Monografia, No. 34, Caracas, Venezuela.

Indap, M. A., R. Y. Ambaye & S. V. Gokhale. 1983. Antitumor and pharmacological effects of the oil from *Semecarpus anacardium*. Indian J. Physiol. Pharmacol. **27(2)**: 83–91.

Jefferson, A. & S. Wangchareontrakul. 1985. Synthesis of urushiol derivatives by the Fries rearrangement. Austral. J. Chem. **38(4)**: 605–614.

Joel, D. M. 1980. Resin ducts in the Mango fruit: A defence system. J. Exp. Bot. **31(125)**: 1707–1718.

——— **& A. Fahn.** 1980a. Ultrastructure of resin ducts of *Mangifera indica* L. (Anacardiaceae). 1. Differentiation and senescence of the shoot ducts. Ann. Bot. **46**: 225–233.

——— & ———. 1980b. Ultrastructure of resin ducts of *Mangifera indica* L. (Anacardiaceae). 2. Resin secretion in the primary stem ducts. Ann. Bot. **46**: 779–783.

Johnson, R., H. Baer, C. Kirkpatrick, C. Dawson & R. Khurana. 1972. Comparison of the contact allergenicity of the four pentadecylcatechols derived from poison ivy urushiol in humans. J. Allergy Clin. Immunol. **49**: 27–35.

Keil, H., D. Wasserman & C. R. Dawson. 1946. Mango dermatitis and its relationship to poison ivy hypersensitivity. Ann. Allergy **4**: 268–281.

Kim, S. W. & J. S. Ju. 1982. The effects of lacquer sap extract on sarcoma 180 cell inoculated Swiss mice. Korea Univ. Med. J. **19(1)**: 23–24. (in Korean).

Kirtikar, K. R., B. D. Basu & I. C. S. An. 1935. Indian medicinal plants. Vol. I, 2nd ed. M/S Bishen Singth Mahendra Pal Singh, Dehra Dun, India.

Kligman, A. M. 1958. Poison ivy (*Rhus*) dermatitis. A. M. A. Arch. Derm. Syph. **77**: 149–180.

Lampe, K. F. & R. Fagerstrom. 1968. Plant toxicity and dermatitis. A manual for physicians. The Williams & Wilkins Co., Baltimore.

—— **& M. A. McCann.** 1985. AMA handbook of poisonous and injurious plants. American Medical Association, Chicago, Illinois.

Lejman, E., T. Stoudemayer, G. Grove & A. M. Kligman. 1984. Age differences in poison ivy dermatitis. Contact Dermatitis **11(3)**: 163–167.

Levy, C. K. & R. B. Primack. 1984. A field guide to poisonous plants and mushrooms of North America. The Stephen Greene Press, Brattleboro, Vermont and Lexington, Massachusetts.

Lewis, W. H. & M. P. F. Elvin-Lewis. 1977. Medical botany. Plants affecting man's health. John Wiley & Sons, Inc., New York.

Lloyd, H. A., C. Denny & G. Krishna. 1980. A simple liquid chromatographic method for analysis and isolation of the unsaturated components of anacardic acid. J. Liq. Chromatogr. **3(10)**: 1497–1504.

Loev, B. 1952. The active constituents of poison ivy and related plants. Structure and synthesis. Unpubl. Ph.D. Thesis. Columbia University, New York.

MacBridge, F. 1951. Flora of Peru. Botanical Series Field Museum of Natural History, Vol. XIII, Part IIIA, No. 1. Field Museum Press, Chicago.

Marcano F., E. de Jesus. 1977. Plantas venenosas en la Republica Dominicana. Publicaciones de la Asociación Medica Dominicana, Santo Domingo, Dominican Republic.

Majima, R. 1922. Uber den Haupbestandteil des Japanlacks. IX Chemische Untersuchung der verschiedenen naturlichen Lackarten, die dem Japanlack nahe verwandt sind. Ber. Dtsch. Chem. Ges. **55**: 191.

Mason, C. T. & P. B. Mason. 1987. A handbook of Mexican roadside flora. Univ. Arizona Press, Tucson.

Mattick, F. 1934. Die Gattung *Astronium*. Notizbl. Bot. Gart. Berlin **110(11)**: 991–1012.

Meyer, T. & F. A. Barkley. 1973. Revision del genero *Schinopsis*. Lilloa **33(11)**: 208–257.

Mitchell, J. & A. Rook. 1979. Botanical dermatology. Greengrass, Vancouver, British Columbia.

Mitchell J. C., J. D. Guin, H. J. Maibach & J. H. Beaman. 1985. Allergenicity of *Toxicodendron sylvestre* (Anacardiaceae). Contact Dermatitis **12(2)**: 113–114.

Mitchell, J. D. & S. A. Mori. 1987. The cashew and its relatives (*Anacardium*: Anacardiaceae). Mem. New York Bot. Gard. **42**: 1–76.

Möbius, M. 1899. Der japanische Lackbaum, eine morphologisch-anatomische Studie. Moritz Diesterweg, Frankfurt.

Morton, J. F. 1971. Plants poisonous to people in Florida and other warm areas. Hurricane House, Miami, Florida.

——. 1976. Herbs and spices. Golden Press, New York.

——. 1981. Atlas of medicinal plants of Middle America: Bahamas to Yucatan. C. C. Thomas, Springfield, Illinois.

Muenscher, W. C. 1972. Poisonous plants of the United States. Revised ed. The Macmillan Company, New York, New York.

Mukherjee, S. K. 1985. Systematic and ecogeographic studies on crop genepools: 1. *Mangifera* L. IBPGR, Rome.

Murthy, B. G. K., M. A. Sirasamban & J. S. Aggarwal. 1968. Identification of some naturally occurring alkyl-substituted phenols in cashew-nut shell by chromatographic techniques. J. Chromatogr. **32(3)**: 519–528.

Nakamura, T. 1985. Contact dermatitis due to *Rhus succedanea.* Contact Dermatitis **12(5)**: 279.

Palgrave, K. C. 1981. Trees of southern Africa, 2nd impression. C. Struik Publishers, Cape Town.

Paviani, T. I. 1965. Contribuição ao Conhecimento do Genero *Schinus* L. Anatomia de quatro especies e uma variedade. Thesis. Santa Maria.

Perry, L. M. & J. Metzger. 1980. Medicinal plants of east and southeast Asia: Attributed properties and uses. The MIT Press, Cambridge, Massachusetts and London, England.

Prasad, F. C., R. Raj & M. Sahu. 1985. Use of Indian medicine as an adjuvant therapy to cancer. Anticancer Res. **5(6)**: 587.

Record, S. J. 1939. American woods of the family Anacardiaceae. Trop. Woods **60**: 11–45.

Rendle, B. J. 1929. 'Paaj' dermatitis produced by red quebracho. Trop. Woods **17**: 7–8.

Robinson, T. 1980. The organic constituents of higher plants. Their chemistry and interrelationships, 4th ed. Cordus Press, North Amherst, Massachusetts.

Roth, I. 1977. Fruits of angiosperms. Handbuch der Pflanzenanatomie **10(1)**. Gebrüder Borntraeger, Berlin and Stuttgart.

Senear, F. E. 1933. Dermatitis due to woods. J. Amer. Med. Assoc. **101**: 1527.

Skopp, G. & G. Schwenker. 1984. Separation of cardanols by reversed-phase high-performance liquid chromatography. Planta Med. **50(6)**: 529.

Sood, S. K., J. H. P. Tyman, A. Durrani & R. A. Johnson. 1986. Practical liquid chromatographic separation of the phenols in technical cashew nutshell liquid from *Anacardium occidentale.* Lipids **21(3)**: 241–246.

Stahl, E., K. Keller & C. Blinn. 1983. Cardanol a skin irritant in pink pepper (*Schinus terebinthifolius*). Planta Med. **48(1)**: 5–9.

Standley, P. C. 1923. Trees and shrubs of Mexico (Oxalidaceae–Turneraceae). Contr. U.S. Nat. Herb. Vol. 23, Part 3. Government Printing Office, Washington, D.C.

Steyermark, J. A. & O. Huber. 1978. Flora del Avila. Bajo los auspicios de Vollmer Foundation y Ministerio del Ambiente y de los Recursos Naturales Renovables, Caracas, Venezuela.

Thomson, R. H. 1971. Naturally occuring quinones, 2nd ed. Academic Press, London.

Tomlinson, P. B. 1986. The biology of trees native to tropical Florida, 2nd printing. Publ. by author, Allston, Massachusetts.

Tyman, J. H. P. & L. J. Morris. 1967. The composition of cashew nut-shell liquid (CNSL). J. Chromatogr. 27(1): 287–288.

Vasistha, S. K. & S. Siddiqui. 1938. Chemical examination of mango 'Chep', the exudation of the fruit of *Mangifera indica.* J. Indian Chem. Soc. 15: 110–117.

Venning, F. D. 1948. The ontogeny of the laticiferous canals in the Anacardiaceae. Amer. J. Bot. 35: 637–644.

Wannan, B. S., J. T. Waterhouse & C. J. Quinn. 1987. A taxonomic reassessment of *Blepharocarya* F. Muell. Bot. J. Linn. Soc. 95(1): 61–72.

Watt, J. M. & M. G. Breyer-Brandwijk. 1962. The medicinal and poisonous plants of southern and eastern Africa. 2nd ed. E. S. Livingstone, Ltd., Edinburgh and London.

Westbrooks, R. G. & J. W. Preacher. 1986. Poisonous plants of eastern North America. University of South Carolina Press, Columbia, South Carolina.

Womersley, J. S. 1983. Blepharocaryaceae. Page 191 *in* B. D. Morley & H. R. Toelken (eds.), Flowering plants in Australia. Rigby Publishers, Adelaide.

Woodruff, J. G. 1979. Tree nuts: Production, processing, products, 2nd ed. Avi Publishing Co., Inc., Westport, Connecticut.

Woods, B. & C. D. Calnan. 1976. Toxic woods. Brit. J. Dermatol. 94(Suppl. 13): 1–97.

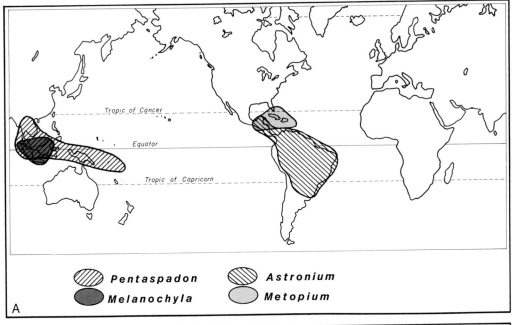

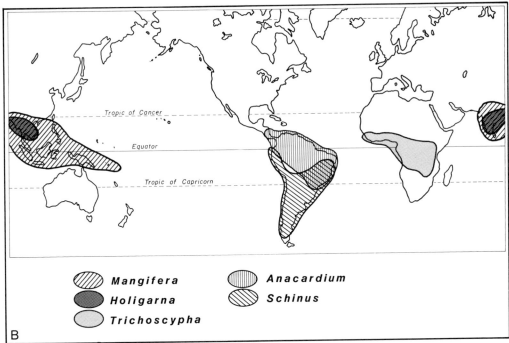

MAP IA. Distributions of *Astronium, Melanochyla, Metopium* and *Pentaspadon.*
MAP IB. Distributions of *Anacardium, Holigarna, Mangifera, Schinus,* and *Trichoscypha.*

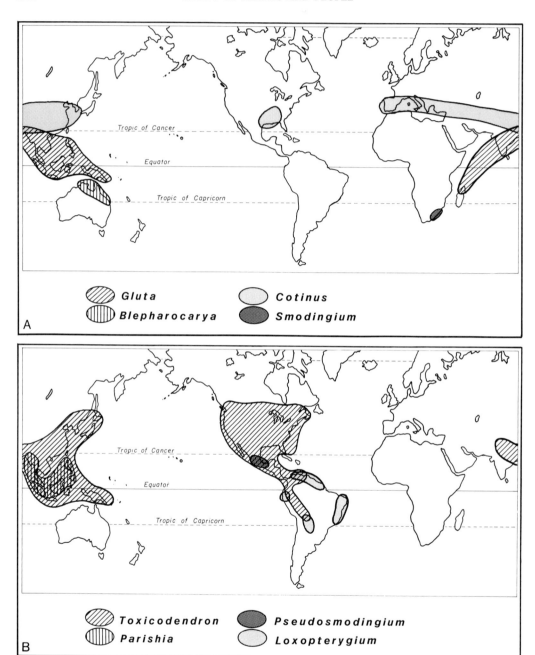

MAP IIA.   Distributions of *Blepharocarya, Cotinus, Gluta* and *Smodingium.*
MAP IIB.   Distributions of *Loxopterygium, Parishia, Pseudosmodingium* and *Toxicodendron.*

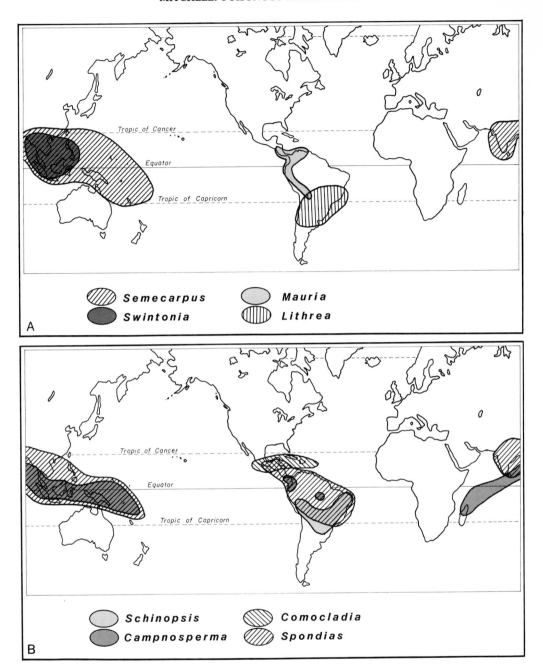

MAP IIIA.   Distributions of *Lithrea, Mauria, Semecarpus* and *Swintonia.*
MAP IIIB.   Distributions of *Campnosperma, Comocladia, Schinopsis* and *Spondias.*

Taxonomy, Ecology, and Economic Botany of the Brazil Nut (*Bertholletia excelsa* Humb. & Bonpl.: Lecythidaceae)

Scott A. Mori and Ghillean T. Prance

Table of Contents

Abstract

MORI, S. A. (Institute of Economic Botany, The New York Botanical Garden, Bronx, New York 10458-5126 U.S.A.) and G. T. PRANCE (Director, Royal Botanic Gardens, Kew, Richmond, Surrey TW9 3AB, United Kingdom). Taxonomy, ecology, and economic botany of the Brazil nut (*Bertholletia excelsa* Humb. & Bonpl.: Lecythidaceae). Advances in Economic Botany 8: 130–150. 1990. The Brazil nut (*Bertholletia excelsa* Humb. & Bonpl.), native to the Amazon basin, is the most economically important species of Lecythidaceae. Its edible seeds are collected almost exclusively from naturally occurring trees. The taxonomy, distribution and ecology, phenology, pollination biology, seed dispersal, economic botany, and cultivation of this rain forest tree are described and related to what is known about other species of New World Lecythidaceae. The technological methods for growing Brazil nuts in plantations are reviewed. Finally, the impact that excessive deforestation is having, and will continue to have, on Brazil nut production is described.

Key words: *Bertholletia excelsa*; Lecythidaceae; Brazil nut; economic botany.

Advances in Economic Botany 8: 130–150, 1990
© 1990 The New York Botanical Garden

Introduction

The Institute of Economic Botany of The New York Botanical Garden approaches the study of economic plants in different ways, ranging from study of the ethnobotany of Amerindians to the development and marketing of under-exploited Amazonian fruits (Institute of Economic Botany, 1986). The approach we have taken is that of undertaking research on the taxonomy and biology of Neotropical members of the Brazil nut family to provide information for its more rational economic exploitation.

The first step in the utilization of any plant is the correct application of its scientific name. Without this step, information about the plant cannot be communicated to other scientists. It is also important that the phylogeny of economic plants be understood so that geneticists know which other species may provide germplasm for breeding programs. Knowledge of geographic distribution and intraspecific variation of economically important traits, such as seed production and quality, is essential for any germplasm collecting program. Moreover, an understanding of pollination and dispersal biology is often critical in deciding if wild populations of under-exploited economic plants, such as the Brazil nut, can be brought into cultivation.

Bertholletia excelsa Humb. & Bonpl. (Figs. 1, 2) is one of a number of economically important plants native to Amazonia and the Guayana Floristic Province (Balick, 1985). Some, e.g., rubber (*Hevea brasiliensis* (Willd. ex Adr. de Juss.) Muell. Arg.) and chocolate (*Theobroma cacao* L.), are of economic importance throughout the world, whereas others, e.g., *guaraná* (*Paullinia cupana* H.B.K.) and numerous medicinal plants have reached only a fraction of their economic potential (Balick, 1985; Farnsworth, 1984). Of the better known economically important plants of the Amazon and Guayana, the Brazil nut is the only one that, until recently, has not been successfully brought into plantation agriculture. In contrast to rubber and chocolate, which are mostly grown in plantations outside of their native ranges, the Brazil nut crop is still mostly gathered from wild trees.

HISTORY OF THE LECYTHIDACEAE

The first reference to a member of Lecythidaceae was made by Gabriel Soares de Sousa who described the *sabucai* in 1587 in his *Noticia do Brasil* (Sousa, 1949). Frei Christovão described the same species, but called it *sapuquaiha*, in his *Historia dos animães e árvores do Maranhão* written between 1624 and 1635 and published in manuscript facsimile in 1967. Both references are to *Lecythis pisonis* Cambess. which grows in eastern extra-Amazonian Brazil, where Soares de Sousa and Christovão observed it, and in Amazonia (Mori & Prance, 1981). Marcgrave illustrated and described the same species in his *Historiae rerum naturalium brasiliae* in 1648. The large size of the fruit of *L. pisonis* and its edible seeds brought this species to the attention of many early explorers.

The first valid scientific publication of a species of Lecythidaceae was that of *Lecythis ollaria* Loefl. in 1758. *Bertholletia* and its single species, *B. excelsa*, were originally described in 1807 by Humboldt and Bonpland from a collection they had made in Venezuela. However, the tree from which they gathered their collection might have been raised from seed introduced from Brazil (Prance, in press). Poiteau (1825) was the first to give familial status to Lecythidaceae when he removed *Bertholletia, Couratari, Couroupita,* and *Gustavia* from the Myrtaceae where they had been traditionally placed.

Neotropical Lecythidaceae have been the subject of one major floristic treatment and three monographs. Berg (1858) treated the family for *Flora Brasiliensis,* Miers (1874) provided the first monograph, Knuth (1939) the second, and Prance and Mori (1979) and Mori and Prance (in press) combined to provide the third. A number of local floristic accounts of Neotropical Lecythidaceae have been published (Barbosa, 1982; Eyma, 1932, 1934; Lemée, 1953; Lindeman & Mennega, 1963; Mori & Prance, 1983; Pittier, 1912, 1927; Reitz, 1981; Woodson, 1958), and a study of the taxonomy and biology of the 27 species of Lecythidaceae found in a proposed national park surrounding Saül, French Guiana has been published (Mori & collaborators, 1987). Because of its economic importance, the Brazil nut has been the object of many studies which are listed in a bibliography of 259 titles by Vaz Pereira and Lima Costa (1981). A great number of these studies were carried out under the auspices of the "Centro de Pesquisas Agropecuário do Trópico Umido" (CPATU) of the "Empresa Brasileira de Pesquisa Agropecuária" (EMBRAPA) at Belém,

FIG. 1.   Remnant Brazil nut tree in disturbed forest in Amapá, Brazil (photo by S. Mori).

$\rightarrow$

FIG. 2.   *Bertholletia excelsa,* A–D (Mori et al. 17503), E, F (Prance et al. 16599), G (Prance blastogeny collection no. 4). **A.** Habit. Note how the petals are tightly appressed to the androecium. **B.** Medial section of the androecium. Note how the hood appendages are swept inwards but do not form a complete coil. **C.** Cross section of the ovary. **D.** Ovary and calyx. Note the 2-parted calyx, truncate ovary base, and long, geniculate style.

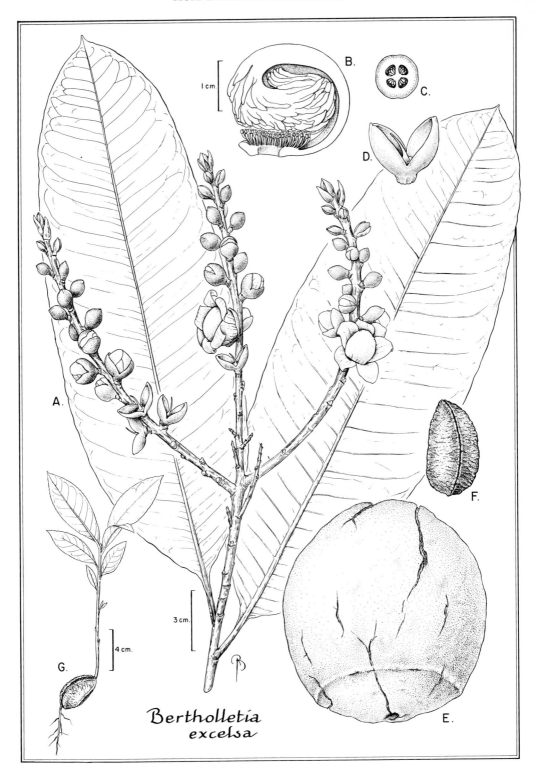

Bertholletia excelsa

Brazil where there is ongoing research on all aspects of Brazil nut biology and cultivation.

USES OF THE LECYTHIDACEAE

The Brazil nut is by far the most economically important Neotropical Lecythidaceae. However, several other species have edible seeds. The *sapucaia* (*L. pisonis*; Mori & Prance, 1981) of eastern extra-Amazonian Brazil and Amazonia possesses seeds equally as delicious as the Brazil nut. However, the fruits open at maturity and the seeds are removed by bats (Greenhall, 1965) thereby making their harvest difficult on a large scale. The three other members of the *sapucaia* group of *Lecythis* (*L. ampla* Miers, *L. lanceolata* Poir., and *L. zabucaja* Aubl.) (Mori & Prance, 1981), also have edible seeds, but their economic exploitation is hindered by the same problems affecting *L. pisonis*. *Lecythis minor* Jacq. of northern Colombia and northwestern Venezuela and *L. ollaria* Loefl. of the llanos of Venezuela, which have edible seeds, unfortunately accumulate toxic levels of selenium when grown on certain soils. Consumption of too many seeds with high selenium may result in nausea and hair and fingernail loss (Kerdell-Vegas, 1966; Mori, 1970). Above-normal accumulations of barium and high incidences of radioactivity have been registered for the seeds of *Bertholletia excelsa* (Souza, 1963). However, no adverse effects on human health have been documented as the result of eating Brazil nut seeds. The pericarps of some species of *Grias* and *Gustavia* are locally consumed, especially in Colombia, Ecuador, and Peru (Prance & Mori, 1979). Other minor economic uses of Neotropical Lecythidaceae are discussed by Prance and Mori (1979).

Exploitation of Brazil nuts began with Amerindians who ate the seeds and extracted the juice to use in flavoring their manioc porridge. The seeds were called *nhá, niá, invia, tacari,* and *tucari* by different groups of Indians. At the end of the 18th and beginning of the 19th centuries, the Dutch invaders of Brazil were sending seeds of Brazil nut back to Europe, therefore commercialization of this species began around 1800. However, it was not until 1866, when Brazilian ports were opened to free trade, that the market in Brazil nuts began to prosper (Almeida, 1963; Souza, 1963).

Taxonomy of the Lecythidaceae

The Lecythidaceae, sensu lato, is a pantropical family of small to very large trees. The family includes four subfamilies: Planchonioideae, with 55 species in six genera distributed through tropical Asia, Malaysia, northern Australia, and the Pacific Islands; Foetidioideae, with 5 species in a single genus distributed in Madagascar, India, and Malaysia; Napoleonaeoideae, with 12 species in two genera in West Africa and one monotypic genus in the upper Rio Negro region of Amazonia; and the Lecythidoideae, with 200 species in ten genera distributed through tropical America from Veracruz, Mexico to southern Brazil (Kowal et al., 1977; Mori & Prance, in press; Prance & Mori, 1979).

The Neotropical Lecythidoideae can be divided into two distinct groups, those with actinomorphic (radially symmetrical) and those with zygomorphic (bilaterally symmetrical) flowers. The genera *Allantoma, Grias,* and *Gustavia* (Fig. 4) are actinomorphic while *Bertholletia* (Figs. 2, 4), *Corythophora, Couratari, Couroupita, Eschweilera,* and *Lecythis* are clearly zygomorphic. *Cariniana* is only slightly zygomorphic. We have discussed and illustrated these basic floral types in detail elsewhere (Mori & Boeke, 1987; Prance & Mori, 1979). *Bertholletia* is readily distinguished from other species of zygomorphic-flowered Lecythidaceae by the features enumerated in the following key.

KEY TO DISTINGUISH *BERTHOLLETIA EXCELSA*

1. Buds enclosed by calyx except for horizontal slit at apex; calyx with 2 lobes at anthesis. Fruits functionally indehiscent, with small, inwardly falling operculum. Seeds with thick, bony testa, remaining inside fruit at maturity (Fig. 2). . . . . *Bertholletia excelsa.*
1. Buds usually not enclosed by calyx, or, if enclosed, then without horizontal slit at apex; calyx entire or with more than 2 lobes at anthesis. Fruits dehiscent, with larger, outwardly falling operculum, infrequently indehiscent but then with no apparent operculum. Seeds without thick, bony testa, usually falling from fruit at maturity. . . . . All other taxa of Lecythidaceae.

Bertholletia excelsa Humboldt & Bonpland, Pl. aequinoct. **1:** 122–127, t. 36. 1807; Hooker, J.

Bot. (Hooker) **2**: 234. 1840; Spruce, J. Bot. Kew Misc. **2**: 74–75. 1850; Berg *in* Mart. Fl. Bras. **14(1)**: 478–479, t. 60–61. 1858; Anonymous, Kew Bull. **27**: 11–13. 1887; Young, Bot. Gaz. (Crawfordsville) **52**: 226–231. 1911; White, Brooklyn Bot. Gard. Leafl. **10(4)**: 1–4. 1922; Petch, Gard. Chron. **77**: 349. 1925; Sands & Sprague, Malayan Agric. J. **14(5)**: 125–126. 1926. Type. Venezuela. Amazonas: Without locality, no date (st), *Humboldt & Bonpland s.n.* (holotype, P; isotypes, P, B-Herb. Willdenow 10107).

Barthollesia excelsa Silva Manso, Enum. subst. braz. p. 45. 1836, sphalm.
Bertholletia nobilis Miers, Trans. Linn. Soc. London **30(2)**: 197–199, t. 37, figs. 4–7. 1874; Young, Bot. Gaz. (Crawfordsville) **52**: 226–231. 1911. Type. Brazil. Pará: Without locality, 1827 (fl), *Martius s.n.* (lectotype, K; isolectotype, E).

Large *trees*, to 50 m tall, unbuttressed, the young branches glabrous, sparsely lenticellate. *Bark* with conspicuous longitudinal fissures. *Leaves* simple, alternate, the blades oblong, 17–36 × 6.5–15.5 cm, coriaceous, glabrous on both surfaces, longitudinally striate, papillate abaxially, the midrib prominulous above, prominent beneath, with 29–45 pairs of lateral veins, prominulous on both surfaces; apex apiculate to mucronate, the acumen 3–10 mm long; base rounded to subcuneate; margins entire or slightly crenulate, undulate; petiole 20–35 mm long, glabrous or sparsely puberulous when young, slightly winged. *Inflorescence* of axillary spikes or of terminal, paniculate arrangements of spikes with one or two orders of branching, the latter with 2–7 lateral branches, the rachises angular and sulcate, sparsely puberulous, the bracts and bracteoles lanceolate, sparsely puberulous abaxially, early caducous; pedicels short and thick, to 1 mm long. *Flowers* ca. 3 cm diam. at anthesis; calyx with 2 lobes, these closed in bud, the lobes oblong-ovate, navicular, 8–14 mm long, puberulous abaxially, often obscurely 3-dentate at apex; petals six, pale yellow to white, oblong-ovate, to 30 mm long, sparsely puberulous abaxially; androecial hood white to yellow, the appendages often brighter yellow, the inside of hood with depression into which fits the staminal ring, the inside of ligule often tinged with pink, the appendages swept inwards but not forming coil;

staminal ring asymmetrical, with 80–135 stamens, the filaments 2.5–3 mm long, swollen and geniculate towards apex, white, the anthers 1 mm long, golden-yellow; hypanthium 3–5 mm long, puberulous; ovary (3–)4(–6)-locular, the 16–25 ovules inserted at base of septum; styles of two lengths, either ca. 8 mm long, geniculate and exserted beyond stamens or less frequently ca. 4 mm long and shorter or equalling stamens in length. *Fruits* round, usually 10–12.5 × 10–12.5 cm but up to 16 × 14 cm, often of varied shape and size even within same tree. *Seeds* 10–25 per fruit, ca. 3.5–5 × 2 cm, the embryo undifferentiated. X = 17.

Taxonomy

The genus was named in honor of the famous chemist L. C. Berthollet (1748–1822), a contemporary of Humboldt and Bonpland. The epithet alludes to the lofty stature of the species.

Miers (1874) described the Brazil nut of commerce as a separate species, *B. nobilis*, separated from *B. excelsa* by a number of minor features now known not to be valid for separating species of Lecythidaceae. Consequently, *B. nobilis* is a synonym of the earlier Humboldt and Bonpland name. Although there is some variation in fruit shape, e.g., most individuals have round fruits while others have fruits longer than broad, there is no justification for recognizing more than one species in the genus.

It is clear that the closest relatives to *B. excelsa* are found in *Lecythis*. In particular, *Lecythis lurida* (Miers) Mori shares a number of features with *B. excelsa*. The abaxial leaf surface of both species possesses papillae of cuticular origin, a feature found in several other species of *Lecythis* but absent in other members of the family. The androecial hood of these two species is also similar, i.e., it bears appendages that are swept inwards but that do not form a complete coil. Finally, the fruits of *L. lurida* also drop to the ground with the seeds inside at maturity. The common name of *L. lurida, jarana,* means false Brazil nut in the Tupi-Guarani language of the Asuruni Indians of the Rio Xingu, Brazil (W. Balée, pers. comm.).

We suggest that these similarites indicate that *B. excelsa* and some species of *Lecythis,* in particular *L. lurida,* share common ancestry. Most of the differences between these genera probably arose as the result of selection for seed dispersal by the agouti (*Dasyprocta* spp.), a topic we will discuss later in this paper. Nevertheless, the two genera have differentiated to such an extent that hybridization between them is probably not possible. Consequently, the search for germplasm for improvement of Brazil nut production will have to come from the variation found within *B. excelsa,* not from closely related species in other genera.

Morphology

A number of morphological features of *B. excelsa* are variable. Moritz (1984), in her detailed study of the floral biology of clones of Brazil nut cultivated on the grounds of CPATU at Belém, Pará, found: (1) stamen number to range from 80 to 130; (2) most flowers to possess four locules (87.8%) but three (0.1%), five (11.8%), and six (0.3%) locules also occur; (3) the presence of long styles in some flowers and short ones in other flowers; (4) fruit size to differ according to their provenance; and (5) seeds per fruit to range from 10 to 21 with a mean of 17–18.

Sampaio (1944) reports considerable variation in Brazil nut seed size. He suggests that the largest seeds come from Lago Abufari along the Rio Purus, and that the smallest originate in the state of Acre. Preference is given to the larger seeds for confectionery while the smaller ones are used for the extraction of oil and the preparation of flour. The smaller seeds are more easily removed from the seed coat (Almeida, 1963).

Bruce Nelson (pers. comm.) reports that seed size and number of seeds per fruit vary considerably within the same population and that intrapopulational variation may be as great as interpopulational variation. He adds that seed size, number of seeds per fruit, and number of fruits from year to year are relatively constant in individual trees. However, these suggestions need to be confirmed by intra- and interpopulational field studies.

Buckley et al. (manuscript), in a study of genetic variation based on isozyme study of 13 enzymes, demonstrated that 54.3% of the sample loci were polymorphic. They also showed that the two populations studied, one from the Careiro-Porto Velho Road in Amazonas (3°23′S, 59°50′W) and the other from Mocambo in Acre (10°45′S, 68°10′W), had more intra- than interpopulation diversity. However, one of their conclusions, that genetic diversity of the Brazil nut can be preserved within one or a few populations, must be taken with extreme caution. In the first place, conclusions based on the comparison of only two populations could easily be modified with increased sample size. In the second place, they have ignored the possibility that one of these populations may have been derived from the other by Amerindian intervention. If one of the populations was established by seed carried by Indians from the other population, then the two would, in effect, be a single population and genetic diversity between them would be expected to be low.

Distribution and Ecology

Bertholletia excelsa is found on non-flooded ground (*terra firme*) in the Guianas, Amazonian Colombia, Venezuela, Peru, Bolivia, and Brazil (Fig. 3) (Prance, in press). It is cultivated in botanical gardens outside of its native range and one relatively successful plantation has been established in Kuala-Lumpur in Malaysia (Müller, 1981).

Trees of *B. excelsa* occur in stands of 50–100 individuals, each stand separated from one another by distances of up to a kilometer. These stands are known as "manchales" in Peru (Sánchez, 1973) and "castanhais" in Brazil (Dias, 1959). Densities as low as one tree per six hectares and as high as 15–20 trees per hectare have been reported (Sánchez, 1973).

Almeida (1963) and Diniz and Bastos (1974) studied climatological data from numerous localities where Brazil nut stands occur. They found that *B. excelsa* trees grow in areas with an annual mean temperature of between 24.3 and 27.2°C, an annual rainfall from 1400 to 2800 mm, and a mean annual relative humidity between 79 and 86%. They also noted that two to seven months of the year are dry, i.e., with less than 60 mm rainfall. Müller (1981) also suggested that two to five months with reduced rainfall are needed for the proper development of *B. excelsa.*

Apparently, Brazil nut trees are light-demanding and need gaps in the forest before they can grow to reproductive size. Pires (1984) found the most frequent size class of Brazil nuts trees in a forest near the Jarí River in Brazil to be 35–40 m × 110 cm DBH. Smaller size classes were rare.

FIG. 3.   Distribution of *Bertholletia excelsa*.

The predominance of larger size classes in Pires' study indicates that Brazil nut trees are gap dependent, as is the case with many other tropical trees (Foster & Brokaw, 1982). However, their occurrence in natural stands may not be explained by gap theory alone because Brazil nut stands have been observed to occur over much larger areas than the area influenced by a single gap (B. Nelson, pers. comm.). It has also been suggested that natural stands of this species are the remnants of plantations started by pre-Colombian Amerindians (Müller et al., 1980). One of us (G.T.P.) has observed that a Brazil nut stand on the Xingu River is near the site of an ancient Indian village. Moreover, it is known

that at least some Amazonian Indians manipulate their environment in such a way as to encourage the growth of useful plants (Balée, 1989; Posey, 1985). Nevertheless, natural stands of Brazil nuts are absent from a large area covered by Tertiary soils around Manaus, an area known to have been heavily populated in the past (Nelson et al., 1985).

Phenology

Our work on phenology and pollination biology of Lecythidaceae has dealt mostly with species other than *B. excelsa* (Mori et al., 1978; Mori & Boeke, 1987; Mori & Prance, 1987b; Prance,

1976; Prance & Mori, 1979). Studies on these aspects of Brazil nut biology have been done mostly by Müller and collaborators at CPATU (Müller, 1981; Müller et al., 1980; Moritz, 1984) and by Nelson et al. (1985) who have studied central Amazonian populations of *Bertholletia*.

Neotropical Lecythidaceae flower mostly in the dry season where a distinct dry season occurs (Mori & Kallunki, 1976; Mori & Prance, 1987a, 1987b; Prance & Mori, 1979). For example, in central French Guiana, where there is a dry season from August to November, 26 of the 27 species of Lecythidaceae flower predominantly in the dry season (Mori & Prance, 1987b). In areas where there is no reliable dry season, other factors may influence flowering. Mori et al. (1980) found that, in southern Bahia where there is no clear cut dry season, *Lecythis pisonis* Cambess. flowered year after year in the spring (September to November), apparently in response to the increasing day length or temperature at that time of the year. Under seasonal climates, Neotropical Lecythidaceae mostly mature seed at the beginning or during the first part of the wet season (Mori & Kallunki, 1976; Mori & Prance, 1987a, 1987b).

Most species of Neotropical Lecythidaceae open their flowers early in the morning and drop their petals and androecia in the same afternoon. However, at least one species, *Lecythis poiteaui* Berg, opens its flowers at dusk and drops its petals and androecia in the morning (0300). In this species, the trees are visited and presumably pollinated by bats (Mori et al., 1978). Other species, *Couratari atrovinosa* Prance (Prance & Anderson, 1976) and *Gustavia augusta* L., open their flowers at night but retain their petals and androecia until late the following afternoon. These species are visited by bees, some of which fly in the early morning or even at night. Finally, some species, e.g., *Eschweilera grandiflora* (Aublet) Sandwith and *E. pedicellata* (Richard) Mori, retain intact flowers for several days (Mori et al., 1978).

Species of Neotropical Lecythidaceae exhibit three flowering patterns in relation to vegetative growth: (1) leaf drop, leaf flush, flowering; (2) leaf drop, flowering, leaf flush; and (3) leaf drop and flowering independent of one another. Members of this family also differ in the number of flowers they produce and in the duration of flowering. Some species produce few flowers daily over long periods ("steady state" flowering); others, abundant flowers daily over short periods ("big bang" flowering); and others, abundant flowers daily over longer periods ("cornucopia" flowering) (Gentry, 1974).

As in most other species of Lecythidaceae, flowering of *Bertholletia excelsa* is related to the dry season. In fact, the species only grows naturally in regions with a three to five month long dry season (Müller, 1981). In eastern Amazonian Brazil, flowering begins at the end of the rainy season in September and extends to February. Peak flowering occurs in October, November, and December (fig. 21 *in* Moritz, 1984).

Toward the end of the rainy season, generally in July, the leaves of Brazil nut trees begin to fall. The new growth flushes from directly below the inflorescences of the previous year, and the new inflorescences are produced at the apex of the current growth flush. This fits the phenological pattern of leaf drop, leaf flush, and flowering described for other species of Lecythidaceae. Large numbers of flowers are produced daily over a relatively long period, and, therefore, *B. excelsa* should be considered a "cornucopia" flowering species.

Anthesis of *Bertholletia excelsa* occurs between 0430 and 0500. However, the anthers start to dehisce within the bud several hours before the flowers open (fig. 7 *in* Müller et al., 1980). The petals and androecia fall in the afternoon of the day that the flowers open. As far as we know, no studies of stigma receptivity have yet been made.

Pollination Biology

The flowers of Neotropical Lecythidaceae are pollinated almost exclusively by bees (Fig. 4) which are rewarded for their visits by fertile pol-

→

FIG. 4. Examples of actinomorphic and zygomorphic-flowered Neotropical Lecythidaceae. **A.** Actinomorphic flower of *Gustavia augusta* L. (Mori & Boom 14701, photo by S. Mori). Note the trigonid bees collecting pollen from the anthers. **B.** Zygomorphic flower of *Bertholletia excelsa* with large bee visiting it (photo by A. Henderson).

len, fodder pollen, or nectar. A wide array of bees visiting Lecythidaceae has been collected, ranging from the meliponine New World social bees to the solitary euglossines (see Mori & Boeke, 1987 and Mori et al., 1978 for reviews). However, it appears that the evolution of zygomorphic-flowered Lecythidaceae has been promoted by the presence of euglossine bees. Both euglossine bees and this kind of Lecythidaceae are restricted to the Neotropics where their ranges nearly overlap (Mori & Boeke, 1987). The only documented exception to bee pollination in Neotropical Lecythidaceae is the bat-pollinated *Lecythis poiteaui* (Mori et al., 1978).

In some species of Neotropical Lecythidaceae with zygomorphic flowers, a differentiated or fodder pollen is produced. In these species, the fodder pollen is always spatially separated from the fertile pollen. It is either restricted to the hood or to a specialized row of stamens in the staminal ring. The two types of pollen can often be distinguished by their different colors. The fodder pollen does not germinate, at least in vitro, in *Couroupita guianensis* Aublet and *Lecythis pisonis*. The fodder pollen, as viewed with the SEM, is morphologically distinct from the fertile pollen in *C. guianensis* but not morphologically distinct in *L. pisonis*. In many other zygomorphic-flowered Neotropical Lecythidaceae, nectar is the principal reward (e.g., all species of *Bertholletia*, *Couratari*, and *Eschweilera* and some species of *Lecythis*). In these species, the staminal ring possesses only fertile pollen while the hood usually provides only nectar (Mori & Prance, 1987a). All actinomorphic species of Neotropical Lecythidaceae possess one kind of pollen, and do not produce nectar. Consequently, the bee pollinators collect the same pollen which also effects fertilization when transferred to the stigma of another flower.

Little information is available on self compatibility and incompatibility in species of Lecythidaceae other than for *Bertholletia excelsa*. The few studies of breeding systems of Neotropical Lecythidaceae indicate that these species are normally outcrossed (Jackson & Salas, 1965; Mori & Kallunki, 1976; Müller et al., 1980; O'Malley et al., manuscript; Ormond et al., 1981; Prance, 1976). It has been demonstrated for *Couroupita guianensis* (Ormond et al., 1981) that some seeds will develop if the flowers are self-pollinated, and it has been suggested that a low level of inbreed-

ing may occur in *B. excelsa* (O'Malley et al., manuscript).

The flowers of *Bertholletia excelsa* are visited and presumably pollinated by a variety of large-bodied bees. Species of *Bombus*, *Centris*, *Epicharis*, *Eulaema*, and *Xylocopa* have been captured visiting Brazil nut trees (Moritz, 1984; Müller et al., 1980; Nelson et al., 1985). These bees have the strength to force open the flower to get at the nectar reward produced in the hood.

Müller et al. (1980) have shown that bees begin to visit trees of *B. excelsa* as early as 0540, and that peak visitation occurs between 0600 and 0700. They point out that *Xylocopa* only visits flowers that have not been previously visited, therefore its visits drop off markedly after 0700 while *Bombus* and *Centris* continue to visit flowers entered by other bees. Individuals of these two genera may visit the canopy throughout the morning. Consequently, they conclude that *Bombus* and *Centris* are the principal pollinators of *B. excelsa* in their study area. They also conclude that individuals of *Epicharis* and *Eulaema* are secondary pollinators because of their less frequent visits. However, we feel that the early visits of *Xylocopa* may also be extremely important for fruit set in *Bertholletia*.

Kitamura and Müller (1984) have suggested that one of the causes of lowered Brazil nut production in the vicinity of Marabá, Brazil has been decreased pollination. According to them, atmospheric smoke caused by the excessive number of slash and burn fires in recent years interferes with the activities of the bees responsible for the pollination of *B. excelsa*.

It should be noted that pollinators have not yet been collected from trees growing in undisturbed forest. Thus, it is not yet possible to say if one or the other group of these large-bodied bees plays a more significant role in pollination of the Brazil nut in undisturbed forests. It is, however, clear that pollination of *B. excelsa* is not restricted to any single group of bees.

Müller et al. (1980) and Moritz (1984) have provided the basic information needed for making crosses between different clones of *B excelsa*. They have shown that the best time for self and cross pollinations is between 0600 and 0900. At this time the pollen still possesses high viability and the stigmas are receptive. They have also demonstrated that *B. excelsa* is mostly self-incompatible. The few fertilized ovules that occurred in their self pollinations could have been caused by thrips or other small insects that may not have been excluded by their bags (Müller et al., 1980). However, O'Malley et al. (manuscript) suggest that, although *B. excelsa* is mostly outcrossed, a significant low level of inbreeding may occur. It has been demonstrated that crosses between different clones have different degrees of seed set (Moritz, 1984).

Seed Dispersal

The seeds of *B. excelsa* are known in English as Brazil nuts. The edible products are the seeds (minus the seed coat or testa) which are produced within a capsular fruit. Botanically, the word nut refers to a fruit, and therefore, Brazil nuts should be more appropriately called Brazil seeds.

The fruits and seeds of *B. excelsa* are unique in the Lecythidaceae (Figs. 5A, 6A). The fruits differ from those of other species of the family which fall to the ground with seeds inside by possessing a thick, woody pericarp and by being functionally indehiscent. At maturity the fruits fall from the tree with the seeds inside. The seeds are retained within because the diameter of the opening of the fruit is smaller than the diameter of the seeds. The seeds of *B. excelsa* differ from other Lecythidaceae in their bony testa and their distinctly triangular shape in cross section. They have massive embryos without differentiated cotyledons similar to those of *Corythophora, Eschweilera,* and *Lecythis* (Prance & Mori, 1979).

Fruit development takes longer in *B. excelsa* than in any other species of Lecythidaceae. Moritz (1984) states that 15 months are needed for the fruits to develop after they have been set. Fruits of *B. excelsa* fall mostly in January and February during the rainy season (Prance, 1986). Other species, such as the large-fruited *Lecythis ampla* Miers and *L. pisonis,* may take from 7 to 10 months to mature (Mori et al., 1980; Prance & Mori, 1979), but most species mature fruit in less than six months. Moreover, the seeds of *B. excelsa,* if not artificially treated, take from 12 to 18 months to germinate (Müller, 1981). The seeds of other species of Lecythidaceae, because of their thinner and softer testas, germinate almost immediately after dispersal.

The seeds of *B. excelsa* are dispersed by agoutis (Huber, 1910), a common rodent in Neotropical forests (Figs. 5B, 6A). The agouti is apparently the only animal that is able to efficiently gnaw through the extremely woody pericarps of the Brazil nut. Agoutis eat some of the seeds immediately and bury others for subsequent use. Although Smythe (1978), in a study of Panamanian agoutis, has shown that this rodent's home range is relatively small, the potential for dispersal is much greater because agoutis in adjacent ranges often remove originally cached seeds and rebury them even further from the parent tree. The seeds are either eaten and destroyed by the agouti or left in a forgotten cache where they eventually germinate.

Economic Botany

Brazil nut trees are of economic importance in Amazonian Brazil, Peru, and Bolivia. The timber is useful for construction. Although it is against the law to fell Brazil nut trees in Brazil, there is still a clandestine market for its timber (Kitamura & Müller, 1984). The bark is used in caulking boats, the fruits are used for fuel and for making an array of souvenirs, and the seeds have a multitude of uses (Almeida, 1963; Souza, 1963). Until recently, Brazil nut seeds were second only to rubber as an export crop in Amazonian Brazil (Almeida, 1963; Knuth, 1939; Dias, 1959). Since 1933, Brazilian production has ranged from 3557 tons in 1944 to 104,487 tons in 1970 (Fig. 7). Low production in the early 1940's was caused by reduced demand by two of the largest consumers, the United States and Great Britain, during World War II. Moreover, many of the resources used to gather Brazil nuts were focused on products essential to the war effort, such as rubber (Kitamura & Müller, 1984). Since 1980, annual production has been around 40,000 tons (Fig. 7). The economic importance of Brazil nuts is not as great in Peru and Bolivia as it is in Amazonian Brazil. Nevertheless, in certain areas, such as Madre de Dios, Peru, where 20% of the department is covered by forests rich in individuals of *B. excelsa,* nearly two-thirds of the population may be engaged in Brazil nut extraction (Sánchez, 1973). Indeed, the economies of many Amazonian towns, such as Puerto Maldonado, Peru (Sánchez, 1973) and Marabá, Brazil (Dias, 1959) still depend heavily on Brazil nut production. In 1986, the total value of shelled and unshelled Brazil nut seeds exported from Manaus alone was $5,773,228.00 (data from the

FIG. 5. Fruit and dispersal agent of *Bertholletia excelsa*. **A.** Fruit in situ. **B.** The agouti (*Dasyprocta* sp.), the principal dispersal agent of Brazil nut seeds (photos by G. T. Prance).

FIG. 6.   Fruits of *Bertholletia excelsa*. **A.** Old fruit with pericarp gnawed through by an agouti. **B.** Removal of seed from fruit with a machete (photos by G. T. Prance).

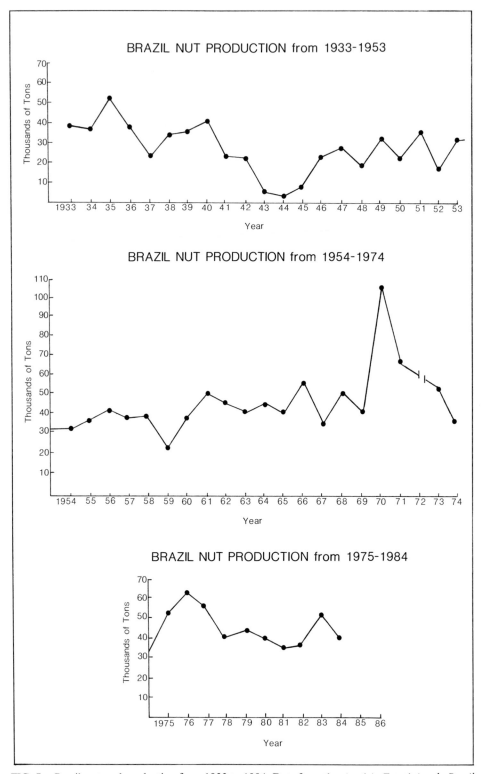

FIG. 7.   Brazil nut seed production from 1933 to 1984. Data from the *Anuário Estatística do Brasil.*

"Associação dos Exportadores da Zona Franca de Manaus"). Most of the seeds exported from Brazil are sent to England, France, the United States, and West Germany. Very little of the Brazilian crop is consumed domestically (Almeida, 1963).

Brazil nuts are eaten raw, roasted, or are used in confectionery. The volatile flavor compounds responsible for the distinctive aroma of the raw seeds have been studied and partially identified by Clark and Nursten (1976). The oil extracted from the seed is bright yellow, nearly odorless and possesses a pleasant nutty taste. It has a specific gravity of 0.9165 at 15°C, solidifies at −4°C, and does not become rancid easily (Knuth, 1939). First extraction yields an excellent cooking oil, and the second extraction is suitable for making soap and burning in lamps. The residue left after extracting the oil can be used as animal feed.

Sánchez (1973) reports Brazil nut seeds to contain 65% oil, 17% protein, 3% ash, 0.9% crude fiber, 4% water, and 10.1% carbohydrates. Zucas et al. (1975) and SUDAM (1976) report that the average protein content in defatted Brazil nut flour is about 46% and includes all of the essential amino acids. However, studies with white rats suggest that the nutritional value of Brazil nut flour is less than that of casein (Zucas et al., 1975). Other detailed summaries of the nutritional properties of Brazil nut seed products are provided by Almeida (1963) and Souza (1963).

The following discussion of Brazil nut harvesting is taken mostly from the works of Almeida (1963) and Souza (1963). These works should be consulted for further details on all aspects of the Brazil nut industry as it occurs in Brazil.

The Brazil nut crop is harvested almost entirely from wild trees during a five to six month period in the rainy season. In contrast, latex from rubber trees is collected during the dry season. The difference in seasonality of these two principal crops therefore allows the same workers to be employed sequentially in their collection.

The seeds of *Bertholletia excelsa* are gathered from December to March throughout Amazonian Brazil. Before the onset of the harvest, Brazil nut gatherers cut trails to the trees and clear the underbrush away from under them. Usually at least two collectors work together in all aspects of the harvest.

At maturity, Brazil nut fruits fall to the ground from where they are gathered by the collectors and placed into wicker baskets made from a species of Araceae (the Arum Family) or from a palm of the genus *Desmoncus*. The fruits are mostly collected in the morning because the weather is generally better, and because there is less danger of accidents from fruit fall, a particularly dangerous part of Brazil nut collection. The fruits, which weigh from 0.5 to 2.5 kg, fall from heights of 40 to 50 meters. They acquire such high velocities when they fall that upon impact they often become embedded in the soil. Severe injuries, and even death, may occur when collectors are struck by falling fruits.

Fruits are gathered as soon after they fall as possible in order to minimize insect and fungal attack of the seeds, and to limit the number of seeds carried away by animals, especially agoutis. A good day of collecting yields 700 to 800 fruits which contain about 200 liters of seeds. Because of the large size of the trees, both in height and diameter, collection from the canopy is not practical. Moreover, the seeds are not ripe until they fall to the ground.

After the fruits are collected, they are split open with a machete (Fig. 6B) or an axe and their 10–25 seeds are removed. The seeds are placed in water to clean them of mud and to determine which are bad. Seeds that sink are good while those that float to the surface are culled out. During storage and transport to the market the seeds are aerated by moving their position. If they become too humid, they are extremely susceptible to fungal attack.

Collection of Brazil nuts is a hazardous occupation. Besides accidents caused by falling fruits, the collectors are exposed to tropical diseases such as leishmaniasis and malaria, other dangers of the forest such as snake bite, and pitfalls encountered in transport of the product, especially navigation of rapids when river transport is employed. In addition, they face the hazards of minor "wars" fought over Brazil nut stands. For example, as recently as 1985, six collectors were killed and another 12 wounded during a series of conflicts over the collection of Brazil nuts on the Fazenda Pau Ferrado in the municipality of Macapá, Brazil (Nóbrega, 1985).

Brazil nut seeds are exported intact or with the seed coat removed. In the latter case, the seeds are soaked for 8–10 hours in a tank of water after which they are submerged in boiling water for 1–2 minutes. The seeds are dried to make the embryo shrink away from the seed coat which facilitates removal of the seed without breakage. Broken seeds have a lower export value. The dried seeds are then placed in a manual press and pressure is applied to both ends to crack the seed coat (Fig. 8A). Women and children are generally employed in this job (Fig. 8B). The shelled seeds are canned before export.

Cultivation

Attempts to propagate Lecythidaceae in general, and *B. excelsa* in particular, from cuttings have generally been unsuccessful (Müller et al., 1980; Sánchez, 1973). The only species known to readily establish from cuttings is *Gustavia augusta* L. Only grafts of *B. excelsa* onto conspecific root stock have been possible. The use of other species of Lecythidaceae as root stock, even the most closely related species of *Lecythis*, have failed (Müller, 1981).

Buds for grafts should be selected from highly productive trees from directly below the inflorescence of the previous year. It is important that the buds come from orthotropic rather than plagiotropic branches. Seedlings are ready to be used as root stock when they are 1–2 cm in diameter at 20 cm from the ground. The grafting method is that of Fokert, often used in grafting rubber trees (Müller, 1981). Brazil nut trees produced from grafts tend to branch near the ground. Consequently, the lower branches should be pruned in order to promote well formed crowns (Müller, 1981).

Most propagation of Brazil nut trees is done from seeds. A major obstacle to plantation cultivation used to be the long period of dormancy of untreated seeds (12 to 18 months). Even after 18 months, only 25% of untreated seeds germinate. Scarification of the seed coat at the apices and along the angles of the seed results in 41% germination after 18 months (Müller, 1981), and slight increases in germination are achieved after treatment with 15% and 25% sulfuric acid (Barbosa et al., 1974). The best method to obtain rapid seed germination is to remove the embryos from the seed coat after they have been soaked in water for 24 hours. With this method, the seeds begin to germinate after 20–30 days, and by the end of five months nearly 78% of them germinate. It is essential to treat the embryos for 90 minutes in a fungicide (e.g., phenyl-mercuric acetate) before planting, and to keep them sus-

pended out of reach of rodents. The seedlings should be planted when they reach about 25 cm in height and possess 16 open leaves (Müller, 1981).

Seedlings should be planted in holes 40 × 40 × 40 cm filled with humus mixed with 10 liters of manure and 50 grams of triple superphosphate. The seedlings should be spaced 10 × 15 m if they are to be planted with other crops, or 20 × 20 m or 10 × 20 m if they are to be planted in pasture (Müller, 1981).

Sánchez (1973) reported that seedlings free from fungal and insect attacks reach 50 cm after a year's growth. According to Knuth (1939), seedlings reach 80 cm after a year, 5 m in 5 years, and 25 m after 60 years. In 1953, a Brazil nut plantation was started on the grounds of CPATU in Belém, Brazil. These trees began production after 10 years and in 1981 they were producing normally. Also at CPATU, lateral buds grafted onto root stock started producing seeds after 3.5 years. Production of the grafted trees increased at six years, and after 11 years each tree was yielding 25 liters of seed. This kind of production, if maintained within a plantation, is better than the 16 to 55 liters of seeds produced per hectare by native stands greater than 50 years old (Müller, 1981). Oliveira (1944) reported a production of 30 to 120 liters of seed per hectare from native stands in the vicinity of Marabá, Brazil.

The areas selected for plantations of *B. excelsa* should possess the ecological features described earlier. In addition, some consideration should be given to preserving habitat for the pollinators, which nest mostly in secondary and undisturbed forests. In order to ensure adequate pollination, patches of forest should be left in the vicinity of the plantation. However, it is not known how important these forest patches are because sufficient pollinators were observed in a 3000 ha plantation in Acre surrounded by pasture (Müller, 1981), and in a small plantation in a disturbed area in the vicinity of Manaus (Nelson et al., 1985).

Another important consideration is the provenance of the seed. Moritz (1984) has shown that fruit production as the result of fertilization between trees of the same clone is low. Consequently, plantations should be started from seed from highly productive trees from several distinct clones. Müller (1981) recommends that, for

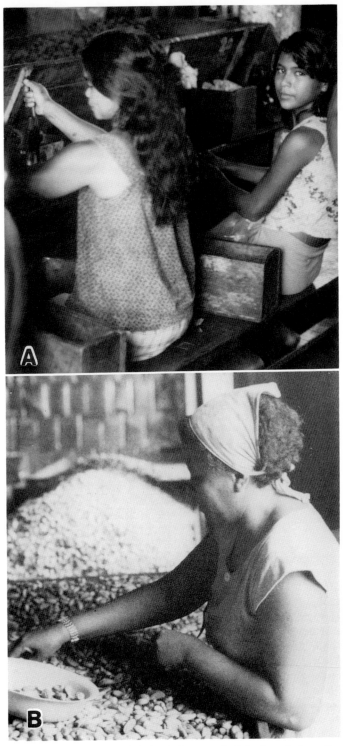

FIG. 8. Processing of the Brazil nut. **A.** Mechanical removal of the kernels from the seeds (photo by S. Mori). **B.** View of factory showing piles of Brazil nut kernels (photo by G. T. Prance.).

grafting, buds be obtained from at least five different trees.

Genetic improvement of Brazil nuts has begun with the work of agronomists and botanists at CPATU. Moritz (1984) has shown that there is considerable variability in fruit size and seed number, and adds that much can be done in the selection of more productive germplasm. She points out that selection for five instead of the normal four locules would increase seed production.

The chromosome number of *B. excelsa* is X = 17 (Kowal et al., 1977; Mangenot & Mangenot, 1958, 1962; Moritz, 1984). This number is found in most species of Neotropical Lecythidaceae, and therefore, chromosome number alone does not limit intergeneric hybridization.

A frequent suggestion has been to introduce fruit indehiscence and fruit fall with seeds intact at maturity from *B. excelsa* into species of dehiscent, large-fruited *Lecythis* with edible seeds. This would make the delicious seeds of these species of *Lecythis* more easily harvested. However, the differences between these genera are so great that successful intergeneric hybridization is highly unlikely.

An especially intriguing problem is the presence of long and short styles in the genus, which has been demonstrated by Moritz (1984). Do these different style lengths indicate a heterostylous breeding system similar to that found in other plants? We hope that CPATU will continue to investigate this and other problems of Brazil nut biology.

Conclusions

In recent years Brazil nut production has declined because of continued deforestation (Anonymous, 1987; Kitamura & Müller, 1984). Many lands that once belonged to the public domain have been or are being deforested, particularly for the extraction of timber and the establishment of pasture land. The custom of burning pasture to control weeds coincident with selective timber harvesting from adjacent forests is resulting in forest fires in some eastern Amazonian forests (Uhl & Buschbacher, 1985). These factors combine to directly destroy Brazil nut stands and to reduce the habitats necessary for maintaining the proper numbers of pollinators (large bees) and dispersal agents (agoutis). The

construction of the Tucuruí dam and subsequent formation of a 100 mile long lake (Prance, 1986) in prime Brazil nut habitat (Moritz, 1984; Oliveira, 1944), and the loss of Brazil nut collectors to more profitable occupations are adding to the diminution of Brazil nut production from wild trees.

Thanks to the research of Peruvian (Sánchez, 1973) and Brazilian (Müller et al., 1980; Müller, 1981; Moritz, 1984) biologists the technology is now available for the cultivation of Brazil nuts in plantations. However, it is critical that the genetic diversity of this important Amazonian crop is collected and preserved before most of it is lost as a result of uncontrolled forest destruction.

Acknowledgments

We are grateful to Douglas Daly, Carol Gracie, H. David Hammond, John D. Mitchell, and Bruce Nelson for their comments on the manuscript. We also thank Bruce Nelson for his help in many other aspects of our study of the taxonomy and biology of the Brazil nut family. We thank Bobbi Angell for the illustration of the Brazil nut, Carol Gracie for preparing the graph of Brazil nut production, and Andrew Henderson for the use of his photo of the Brazil nut flower. We thank the Fund for Neotropical Plant Research of The New York Botanical Garden for financing the preparation of some of the illustrations. Our work on Neotropical Lecythidaceae was supported by National Science Foundation Grants BMS 75-0324 A02 (three years) and DEB-8020920 (two years). This is publication number 53 in the series of the Institute of Economic Botany of The New York Botanical Garden.

Literature Cited

Almeida, C. P. de. 1963. Castanha do Pará: Sua exportação e importância na economia Amazonica. Edições S. I. A. Estudos Brasileiros **19**: 1–86.

Anonymous. 1987 Ocupação devasta os castanhais. Jornal do Brasil. 19 Jan 1987.

Balée, W. 1989. The adaptation to culture in Amazonia. *In* D. A. Posey & W. Balée (eds.), Resource management in Amazonia: Indigenous and folk strategies. Adv. Econ. Bot. **7**: 1–21.

Balick, M. J. 1985. Useful plants of Amazonia: A resource of global importance. Pages 339–368 *in* G. T. Prance & T. E. Lovejoy (eds.), Amazonia. Pergamon Press, New York.

Barbosa, M. M. S., W. T. Lellis & A. F. de Souza Pinho. 1974. Ensaio sobre germinação da castanha do Pará. Bol. Inst. Biol. Bahia **13**: 100–106.

Barbosa, M. R. de V. 1982. Lecythidaceae do Parque Nacional da Tijuca. Cadernos Feema Flora, Alguns Estudos **1**: 9–16.

Berg, O. C. 1858. Myrtaceae, tribus II. Barringtonieae, tribus III. Lecythideae. *In* C. F. P. von Martius. Fl. bras. **14(1)**: 469–526.

Buckley, D. P., D. M. O'Malley, V. Apsit, G. T. Prance & K. S. Bawa. (Manuscript). Genetics of Brazil "Nut" (*Bertholletia excelsa*): I. Genetic variation in natural populations. For submission to Theoretical and Applied Genetics.

Christovão, Frei. 1967. Histora dos animães e árvores de Maranhão. Published in 1967 by Archivo Histórico Ultramarino, Lisboa; and the botanical part in 1968, with editorial and background comments by several authors, by the Universidade Federal do Paraná, Conselho de Pesquisas, Curitiba.

Clark, R. G. & H. E. Nursten. 1976. Volatile flavour compounds of Brazil nuts (*Bertholletia excelsa* Humb. & Bonpl.). J. Sci. Food Agric. **27**: 713–720.

Dias, C. V. 1959. Aspectos geográficos do comércio da castanha no médio Tocantins. Revista Brasil. Geogr. **21(4)**: 77–91.

Diniz, T. D. de A. S. & T. X. Bastos. 1974. Contribuição ao conhecimento do clima típico da castanha do Brasil. Bol. Técn. IPEAN **64**: 59–71.

Eyma, P. J. 1932. The Polygonaceae, Guttiferae and Lecythidaceae of Surinam. Meded. Bot. Mus. Herb. Rijks Univ. Utrecht **4**: 1–77.

———. 1934. Lecythidaceae. *In* A. Pulle (ed.), Flora of Suriname **3(1)**: 119–155.

Farnsworth, N. R. 1984. How can the well be dry when it is filled with water? Econ. Bot. **38**: 4–13.

Foster, R. B. & N. V. Brokaw. 1982. Structure and history of the vegetation of Barro Colorado Island. Pages 67–81 *in* E. G. Leigh, Jr., A. S. Rand & D. M. Windsor, The ecology of a tropical forest: Seasonal rhythms and long-term changes. Smithsonian Institution Press, Washington, D.C.

Gentry, A. H. 1974. Coevolutionary patterns in Central American Bignoniaceae. Ann. Missouri Bot. Gard. **61**: 728–759.

Greenhall, A. M. 1965. Sapucaia nut dispersal by greater spear-nosed bats in Trinidad. Carib. J. Sci. **5**: 167–171.

Huber, J. 1910. Mattas e madeiras amazônicas. Bol. Mus. Paraense Hist. Nat. **6**: 91–225.

Humboldt, H. & A. Bonpland. 1807. Plantae aequinoctiales. Lutetiae Parisiorum, Paris.

Institute of Economic Botany. 1986. Discoveries in economic botany 1981–1986. The New York Botanical Garden, Bronx, New York.

Jackson, G. C. & J. B. Salas. 1965. Insect visitors of *Lecythis elliptica* H.B.K. J. Agric. Univ. Puerto Rico **49**: 133–140.

Kerdell-Vegas, F. 1966. The depilatory and cytotoxic action of coco de mono (*Lecythis ollaria*) and its relationship to chronic seleniosis. Econ. Bot. **20**: 187–195.

Kitamura, P. C. & C. H. Müller. 1984. Castanhais nativos de Marabá-PA: Fatores de depredação e bases para a sua preservação. EMBRAPA, Centro de Pesquisas Agropecuário do Trópico Umido. Documentos **30**: 1–32.

Knuth, R. 1939. Lecythidaceae. *In* A. Engler, Pflanzenreich IV. **219a**: 1–146.

Kowal, R. R., S. A. Mori & J. A. Kallunki. 1977. Chromosome numbers of Panamanian Lecythidaceae and their use in subfamilial classification. Brittonia **39**: 399–410.

Lemée, A. M. V. 1953. Flore de la Guyane Française. Vol. 3. Lechevalier, Paris.

Lindeman, J. C. & A. M. W. Mennega. 1963. Bomenboek voor Suriname. Uitgave Dienst's Lands Bosbeheer Suriname, Paramaribo.

Loefling, P. 1758. Iter hispanicum. Stockholm.

Mangenot, S. & G. Mangenot. 1958. Deuxième liste de nombres nouveaux chez diverse Dicotylédonées et Monocotylédonées d'Afrique occidentale. Bull. Jard. Bot. Etat **28**: 315–329.

——— & ———. 1962. Enquête sur les nombres chromosomiques dans une collection d'espèces tropicales. Rev. Cytol. Biol. Vég. **25**: 411–447.

Marcgrave, G. 1648. Historiae rerum naturalium brasiliae. Pages 128, 293 *in* G. Piso & G. Margrave, Historiae naturalis brasiliae, auspicio et beneficio illustris I. Mauritii Com. Nassau, Leiden.

Miers, J. 1874. On the Lecythidaceae. Trans. Linn. Soc. London **30(2)**: 157–318.

Mori, S. A. 1970. The ecology and uses of *Lecythis* in Central America. Turrialba **20(3)**: 344–350.

——— & J. Boeke. 1987. Chapter XII. Pollination. *In* S. A. Mori & collaborators, The Lecythidaceae of a lowland Neotropical forest: La Fumée Mountain, French Guiana. Mem. New York Bot. Gard. **44**: 137–155.

——— & collaborators. 1987. The Lecythidaceae of a lowland Neotropical forest: La Fumée Mountain, French Guiana. Mem. New York Bot. Gard. **44**: 1–190.

——— & J. A. Kallunki. 1976. Phenology and floral biology of *Gustavia superba* (Lecythidaceae) in central Panama. Biotropica **8**: 184–192.

———, L. A. Mattos Silva & T. S. dos Santos. 1980. Observações sobre a fenologia e biologia floral de *Lecythis pisonis* Cambess. (Lecythidaceae). Revista Theobroma **10**: 103–111.

——— & G. T. Prance. 1981. The "sapucaia" group of *Lecythis* (Lecythidaceae). Brittonia **33**: 70–80.

——— & ———. 1983. Lecitidáceas: Família da castanha-do-Pará. Centro de Pesquisas do Cacau. Bol. Téc. **116**: 1–35.

——— & ———. 1987a. Species diversity, phenology, plant-animal interactions, and their correlation with climate, as illustrated by the Brazil nut family (Lecythidaceae). Pages 69–89 *in* R. E. Dickinson (ed.), The geophysiology of Amazonia. John Wiley & Sons, New York.

——— & ———. 1987b. Chapter XI. Phenology. *In* S. A. Mori & collaborators, The Lecythidaceae of a lowland Neotropical forest: La Fumée Mountain, French Guiana. Mem. New York Bot. Gard. **44**: 124–136.

——— & ———. (In press). Lecythidaceae Part—II. The zygomorphic-flowered New World Genera (*Bertholletia, Corythophora, Couratari, Couroupita, Eschweilera,* & *Lecythis*). Fl. Neotrop. Monogr. **21, Pt. II:** 1–000.

———, ——— & **A. B. Bolten.** 1978. Additional notes on the floral biology of Neotropical Lecythidaceae. Brittonia **30**: 113–130.

Moritz, A. 1984. Estudos biológicos da castanha-do-Brasil (*Bertholletia excelsa* H.B.K.). EMBRAPA, Centro de Pesquisa Agropecuário do Trópico Umido. Documentos **29**: 1–82.

Müller, C. H. 1981. Castanha-do-Brasil; estudos agronômicos. EMBRAPA, Centro de Pesquisas Agropecuário do Trópico Umido. Documentos **1**: 1–25.

———, **I. A. Rodrigues, A. A. Müller & N. R. M. Müller.** 1980. Castanha-do-Brasil. Resultados de pesquisa. EMBRAPA, Centro de Pesquisas Agropecuário do Trópico Umido. Miscelânea **2**: 1–25.

Nelson, B. W., M. L. Absy, E. M. Barbosa & G. T. Prance. 1985. Observations on flower visitors to *Bertholletia excelsa* H.B.K. and *Couratari tenuicarpa* A. C. Sm. (Lecythidaceae). Acta Amazônica **15**(1/2) supl.: 225–234.

Nóbrega, E. C. Castor da. 1985. Guerra declarada no Araguaia: A "Pau Ferrado" ninguém toma. Jornal do Belém, 13 Jan 1985.

Oliveira, A. B. de. 1985. Considerações sobre a exploração da castanha no baixo e médio Tocantins. Pages 278–283 *in* Excerptos da Revista Brasileira de Geografia. Amazonia Brasileira. Instituto Brasileira de Geografia e Estatística, Rio de Janeiro.

O'Malley, D. M., D. P. Buckley, G. T. Prance & K. S. Bawa. (Manuscript). Genetics of Brazil "Nuts" (*Bertholletia excelsa*) II. Mating system. To be submitted to Theoretical and Applied Genetics.

Ormond, W. T., M. C. B. Pinheiro & A. R. Cortella de Castells. 1981. A contribution to the floral biology and reproductive system of *Couroupita guianensis* Aubl. (Lecythidaceae). Ann. Missouri Bot. Gard. **68**: 514–523.

Pires, J. Murça. 1984. The Amazonian forest. Pages 581–602 *in* H. Sioli (ed.), The Amazon. Limnology and landscape ecology of a mighty tropical river and its basin. Dr. W. Junk Publishers, Dordrecht, Netherlands.

Pittier, H. 1912. The Lecythidaceae of Costa Rica. Contr. U.S. Natl. Herb. **12**: 95–101.

———. 1927. The Lecythidaceae of Central America. Contr. U.S. Natl. Herb. **26**: 1–14.

Poiteau, M. A. 1825. Mémoire sur Lecythidées. Mém. Mus. Hist. Nat. **13**: 141–165.

Posey, D. A. 1985. Indigenous management of tropical forest ecosystems: The case of the Kayapó Indians of the Brazilian Amazon. Agroforestry Systems **3**: 139–158.

Prance, G. T. 1976. The pollination and androphore structure of some Amazonian Lecythidaceae. Biotropica **8**: 235–241.

———. 1986. The Amazon: Paradise lost? Pages 62–106 *in* L. Kaufman & K. Mallory (eds.), The last extinction. MIT Press, Cambridge, Massachusetts.

———. (In press). *Bertholletia. In* S. A. Mori & G. T. Prance, Lecythidaceae—Part II. The zygomorphic-flowered New World genera (*Bertholletia, Corythophora, Couratari, Couroupita, Eschweilera,* & *Lecythis*). Fl. Neotrop. Monogr. **21, Pt. II:** 1–000.

——— & **A. B. Anderson.** 1976. Two new species of Amazonian Lecythidaceae. Brittonia **28**: 298–302

——— & **S. A. Mori.** 1979. Lecythidaceae—Part I. The actinomorphic-flowered New World Lecythidaceae (*Asteranthos, Gustavia, Grias, Allantoma,* & *Cariniana*). Fl. Neotrop. Monogr. **21**: 1–270.

Reitz, R. 1981. Lecítidáceas. Flora Illustrada Catarinense **1**(LECI): 1–32.

Sampajo, A. J. 1944. A flora amazônica. Pages 92–102 *in* Excerptos da Revista Brasileira de Geografia. Amazonia Brasileira. Instituto Brasileira de Geografia e Estatística, Rio de Janeiro.

Sánchez, J. S. 1973. Explotación y comercialización de la castaña en Madre de Dios. Ministerio de Agricultura, Dirección General de Forestal y Caza, Informe No. 30. Lima, Peru.

Smythe, N. 1978. The natural history of the Central American agouti (*Dasyprocta punctata*). Smithsonian Contr. Zool. **257**: 1–51.

Sousa, Gabriel Soares de. 1949. Noticia do Brasil. Introdução, comentários e notas pelo Prof. Pirajá da Silva. Martins, Bibliotêca Histórica Brasileira, São Paulo.

Souza, A. H. 1963. Castanha do Pará: Estudo botánico, químico e tecnológico. Edições S.I.A., Estudos Técnicos **23**: 1–69.

SUDAM. 1976. Estudos e pesquisas sobre a castanha-do-Pará. Ministerio do Interior, Superintendência do Desenvolvimento da Amazonia, Departamento de Recursos Naturais, Belém.

Uhl, C. & R. Buschbacher. 1985. A disturbing synergism between cattle ranch burning practices and selective tree harvesting in the eastern Amazon. Biotropica **17**: 265–268.

Vaz Pereira, I. C. & S. L. Lima Costa. 1981. Bibliografia de Castanha-do-Pará (*Bertholletia excelsa* H.B.K.). EMBRAPA, Centro de Pesquisas Agropecuário do Trópico Umido. Belém, Pará.

Woodson, R. E. 1958. Lecythidaceae. *In* R. E. Woodson & R. W. Schery (eds.), Flora of Panama. Ann. Missouri Bot. Gard. **45**: 115–136.

Zucas, S. M., E. C. V. Silva & M. I. Fernandes. 1975. Farinha de castanha do Pará. Valor de sua proteína. Revista Far. Bioquim. Univ. São Paulo **13**: 133–143.

Santa Rosa: The Impact of the Forest Products Trade on an Amazonian Place and Population

Christine Padoch and Wil de Jong

Table of Contents

Abstract

PADOCH, C. and W. DE JONG (Institute of Economic Botany, New York Botanical Garden, Bronx, New York 10458). Santa Rosa: The Impact of the Forest Products Trade on an Amazonian Place and Population. Advances in Economic Botany **8**: 151–158. 1990. The history of many rural communities in the lowland Peruvian Amazon is closely linked to the history of minor forest product exploitation in the region. This article recounts the history and travels of an Amazonian population that eventually established the village of Santa Rosa on the lower Ucayali River. Prior to the founding of the settlement, members of this group of people with varied tribal and non-tribal backgrounds, were taken by a *patron* to several farflung sites in the Peruvian Amazon to collect forest products. The very heterogeneous ethnic composition of Santa Rosa is largely the result of the vagaries of markets in minor forest products in past decades and of the economic system involved in their exploitation.

Key words: Santa Rosa; Peruvian Amazon; Ucayali River; forest products; rubber; leche caspi; rosewood; aguaje; barbasco.

Resumen

La historia de numerosas comunidades rurales en la Selva Baja de la Amazonía Peruana está estrechamente ligada á la historia de la explotación de productos forestales menores en la región. El presente artículo narra la historia y los viajes de una población Amazónica que al final fundó el caserío de Santa Rosa en el margen del Río Ucayali. Previo a la fundación del asentamiento, miembros del grupo con antecedentes tribales y no tribales fueron llevados

por un patrón a sitios alejados a colectar productos forestales. La composición étnica heterogénea de Santa Rosa es principalmente el resultado de los cambios de mercados de
productos forestales menores en décadas pasadas y de los sistemas económicos relacionados
con su explotación.

Introduction

For about a third of a century, ending in 1914,
the economy and life of the lowland Peruvian
Amazon were intimately tied to one forest product: rubber. Other plant products of the tropical
forests had been extracted and sold before rubber; vanilla, sarsaparilla, waxes, and copal had
all been important (Villarejo, 1979: 128–130).
But the extremely high prices that rubber offered
changed the economy of the area drastically, and
left a social and cultural mark on the area which
persists until this day (San Roman, 1975).

The Rubber Boom's demographic alteration
of the tropical lowland area may be the most
dramatic change that can be discerned. Massive
migration flows from neighboring Departments
of Peru, particularly San Martin and Amazonas,
swept the lowlands (San Roman, 1975: 139); other prospective rubber tappers, traders, and entrepreneurs came from Brazil and Colombia, as
well as from Europe and North America. While
the European and *mestizo* populations boomed,
many of the native Amerindian societies that had
survived earlier European incursions were destroyed, some physically, and most culturally.
Rapid deculturation was the consequence of the
immigration for many tribal peoples (cf. Murphy, 1954, for a Brazilian example). Other groups,
especially the Witoto of the Putumayo River,
suffered atrocities that resulted in the deaths of
many indigenous people, arousing international
attention and protest (Collier, 1968; Hardenburg, 1912). Those native groups that persisted
were driven or escaped far into the forests that
labor-hungry rubber traders found difficult to
penetrate on their *correrías* in search of Indian
slaves.

Villages and towns were created and disappeared all within the few rubber decades (Larrabure i Correa, 1909: 206, 213). Family and
other social units were completely transformed.
After the Rubber Boom virtually no one in the
lowland Peruvian Amazon was as he or she had
been before: everyone had moved, everyone had
changed.

The enormous impact that rubber had on Amazonian towns and villages, especially in Brazil
has been discussed in several works (Parker, 1985;
Wagley, 1976; Weinstein, 1983). In a history of
the caboclos of the Brazilian Amazon, Parker has
stated that the influx of northeasterners was the
last "significant direct infusion of new peoples"
into the "caboclo" class of Amazonia (Parker,
1985: xxiv). While in large measure this was also
true in the Peruvian Amazon, many events after
the end of the Rubber Boom changed the demographic makeup of the region. The smaller
"booms" and "busts" of subsequent years—
mostly built on commerce in other forest products—also resulted in migrations, ethnic changes,
and the creation and demise of numerous settlements. Yet the impact of these later small
"booms," other than the revival of rubber tapping during World War II, has rarely been mentioned in the literature. While several works on
the Amazon region do provide a general outline
of the history of the forest products trade in the
Peruvian lowlands (Padoch, 1988; San Roman,
1975; Villarejo, 1979), only a very few detail the
social consequences of these market fluctuations
on particular human settlements of the region
(Stocks, 1978).

In this article we shall briefly sketch the history
of a place and a population, including the story
of the founding and dissolution of a large agricultural estate, the extensive migrations of a
mixed tribal and mestizo group, and the final
founding of the small village of Santa Rosa along
the lower Ucayali River in Peru. Throughout we
shall note how the vagaries of international trade
in forest products affected the region and its peoples, particularly in the decade of the 1950's. We
cannot prove nor do we wish to suggest that this
specific case is either typical of a large class of
settlements in the Peruvian Amazon, or that it
is exceptional.

The few works that trace the history of human
settlements in the Peruvian Amazon, tend to recount the fates of major cities, towns, and early
mission settlements (Izaguirre, 1923–1929;
Rumrill, 1983; Stocks, 1978). Apart from a few

brief notes (e.g., Hiraoka, 1985; IPA, 1974), virtually nothing has been published about the formation and change of *caseríos,* the non-tribal villages that are by far the most numerous human settlements in the region. The short history of the northeastern Peruvian Amazon written by San Roman (1975) offers the best general outline of social and economic changes in the area. But no equivalent of Wagley's *Amazon Town* exists for the Peruvian lowlands.

The history that will be sketched here remains an incomplete one. In the absence of retrievable written documents it is based almost entirely on informally collected life histories of a large number of the village's present residents. In addition, interviews were conducted with former residents now living in the neighboring village of Yanallpa as well as in the city of Iquitos. These interviews, during which we asked people to recall events of several decades past, contain considerable confusion over dates, names, and places. These and other apparent contradictions were not easily resolved. We have, however, put together the available material in the manner we thought most consistent and plausible.

Several can speak native Amazonian languages, and, when specifically questioned, a few will acknowledge descendence from a particular tribal Amazonian group. However, no one identifies him- or herself as a current member of a native group. The village includes descendents of at least four and probably five different tribal groups, and others that can be best classed as mestizos. These include: a predominant population of descendents of the Cocamas, once the dominant tribal group of this and neighboring floodplain areas; several former members of Campa or Ashaninka tribal groups from the Upper Ucayali, and their descendents; a few Quichuas from the upper Napo River; a number of women who previously were members of Yagua tribal communities, one older resident who is said by his neighbors to have been a Cashibo (but who refuses to identify his tribal ancestry), offspring of unions among members of all the above groups, as well as several persons who both physically and culturally appear descended largely from European immigrants. The reason for such extreme ethnic diversity within this small community lies in the recent history of the region.

Santa Rosa Today

The village or *caserío* of Santa Rosa lies on the true right bank of the Ucayali River, about 15 kilometers downstream from the town of Requena (Fig. 1). While the majority of the houses comprising the village occupy a small rise which lies above the Ucayali's floodplain, other houses as well as a large part of the village's agricultural fields are within the zone that experiences annual or periodic flooding.

In May of 1986, Santa Rosa had a population of 335 persons divided among 46 households. All the households engage in a complex mix of farming, fishing, hunting, and gathering of forest products for their own subsistence and for the market. The exact population changes frequently. A few households maintain second residences in Requena. When secondary schools are in session or when little farm work needs to be done in Santa Rosa, several or all members of each of these families can be found in the larger town.

All Santa Rosinos speak Spanish and consider themselves *ribereños,* a term which denotes non-tribal, rural residents of the Peruvian Amazon.

A History of Santa Rosa

We have been unable to find specific information concerning the pre-Rubber Boom inhabitants of the area where Santa Rosa now stands. It should be kept in mind that the small hillock on which the village is located today doubtless had a very different relationship to the meandering Ucayali in years or centuries past. The hillock is now separated from the river by a very narrow strip of palm swamp. According to older Santa Rosinos, within the last thirty years or so, several hundred meters of that swamp have eroded. One could surmise that earlier still, the present swamp which fronts on the river was not only much wider, but also was separated from the Ucayali by a wide levee, and perhaps a seasonal beach or mudflat. Thus the exact location of Santa Rosa may once have been far back from the river.

The general region of the lower Ucayali is said by archeologists to have been the domain of the once large Tupi-speaking Cocama group (De-Boer, 1981; T. P. Myers, 1974, 1978). This accessible, floodplain dwelling population experi-

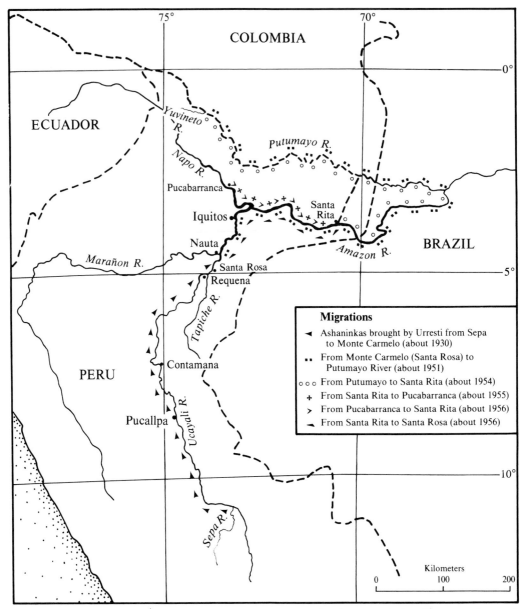

FIG. 1.   Map showing location of modern-day Santa Rosa and migrations over the last 60 years as discussed in the text.

enced contact with Europeans early and beginning in the latter half of the 16th century was repeatedly devastated by European diseases (T. P. Myers, 1978; Stocks, 1978: 139). That the chronicles of missionaries and other travellers in the 18th and 19th centuries do not mention seeing permanent settlements in the lowest reaches of the Ucayali where Santa Rosa now stands, may well be the result of the near annihilation of the Cocama nation by disease and raiding. Because of this devastation as well as the effects of missionization, economic exploitation, and other forms of domination by Europeans and their descendants, the Cocama are no longer distinguishable as a tribal group. However, their language and many cultural traits persist throughout many small settlements of the lower Ucayali and neighboring regions.

While none of our informants was old enough to remember the Rubber Boom well, several did suggest that during the heyday of rubber there were rubber collectors operating in the area where Santa Rosa stands today. The nearby Tapiche River was an especially rich source of *Hevea brasiliensis* latex. Just after the turn of the century (about 1907) the town of Requena was established by the Franciscan missionary, Augustin Lopez, who noted in his journals the existence of numerous "fundos" (agricultural estates) in the general area (Izaguirre, 1923–1929: v. 12: 237). Presumably, these turn-of-the-century estates on the Ucayali, like those on the Napo (San Roman, 1975: 137), combined extraction of rubber and other forest products, with some minor agricultural and livestock production.

Although we lack exact data, we assume that when the price of rubber began a steep decline after 1912, many of the *fundos* and much of the population disappeared. Many rubber tappers went to the new but growing town of Iquitos, while other immigrants brought to the Amazon by the Rubber Boom, left the region altogether.

After the Rubber Boom

The history that the present residents of Santa Rosa recount largely begins with what they can remember of their own lifetimes, and thus commences around 1930. By that time the Rubber Boom had long been over, and the economy based on rubber exploitation had long decayed. However, some commercial enterprises had survived in other forms and new ones had been started.

Just upriver from the present village site, occupying an extensive area of high ground or *terra firme*, a large *fundo* operated, producing sugar cane and cane-based *aguardiente*, cattle, pigs, and dealing in a variety of forest products. Among those products were such exportable ones as the skins of forest animals and fine construction and cabinetry timbers, and other locally used products such as the fruit of the naturally occurring *aguaje* palm (*Mauritia flexuosa*) which was both eaten by local populations and gathered to feed the *fundo's* pigs. The large, diversified, and financially successful enterprise, named Monte Carmelo, was founded around 1930 by the Urresti family, which also had holdings on the upper Ucayali.

The Urrestis brought their own labor force with them when they descended the river. The laborers they moved were mostly Ashaninkas (Campas) recruited by persuasion and by force from the vicinity of their other holdings on the Sepa River, a tributary of the Ucayali. Tied to the *patrón* and the *fundo* by debt peonage, this tribal group was apparently considered especially desirable for the Monte Carmelo site because they were said to be fierce. Until well into the 1960's groups of Matses (Mayoruna) tribesmen who lived on nearby uplands were greatly feared, and indeed some raiding for women had occurred in the area. Monte Carmelo's lack of problems with Matses incursions was attributed to the presence of the warlike Campa.

Apart from the Campa families, Monte Carmelo also employed Cocama and mestizo laborers. The total population of the fundo is very difficult to ascertain, but appears to have been approximately 50 families. The positions of administrator and *capatazes* (foremen) were in the hands of local mestizo men.

In 1934, a campaign to increase production of *barbasco* (*Lonchocarpus* sp.), a plant containing rotenone, a natural insecticide, got underway in the Peruvian Amazon (Villarejo, 1979: 136). The plant had long been used by local native and ribereno groups for capturing fish. The Astoria Company, an American company based in Iquitos, exported the roots of the plant to North America; in 1946 exports from the Peruvian Amazon reached a high of almost 5.5 tons per annum (Villarejo, 1979: 136).

In the late 1930's and '40's much of Monte Carmelo's land was planted to barbasco. The boom in barbasco production also had a strong impact on other villages, as small entrepreneurs and others struggled to clear more forest to plant and sell the valuable product (Stocks, 1978: 196–199). However, less than fifteen years after barbasco fever gripped the region, the price had begun to fall. The development of DDT and other synthetic insecticides replaced the product of the Amazon's forests and farms, and exports plummetted. While the plant, in powdered form, continues to be exported from the Peruvian Amazon to this day, quantities and profits are low.

With the collapse of barbasco's high prices,

Monte Carmelo which had a substantial investment in the plant, experienced considerable financial difficulties. While the exact reasons for the demise of the large *fundo* are complex and difficult to ascertain, the collapse in barbasco prices was doubtless a major factor. Also contributing to the estate's troubles were the death of Luis Enrique Urresti, the founder of the *fundo*, as well as the subsequent quarrels between his various sons. Changes in transportation facilities, and therefore in the countryside's access to Iquitos and other markets, as well as increased activity among traveling traders, also threatened the hold Monte Carmelo and other *fundos* had over their workforce.

Travels in Search of Forest Products

With his fortunes and perhaps his interest in Monte Carmelo failing, Victor Urresti, the *fundo's* new owner, searched for other opportunities that the Amazonian environment and the world market for forest products offered. Apart from barbasco, a product that was considered capable of "saving the region from the marasmus that resulted from the fall of rubber" (Villarejo, 1979: 109) was *leche caspi* (*Couma macrocarpa*). The native tree could be tapped, much like rubber, for its milky-white resin, which was exported for use in the manufacture of chewing gum, paints, and varnishes. But while *leche caspi* trees did grow in the vicinity of Monte Carmelo, they were scarce. Thus, Victor Urresti left the *fundo*, and taking 60 of his male laborers with him, travelled to some promising extraction sites on the Putumayo River. The trip itself, by slow steamer, took a month and a half. The women and children of Monte Carmelo were left behind to carry on with subsistence farm production. This disruption of family and community life was reminiscent of what had happened during the great Rubber Boom.

The men stayed on the Putumayo for approximately three years, tapping leche caspi, and hunting forest animals for their valuable skins and meat. While some profits came in, apparently the prices received were not high enough to warrant a move to new tapping sites.

By the mid 1950's the market in forest products was generally waning, and agriculture was being encouraged. In 1954, with funds advanced by one of Iquitos' largest commercial trading houses, the Kahn Company, Urresti went back to managing a largely agricultural enterprise. He established a new *fundo* at Santa Rita, near the Colombian border. The workers who had accompanied him from Monte Carmelo to the Putumayo were taken there, and their wives and children were at last brought down from the Ucayali to join them.

Santa Rita was again a large enterprise. Reflecting a new agricultural development in the region, the *fundo* operated largely as a rice processing center; rice had begun to be planted in some quantity by smallholders in the lower Peruvian Amazon. The estate also produced the more traditional sugar cane (for aguardiente) and a few cattle. A large mill was installed and local farmers were contracted to grow rice.

But this enterprise did not last long. Victor Urresti apparently had a falling out with a powerful neighboring *fundero*—the matter concerned the other's daughter—and fled the region, never to be seen again in the Peruvian Amazon.

Urresti's workers, sixty or more families, found themselves abandoned on a leaderless fundo. Most stayed in the area and engaged in small-scale agricultural production or worked on neighboring estates. A number of the men who had originally come from Monte Carmelo took wives from among the Yagua Indians who were employed on a nearby *fundo*. However, a group of approximately ten families were recruited again to move by Andres Rodriguez, Urresti's former administrator.

In 1955, Rodriguez was hired as an administrator by the Abensur Company, another of Iquitos' major commercial houses to run its operations at Pucabarranca on the Napo River, a major tributary of the Amazon. The large operation, well up the Napo River, was then focussed on the cutting of rosewood (*Aniba rosaeodora*) trees for the extraction of their precious essential oil. Villarejo (1979: 137) reports that the prices for the oil, used in the manufacture of soaps and perfumes, experienced a "meteoric rise" in the 1950's, and the entrepreneurs of the Peruvian Amazon were quick to take notice and advantage.

At Pucabarranca therefore, the much displaced Monte Carmelinos returned to the forest to fell rosewood trees and cut the logs into small pieces which would then be ground up, and the oil distilled. As no thought was given to replanting the species, and the profits to be made from the cutting were high, accessible forests were quickly stripped of the valuable trees.

Some of the workers at the great fundo stayed on the plantation and were occupied in cattle raising. The mixed group from the Ucayali were exposed on the Napo to still other alien cultures. The tribal group native to the region and the majority of the workers on the fundo were Quichuas. Many of the Monte Carmelo group, especially the women, still recall their difficulties in communicating and getting along with these speakers of an unrelated language.

Apparently Rodriguez himself had difficulty adapting, because a year later he quit and left, again followed by most of the faithful group. The folks returned to Santa Rita for a time, but eventually most of those who had gone up the Napo decided to return to what was left to Monte Carmelo. However, because of fear of raids by the Matses, they decided to settle in Santa Rosa, a small property or *puesto* owned by Andres Rodriguez, just downriver from Monte Carmelo. As Santa Rosa was located on a natural levee and separated from the *terra firme* territory of the Matses by a broad swath of seasonally inundated, swampy ground, it was considered a safer place to live.

The group that returned to the area on the Ucayali, however, was not the same one that had left. Many of the original workers had decided to stay in Santa Rita, on the Putumayo, or along the Napo. Other families had attached themselves to the Monte Carmelinos. Some Quichuas from Pucabarranca, as well as some mestizos from the border area of Santa Rita had joined the group. Many of the men had married during their travels and thus a few Yagua wives entered the community. Many families had also changed or ceased to exist because of the major disruptions of distant travels and lengthy separations. And the Ashaninkas and other former tribal members had abandoned much of the languages and folkways that they had maintained when living in large groups at Monte Carmelo or other *fundos*; the residents of the new village of Santa Rosa were now all ribereños.

The population of Santa Rosa has swelled over the years with immigrants from many other places. However, the group that came with Andres Rodriguez around 1956 formed the core of the present population of Santa Rosa, and the diverse ethnic backgrounds of today's Santa Rosinos reflect the extensive travels of Monte Carmelo's workers in the 1950's.

Summary and Conclusions

The impact of international trade in forest products on the people and settlements of the Peruvian Amazon has been not only pronounced, but also continuing and complex. However, while the effects of the Rubber Boom have justly received a good deal of attention, little has been written about the numerous, short-lived smaller "booms" in Amazonian forest products.

The brief history presented above, of the geographical site and the population that founded the Ucayali *caserío* of Santa Rosa, serves as an example of the effects on Amazonian populations of abrupt shifts in the prices of forest products in the late 1940's and '50's. There are surely more dramatic histories to be told, and many other forest products—tagua (*Phytelephas macrocarpa,* balata (*Manilkara* spp.), and many timbers, among others—that fetched high prices for a short time and caused major disruptions in the lives of communities. But while not demonstrably typical, Santa Rosa's history should not be considered unusual. Most of today's *caseríos* were *fundos* until several decades ago (Hiraoka, 1985) and *mestizaje* or detribalization of native peoples in the Peruvian Amazon has in many cases occurred or was greatly accelerated within the last half century. The wholesale movement of people from one location to another many kilometers away, by a *patrón* or *fundero* in search of profits from forest products has been reported by other researchers (Chaumeil, 1984; Denevan & Padoch, 1988) for native groups.

Ephemeral economic booms and busts did not end in the Peruvian Amazon with the 1950's. Timber, petroleum, and drug trafficking have more recently led to temporary shifts in the area's economic fortunes and the disruption of families and communities. The *ribereño* populations of the lowland Amazon continue to be extremely flexible and mobile groups. The expectation of another "boom," of another forest product that will lift the region from economic doldrums, persists. The history of the social effects of these booms should, however, not be forgotten. Although not as overwhelming as the Rubber Boom, among the results of these shorter, minor episodes were the erosion of subsistence production, disruption of family life, destruction of communities, and rapid cultural change. Little permanent economic development in the lowland Peruvian Amazon has resulted from those brief moments of affluence.

The patron-client relationships that were prevalent in the 1950's in the lower Ucayali, which allowed *funderos* to move populations from place to place, and to maintain large workforces in virtual slavery, exists today only in the more remote areas of the Peruvian Amazon (Chaumeil, 1984). The exact story of Santa Rosa is not apt to repeat itself. However, the calls for greater future exploitation of Amazonian forest products which abound in the recent professional and popular literature (Balick, 1985; N. Myers, 1983, 1984) should be informed and cautioned by the history of forest product exploitation in the past.

Acknowledgments

This article is based on work done as part of a research agreement between the Institute of Economic Botany of The New York Botanical Garden and the Instituto de Investigaciones de la Amazonía Peruana. This is publication number 136 in the series of the Institute of Economic Botany of the New York Botanical Garden.

Literature Cited

Balick, M. J. 1985. Useful plants of Amazonia: A resource of global importance. Pages 339–368 *in* G. T. Prance & T. E. Lovejoy, eds., Amazonia. Pergamon Press, New York.

Chaumeil, J. P. 1984. Between zoo and slavery: The Yagua of eastern Peru in their present situation. IWGIA 49.

Collier, Richard. 1968. The river that God forgot. Dutton, New York.

DeBoer, W. R. 1981. Buffer zones in the cultural ecology of aboriginal Amazonia: An ethnohistorical approach. Amer. Antiquity 46(2): 364–377.

Denevan, W. M. & C. Padoch. 1988. Introduction: The Bora agroforestry project. Pages 1–7 *in* W. M. Denevan & C. Padoch (eds.), Swidden-fallow agroforestry in the Peruvian Amazon. Adv. Econ. Bot. 5: 1–107.

Hardenburg, W. E. 1912. The Putumayo; the devil's paradise. R. Enock (ed.). London.

Hiraoka, M. 1985. Floodplain farming in the Peruvian Amazon. Geogr. Rev. Japan 58(1), ser. B: 1–23.

IPA (Investigación y Promoción de la Amazonia). 1974. Estudios socio-económicos de los rios Amazonas y Napo. 2 vols. Iquitos.

Izaguirre, Fr. B. 1923–1929. Historia de las misiones franciscanas. Talleres Tipograficos de la Penitenciaria, Lima.

Larrabure i Correa, C. 1909. Collección de leyes, decretos, resoluciones i otros documentos oficiales referentes al Departamento de Loreto. Imprenta "La Opinion Nacional," Lima.

Murphy, R. F. 1954. The rubber trade and the Mundurucu village. Unpubl. Ph.D. Dissertation. Columbia University.

Myers, N. 1983. A wealth of wild species: Storehouse for human welfare. Westview Press, Boulder, Colorado.

———. 1984. The primary source: Tropical forests and our future. Norton, New York.

Myers, T. P. 1974. Spanish contacts and social change on the Ucayali River, Peru. Ethnohistory 21(2): 135–158.

———. 1978. The impact of disease on the upper Amazon. Unpubl. paper read at the 26th annual meeting of the American Society for Ethnohistory, Austin, Texas.

Padoch, C. 1988. The economic importance and marketing of forest and fallow products in the Iquitos region. Pages 74–89 *in* W. M. Denevan & C. Padoch (eds.), Swidden-fallow agroforestry in the Peruvian Amazon. Adv. Econ. Bot. 5: 1–107.

Parker, E. P. 1985. The Amazon cabolco: An introduction and overview. Pages xvii–li *in* E. P. Parker (ed.), The Amazon cabolco: Historical and contemporary perspectives. Studies in Third World Societies 32.

Rumrill, R. 1983. Iquitos: Capital de la Amazonía. Published by the author, Iquitos, Peru.

San Roman, J. 1975. Perfiles históricos de la Amazonía peruana. Ediciones Paulinas, Lima.

Stocks, A. W. 1978. The invisible Indians: A history and analysis of the relations of the Cocamilla Indians of Loreto, Peru, to the state. Unpubl. Ph.D. Dissertation. Univ. Florida.

Villarejo, A. 1979. Así es la selva, 3rd ed. CETA, Iquitos, Peru.

Wagley, C. 1976. Amazon town: A study of man in the tropics, 2nd ed. Oxford University Press, London.

Weinstein, B. 1983. The Amazon rubber boom, 1850–1920. Stanford University Press, Stanford.

Fruits from the Flooded Forests of Peruvian Amazonia: Yield Estimates for Natural Populations of Three Promising Species

Charles M. Peters and Elysa J. Hammond

Table of Contents

Abstract

PETERS, C. M. (Institute of Economic Botany, The New York Botanical Garden, Bronx, NY 10458-5126, U.S.A.) and E. J. HAMMOND (School of Forestry and Environmental Studies, Yale University, New Haven, CT 06511, U.S.A.). Fruits from the flooded forests of Peruvian Amazonia: Yield estimates for natural populations of three promising species.

Advances in Economic Botany 8: 159–176, 1990

Advances in Economic Botany **8**: 159–176. 1990. *Myrciaria dubia* (Myrtaceae), *Grias peruviana* (Lecythidaceae) and *Spondias mombin* (Anacardiaceae) are common species in the flooded forests of Peruvian Amazonia. The fruits produced by these trees are very popular in the region, and considerable quantities are collected from the forest for local consumption or sale. Given the ecological importance and economic potential of these wild fruit trees, natural populations of *M. dubia*, *G. peruviana*, and *S. mombin* were inventoried and mapped, and size-specific fruit production rates were quantified for each species. The *M. dubia*, *G. peruviana*, and *S. mombin* study populations contained 8714, 508, and 17 individuals/ha ≥1.0 cm DBH, respectively. Fruit production was significantly related to tree diameter for *M. dubia* ($r^2 = 0.98$) and *G. peruviana* ($r^2 = 0.86$); no statistical relationship between these parameters was detected for *S. mombin*. Total annual fruit production by *M. dubia* was estimated at 9.5 to 12.7 t/ha, *G. peruviana* produced 2.3 t/ha of fruit, and *S. mombin* produced 0.6 t/ha. The density and productivity of natural populations of *M. dubia*, *G. peruviana*, and *S. mombin* suggest that in situ management is a viable development alternative for these three native fruits.

Key words: forest fruits; population ecology; yield; *Myrciaria dubia*; *Grias peruviana*; *Spondias mombin*; Peruvian Amazonia.

Resumen

Myrciaria dubia (Myrtaceae), *Grias peruviana* (Lecythidaceae), y *Spondias mombin* (Anacardiaceae) son especies comunes de los bosques inundables de la Amazonía Peruana. Las frutas producidas por estos arboles son muy apreciadas en la región, y se colectan cantidades significativas de ellas para el consumo o la venta local. El objetivo de este estudio fue describir la estructura y la abundancia de poblaciones naturales de cada especie, y cuantificar el número de frutas producidas por cada población. Las poblaciones de *M. dubia*, *G. peruviana* y *S. mombin* contienen 8714, 508 y 17 individuos/ha ≥1.0 cm DAP. La producción de frutas fue significativamente relacionada con el diámetro del arbol en el caso de *M. dubia* y *G. peruviana*, pero no se pudo comprobar ninguna relación estadística entre estos dos parámetros para *S. mombin*. La producción total anual de frutas estimada para *M. dubia* es 9.5 a 12.7 t/ha, *G. peruviana* produjo 2.3 t/ha de frutas, y *S. mombin* produjo 0.6 t/ha de frutas. La densidad y productividad de estas poblaciones naturales sugieren que el manejo forestal es una alternativa viable para fomentar el desarrollo de estos frutales nativos.

Introduction

The documentation and description of useful tropical plants has been a major focus of economic botany in recent years. Inventories, market surveys, and ethnobotanical studies conducted in the tropics have produced a growing list of species which represent promising new sources of food, fuel, fiber, forage, oil, medicine, and chemical compounds. These studies underscore the great economic potential of tropical forests and provide a strong argument for the rational use and conservation of these important ecosystems. However, before the increased exploitation or management of any of these wild plants can be seriously considered, several fundamental questions concerning resource availability need to be answered. For example, how abundant is the species in the forest? Is the plant regenerating itself *in situ*? When does the species flower and fruit? How much of the desired resource is produced by natural populations? Although of fundamental importance in the overall evaluation of a plant resource, very few studies in economic botany have been concerned with quantifying these parameters.

Amazonian forests contain a large variety of wild fruit trees. Native fruits play an important role in the diet of rural populations, and a large number of them are collected and sold in local markets. Many forest fruits are exceptionally rich in vitamins. The general characteristics of the more common native fruits in Amazonia are described in the classic works of Le Cointe (1934),

Romero-Castaneda (1961), and Cavalcante (1976, 1978, 1980), yet very little has been written about the density, phenology or productivity of natural populations of these important tropical forest resources.

Fruit production data is reported here for three promising fruit trees from Peruvian Amazonia: *Myrciaria dubia, Grias peruviana,* and *Spondias mombin.* All of these trees grow in the seasonally flooded forests of the upper Amazon, and each species occurs naturally in relatively dense stands. All three species are exploited commercially to varying degrees. Detailed information on the density, size structure, and reproductive phenology of natural populations of each species is presented, and the potential for increased exploitation is also evaluated.

Species Descriptions

MYRCIARIA DUBIA

Myrciaria dubia (HBK) McVaugh (Myrtaceae) is a shrub or small tree which grows along the seasonally flooded banks of the tributaries and ox-bow lakes of the Amazon river. It is an important component of the riparian vegetation in Peru, Brazil, and possibly Colombia (McVaugh, 1963). The species is especially abundant in Peruvian Amazonia where it forms dense aggregations along the watercourses associated with the Napo, Nanay, Ucayali, Marañon, and Tigre rivers. It is known locally in Peru as "camu-camu" and in Brazil as "caçari."

Individual plants of *M. dubia* may attain a height of 6–8 m and a diameter of 10–15 cm. Basal sprouting is extremely profuse and branches are thin and slightly pendent giving the plant a sprawling appearance. The bark, which is shed periodically in thin plates, is smooth and of a light, grayish-brown color. Leaves are simple, opposite, 6–10 cm in length, with an acute tip and conspicuous glands on both surfaces. Flowers are perfect, subsessile, 1.0–1.2 cm in diam., with four white petals. The fruit is glabrous, 1.5–2.0 cm in diam. and dark reddish-purple in color on maturity (Fig. 1).

Although rarely cultivated, camu-camu is a very popular fruit in Peruvian Amazonia and a growing market for the species exists in Iquitos. The fruit is used in juice drinks, ice creams, and pastries, and a homemade liqueur, "camu-camuchada," is prepared by mixing the fruit pulp with cane alcohol. The pericarp is occasionally eaten raw with salt. Chemical analyses have shown that camu-camu is exceptionally rich in vitamin C, its fruit containing 2000–2994 mg of ascorbic acid per 100 g of pulp (Ferreyra, 1959; Roca, 1965). A chewable vitamin C tablet produced from camu-camu is marketed on a small scale by American Health Products, Inc. under the name "Camu-Plus."

GRIAS PERUVIANA

Grias peruviana Miers (Lecythidaceae) is a medium-sized tree of the seasonally flooded forests of northwest Amazonia. The species is commonly encountered on well-drained alluvial soils subject to inundation by white water, and may form high-density populations on extremely favorable sites. It is known as "sacha mangua" in Peru and "llanero," "apai," or "piton" in Ecuador (Prance & Mori, 1979). The common name for the species in Peru apparently results from the color of the fruit, which is similar to that of a mango (from Quechua, "sacha" = pseudo or bearing a resemblance to; from Spanish, "mangua" = mango).

The species has a distinctive form which is easily recognized in the forest. Leaves are simple, oblanceolate, very large (100 cm–150 cm in length) and in terminal clusters; branching is minimal; flowers and fruits grow directly from the trunk. Adult trees reach heights of 20–25 m and diameters of 30–40 cm. Flowers are perfect, 3.5–7.0 cm in diam., with four white petals and a yellow androecium. Fruits are brown, elliptical, indehiscent, 8–13 cm long with a single seed; the pulp turns yellowish-orange on maturity (Fig. 2).

The fruits of sacha mangua are collected and consumed in many areas of Peruvian Amazonia, and there is a small commercial demand for the species in the Iquitos market. The mature fruit is peeled, and slices of the mesocarp are separated from the seed and eaten raw with fariña (granules of toasted manioc) or roasted. An oil is also extracted by boiling the fruit. The mesocarp is rich in vitamin A, proximate analyses indicating a carotene content of 2.2 mg per 100 g of pulp (INDDA, 1984). The species is occasionally cultivated in house gardens both as a fruit tree and an ornamental.

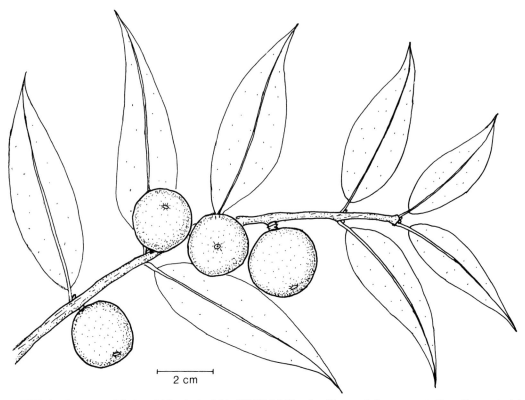

FIG. 1.   Leaves and fruits of *Myrciaria dubia* (HBK) McVaugh.   Figures 1–3 were made from live material collected in Jenaro Herrera region of lower Ucayali river, Loreto, Peru.

Spondias Mombin

Spondias mombin L. (Anacardiaceae) is a large forest tree which is widely distributed throughout the neotropics. Although the species shows an affinity for dry microsites in the evergreen and semi-evergreen forests of Mexico, Costa Rica and Panama (Janzen, 1985), in Peru it is most abundant in forests subject to seasonal flooding. Known locally as "uvos," other common names for *S. mombin* include "jobo," "ciruela amarilla" (Mexico, Central America, Colombia, Venezuela), "caja," tapereba (Brazil), hogplum and yellow mombin (Caribbean).

Adult trees reach heights of 40 m and diameters of up to 110 cm, the trunks of large trees often presenting small plank buttresses. The leaves are alternate, imparipinnately compound, up to 60 cm long with 3 to 17 pairs of leaflets (Fig. 3). Flowers, born in terminal panicles, are small, yellowish-white with five petals, 10 sub-

equal stamens and 4–5 styles (Croat, 1974; Macbride, 1951). The fruits are ovoid, 2.5–4.0 cm in length, with a hard, fibrous endocarp containing five seeds. On maturity the pericarp turns bright orange. Two varieties of the tree are recognized in the Peruvian Amazon based on bark texture and color. The red variety, "uvos rojo," has thick, coarsely-fissured bark with a pink to red periderm marked with prominent white striations. The white variety, "uvos blanco," has thinner, smoother bark with a light pink periderm lacking striations.

Almost every part of *S. mombin* is used in some manner by local people. The fresh fruits are eaten raw or made into juice drinks, ice creams, jams, or jellies. A tea made from the leaves and bark is used medicinally to treat diarrhea, stomach and vaginal infections, and dermititis. A tonic made from the bark is claimed to be an effective contraceptive. The wood is a medium grade sawtimber and is used in light

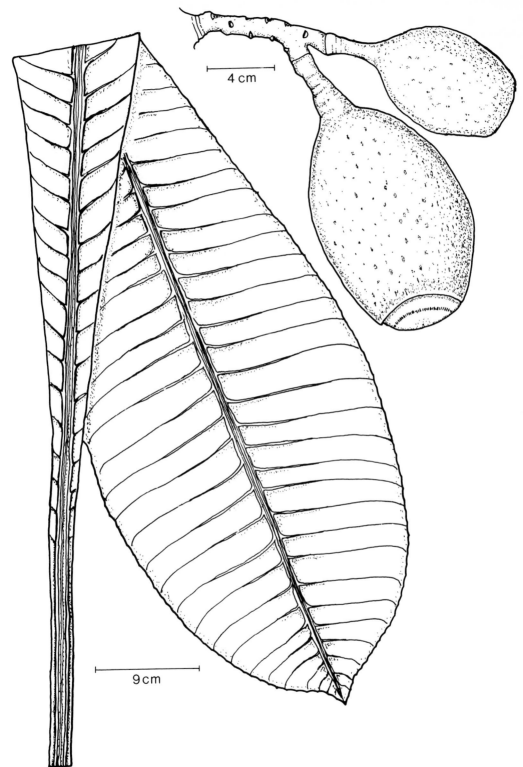

FIG. 2. Leaves and fruits of *Grias peruviana* Miers.

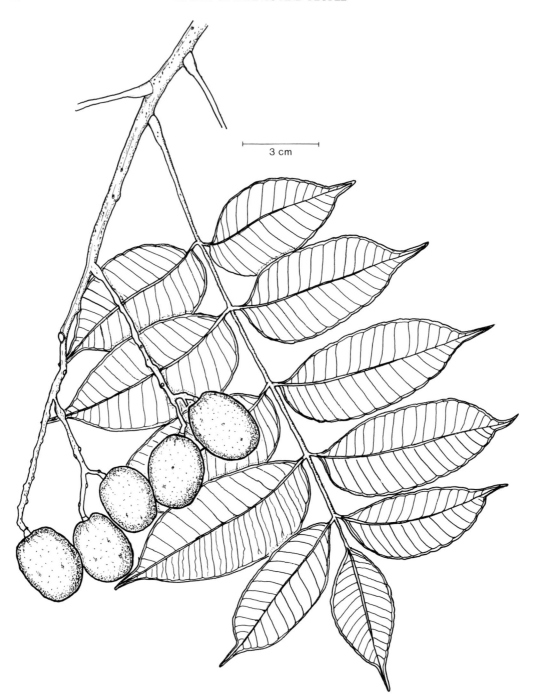

FIG. 3. Leaves and fruits of *Spondias mombin* L.

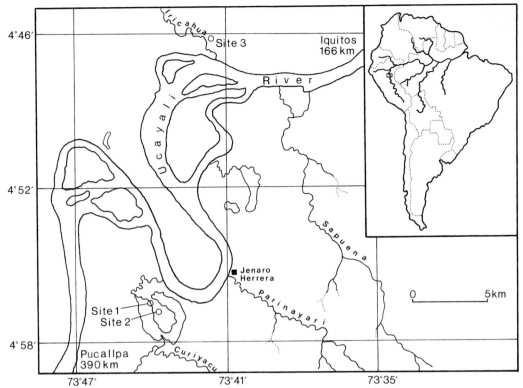

FIG. 4. Map of lower Ucayali river, Loreto, Peru showing Jenaro Herrera and location of *Myrciaria dubia* (Site 1), *Spondias mombin* (Site 2), and *Grias peruviana* (Site 3) study populations.

construction. The species is easily propagated from cuttings and is frequently cultivated in house gardens or used as a living fence.

Study Sites

Fieldwork on all species was based at the Instituto de Investigaciones de la Amazonia Peruana (I.I.A.P.) research station at Jenaro Herrera (73°40'W, 4°55'S), Provincia de Requena, Departmento de Loreto, Peru. The small village of Jenaro Herrera is located on the eastern bank of the lower Ucayali river, approximately 165 km upriver from the city of Iquitos (Fig. 4). Annual rainfall in this region averages 2889 mm; mean annual temperature is 29.9°C (unpubl. I.I.A.P. records). The local vegetation is classified as wet tropical forest (Holdridge et al., 1971), and contains both unflooded upland forests and lowland forests seasonally flooded by either black or white water (Encarnación, 1985).

Natural populations of *M. dubia, G. peruviana*, and *S. mombin* were located by interviewing local people and by exploring the lowland forests within a 30 km radius of Jenaro Herrera. From the numerous populations surveyed, one permanent study site was selected for each species. Site selection was based on the distribution and abundance of the study species, the distance to Jenaro Herrera, the accessibility during low and high water levels, and the probability that the forest would be logged or cut in the near future. All three study sites are exploited to varying degrees by local fruit collectors.

MYRCIARIA DUBIA

The *M. dubia* study population was located on the eastern margin of Sahua Cocha, an 80 ha oxbow lake of the Ucayali river (Site 1 in Fig. 4). The lake is continually fed by two black water rivers, but white water entering from the Ucayali predominates during peak flooding. The riparian vegetation bordering the lake is flooded for 6–7 months each year. Associated species on the site include *Eugenia inundata* DC. (Myrtaceae), *Laetia americana* L. (Flacourtiaceae), and *Symmeria paniculata* Benth. (Polygonaceae).

GRIAS PERUVIANA

Grias peruviana was studied in a tract of flooded forest located 1.5 km from the mouth of the Iricahua river, a small black water tributary of the Ucayali (Site 3 in Fig. 4). Representative canopy trees in the forest include *Ceiba samauma* K. Schum. (Bombacaceae), *Hura crepitans,* and *Maquira coriacea*; palms of the genera *Euterpe, Astrocaryum,* and *Scheelea* are especially prevalent. The forest floods to a depth of 1.5 m each year.

SPONDIAS MOMBIN

The *S. mombin* study site (Site 2 in Fig. 4) was located in the flooded forest immediately behind Site 1. The forest occupies a low "restinga" or levee which contours the entire eastern shore of Sahua Cocha. In addition to *S. mombin,* dominant canopy species on the site include *Calycophyllum spruceanum* (Benth.) K. Schum. (Rubiaceae), *Couroupita guianensis* Aubl. (Lecythidaceae), *Hura crepitans* L. (Euphorbiaceae), and *Maquira coriacea* (Karsten) C. C. Berg (Moraceae). The forest is flooded each year from March to May, water levels reaching 2.0 m during severe floods.

Methods

The basic procedure used to estimate fruit yield was identical for all three species. The size-class distribution of individuals within the population was first determined, and then size-specific rates of fruit production were measured for a subsample of individuals of differing size. Exact sampling methodologies, however, necessarily varied with each species in response to differences in plant size, abundance, and reproductive phenology.

MYRCIARIA DUBIA

The data for *M. dubia* was collected as part of a long-term demographic study of the species. In

September 1984 during low water level, ten 10 × 10 m contiguous plots (1000 m²) were established in the riparian vegetation growing along the bank of Sahua Cocha. All *M. dubia* plants in each plot were measured for height and basal diameter, and the exact location of each plant was mapped to the nearest 0.5 m using a system of cartesian coordinates. Each individual was permanently numbered with an embossed metal tag. Fruit production was measured on 25 adult trees selected to represent a range of basal diameters from 2.0 to 14.0 cm. Using a small boat, all the mature fruits produced by each sample tree were harvested and counted. Fruit collections were continued over two fruiting seasons (1984 and 1985), the same 25 individuals being measured each year.

GRIAS PERUVIANA

The size-structure of the *G. peruviana* population was determined using eight 20 × 20 m plots (3200 m²). All the *G. peruviana* individuals within each plot were counted, mapped, and permanently numbered. Seedlings were measured for height; both height and diameter were measured for juveniles and adults. Bi-weekly censuses of 15 adult trees ranging in diameter from 8.0 to 24.0 cm DBH were used to quantify size-specific rates of fruit production. At each census, the total number of mature fruits present on each sample tree was counted, and all fruits were marked with paint to avoid duplication in subsequent counts. The censuses were continued for 12 months (Feb 1985–Feb 1986).

SPONDIAS MOMBIN

To determine the density and size-class distribution of the *S. mombin* population, 125 contiguous 20 × 20 m plots (5.0 ha) were established in the flooded forest at Site 2. The plots were arranged in a rectangular configuration (100 × 500 m) to conform to the natural boundaries of the levee on which the forest was growing. All the *S. mombin* individuals ≥ 1.0 cm DBH in each plot were counted, measured, and mapped to the nearest 0.5 m. The variety ("rojo" or "blanco") and reproductive status of each tree were also recorded. Litter traps were used to estimate fruit production. Eight trees ranging in diameter from 36.0 to 85.0 cm DBH were sampled, tree selection being limited to healthy individuals whose crowns did not overlap with other conspecific adults. The vertical projection of the crown of each sample tree was determined by measuring out from the trunk to the outermost branches along at least four radii. Eight litter traps were then positioned within this area under the crown using random bearings and distances from the trunk. Each trap provided a sample area of 0.5 m^2 and consisted of a 79.8 cm diam. hoop of stiff metal wire supported by three 1.0 m stakes. Cloth mesh bags were placed inside each hoop to collect any material falling from the crown. All litter traps (n = 64) were emptied weekly during the 1986–1987 fruiting season, this interval being reduced to every three days during peak fruit production. Total fruit production for each tree was estimated by summing the number of fruits collected during each sampling period and expanding the result by the percent of the crown area sampled.

The spatial arrangement of individuals within a plant population can provide useful information about seed dispersal and the regeneration and growth requirements of a species (Greig-Smith, 1983). The distance between conspecific trees is also a key factor in determining the relative ease with which a forest fruit can be harvested. The spatial distribution of M. dubia, G. peruviana, and S. mombin was examined using Morisitas's (1959) Index of Dispersion (I_δ). This measure is given by:

$$I_\delta = \frac{\Sigma n_i(n_i - 1)}{N(N - 1)} q$$

where q = the number of samples, n_i = the number of individuals in the ith sample, and N = the total number of individuals in all samples. The I_δ value equals 1.0 when the individuals are randomly dispersed, i.e., independently assorted among the samples with equal probability. If the individuals are aggregated, then $I_\delta > 1.0$, and if uniformly or hyperdispersed, then $I_\delta < 1.0$. The index increases monotonically as clumping becomes more pronounced in the sample populations. To estimate the actual size of individual clumps of adult trees, the stem maps drawn for each species were sampled repeatedly using contiguous square plots ranging in size from 1–1600 m^2 and I_δ values were calculated for each plot size. An F statistic was computed for each value to test for significant differences from unity.

Results

SIZE STRUCTURE AND PATTERN OF TREE POPULATIONS

The three tree species differed greatly in terms of population density. The M. dubia population had the highest density with 8714 individuals/ha, followed by the G. peruviana and S. mombin populations with 508 and 17 individuals/ha, respectively. All density data refer to individuals greater than 1.0 cm in diameter.

The distribution by diameter class of the individuals in each population is shown in Figure 5. The data for M. dubia and G. peruviana are standardized to represent 1.0 ha, while the size-class distribution for S. mombin is based on 5.0 ha given the relatively low density of this species in the forest. The number of individuals in each size class is plotted on a $\log_{10}$ scale. Although the exact shape of the frequency histograms varies from species to species, all of the diameter distributions are characterized by a greater number of individuals in the smaller size classes than in the larger ones. Regression analyses revealed that the size structure of the M. dubia ($r^2 = 0.96$; $P < 0.001$), G. peruviana ($r^2 = 0.95$; $P < 0.001$), and S. mombin ($r^2 = 0.82$; $P < 0.05$) populations closely approximate a negative exponential distribution, the reduction in numbers from one diameter class to the next being relatively constant. Several authors (e.g., Leak, 1965; Meyer, 1952) have reported that diameter distributions conforming to a negative exponential are characteristic of stable, self-maintaining plant populations.

The dotted vertical line included in the histograms of Figure 5 represents the division between juvenile and adult trees in each population. Estimates of minimum reproductive size are based on the diameter of the smallest individual in each population which was observed with flowers or fruits. The diameter at which reproduction first occurs is directly related to the maximum size obtained by each species. The largest tree, S. mombin, does not begin to flower until approximately 20.0 cm DBH; G. peruviana exhibits a minimum reproductive size of 8.0 cm DBH under forest conditions; and M. dubia, a small shrub, initiates flowering soon after attaining a basal diameter of 2.0 cm.

The results of the spatial pattern analysis are presented in Table I. Indices of dispersion (I_δ)

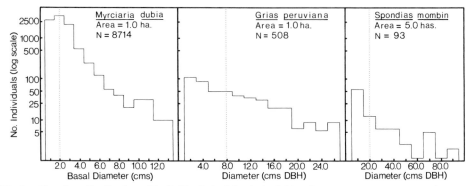

FIG. 5. Size-class distribution of individuals in *Myrciaria dubia, Grias peruviana,* and *Spondias mombin* study populations. Note $\log_{10}$ scale on y-axis. N represents the total number of individuals ≥ 1.0 cm in diam. recorded in each population; dotted vertical line in each histogram indicates the division between juvenile and adult trees.

calculated for all the adult trees in each population are shown for plots ranging in size from 1 m² to 1600 m². I_δ values determined to be significantly different ($P < 0.001$; one-tailed F-test) from 1.0 are marked with an asterisk. The distribution of adult *G. peruviana* trees could not be distinguished from random, although there is a slight tendency towards hyper-dispersion at the smaller plot sizes. An aggregated spatial pattern was detected for both *M. dubia* and *S. mombin.* The high I_δ values calculated for the latter species reflect the distinct clumps of adult trees which were noted during the inventory of Site 2.

REPRODUCTIVE PHENOLOGY

A graphic representation of flowering, fruiting, and leaf flushing phenology by *M. dubia, G. peruviana,* and *S. mombin* is presented in Figure 6. The patterns illustrated are based on periodic observations of marked individuals in each population. Monthly precipitation and river level data for Jenaro Herrera are shown in the lower half of Figure 6.

All three species flower and fruit on an annual basis, yet there are notable differences between species in terms of the timing and duration of each reproductive event. *Myrciaria dubia* initiates flowering at the end of August when the water level of the Ucayali river is at its minimum and all plants are on dry land. Flowering continues in distinct pulses, i.e., a large floral display followed by a 7–10 day period of inactivity, until all plants are completely covered by water, usually some time at the beginning of February. Mature fruits start to appear during the middle of October, and continue to be produced until flooding occurs. The later the flood peak, the longer the plants continue to produce fruit. New leaves are produced continually by *M. dubia* for as long as the plants remain unflooded.

The reproductive phenology of *Grias peruviana,* however, is notably aseasonal, with flowers and fruits being produced in small quantities

Table I

Spatial distribution of *Myrciaria dubia, Grias peruviana,* and *Spondias mombin* individuals ≥ 1.0 cm in diameter as determined by Morisita's Dispersion Index (I_δ). I_δ values shown for a range of plot sizes; values >1.0 indicate aggregation of individuals

Species	Plot size (m²)					
	1	6.25	25	100	400	1600
Myrciaria dubia	1.46*	4.50*	1.75*	1.37*	1.01	
Grias peruviana		0.80	0.86	0.95	1.03	1.05
Spondias mombin			7.25*	3.62*	1.81*	0.90

* Significantly different ($P < 0.001$) from a random pattern as determined by a one-tailed F-test.

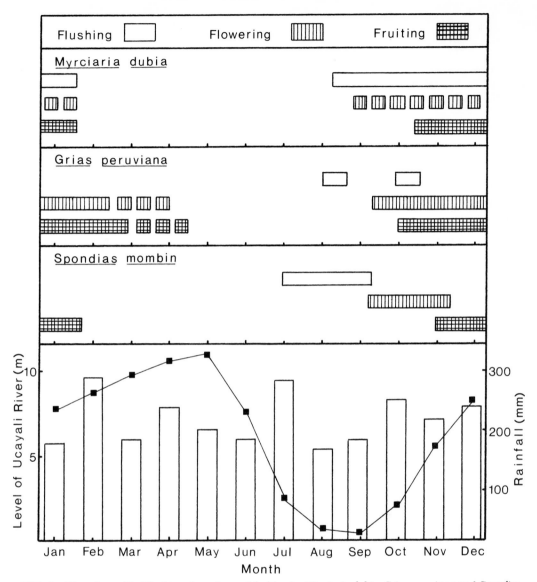

FIG. 6. Phenology of leaf flushing, flowering and fruiting by *Myrciaria dubia*, *Grias peruviana*, and *Spondias mombin* growing near Jenaro Herrera, Loreto, Peru. Monthly rainfall (bars) and river level (squares) data during 1985 (unpubl. I.I.A.P. records) are shown in lower half of figure.

over an eight month period (early September–late April). There is a slight peak in fruit production from December to February. Most trees produce at least two flushes of leaves each year, one in August after flooding subsides, the other in late October. Vigorously growing trees were observed to produce up to four flushes of leaves a year.

The production of flowers and fruits by *Spondias mombin*, in contrast, is also markedly sea-

sonal, but the reproductive phenologies of different individuals in the population are not completely synchronized. Trees drop their leaves in late April or early May during the flood peak, and remain leafless for two or three months. Flushing of new leaves occurs in late June; flowering is initiated in late August concurrent with the expansion of new foliage. This behavior is in contrast to that observed for *S. mombin* in Central America where the species flowers while leaf-

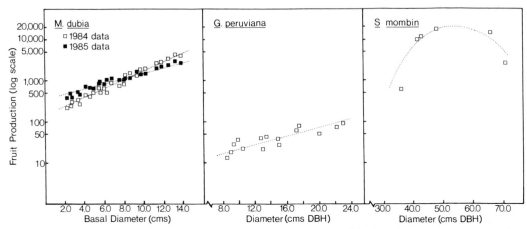

FIG. 7. Fruit production as related to tree diameter for *Myrciaria dubia, Grias peruviana*, and *Spondias mombin*. Fruit production data are shown on $\log_{10}$ scale; regression lines based on general linear model, i.e., $\log_{10}$ fruit production = a +b(diameter), are fitted to *M. dubia* and *G. peruviana* data; parameter values and coefficients of determination are as follows: *M. dubia* 1984 data, a = 2.17, b = 0.11, r^2 = 0.97; *M. dubia* 1985 data, a = 2.38, b = 0.09, r^2 = 0.99; *G. peruviana*, a = 0.93, b = 0.048, r^2 = 0.86. No significant linear relationship was detected for *S. mombin* (r^2 = 0.025); the curve shown for this species was fitted by eye.

less during the dry season (Croat, 1974; Janzen, 1985). Individual trees may differ by as much as six to eight weeks in the timing of floral initiation. The first mature fruits begin to appear in early November; fruiting usually continues until the end of February.

FECUNDITY OF INDIVIDUAL TREES

Size-specific fruit production rates for *M. dubia, G. peruviana*, and *S. mombin* are shown in Figure 7. Given the large range in values and the exponential nature of the relationship between size and fecundity in most cases, the fruit production data are plotted on a $\log_{10}$ scale. The results from both the 1984 and 1985 collections are shown for *M. dubia*. Least-squares linear regression using log-transformed fruit production data showed that fecundity was significantly related to plant size for *M. dubia* (1984 data r^2 = 0.97, $P < 0.01$; 1985 data r^2 = 0.99, $P < 0.001$) and *G. peruviana* (r^2 = 0.86, $P < 0.05$). In both species, an increase in diameter results in an exponential increase in the number of fruits produced.

Fruit production by *S. mombin* displays a different pattern, and regression analysis detected no significant relationship between fecundity and tree diameter in this species (r^2 = 0.025, N.S.). Complete data sets, however, were collected from

only six sample trees, a rapid increase in the water level of Sahua Cocha totally flooding the litter traps under the remaining two trees during peak fruit production. These two trees, both of large diameter (67.5 and 85.5 cm DBH), were the last adults in the population to initiate flowering. In spite of the small sample size, the data in Figure 7 do suggest that fruit production by *S. mombin* is not a linear function of size.

TOTAL POPULATION FRUIT YIELD

Estimates of the total number of fruits produced in each plant population were calculated using the data collected on population structure and size-specific fecundity. The study populations were divided into size classes, and, as none of the species is dioecious, the number of individuals in each size class was simply multiplied by the fecundity value for that class. The fruit production totals for each class were then summed.

The results calculated for *M. dubia* using this procedure are presented in Table II. Estimates of total population fruit production are shown for both 1984 and 1985; the fecundity estimates for each class are based on the regression equations calculated for each year (Fig. 7). Natural populations of *M. dubia* yield between 1,202,519 and 1,611,336 fruits/ha/year. The great majority

Table II

Total annual fruit production by a 1.0 ha population of *Myrciaria dubia* growing near Sahua Cocha, Loreto, Peru during 1984 and 1985. N is equal to the number of adult trees/size class; class total represents fruit yield/size class/year

Basal diameter class (cm)	N	1984		1985	
		Size-specific fecundity	Class total	Size-specific fecundity	Class total
2.1–3.0	1800	278	500,000	403	725,400
3.1–4.0	600	359	215,400	495	297,000
4.1–5.0	310	462	143,220	609	188,790
5.1–6.0	150	595	89,250	750	112,500
6.1–7.0	50	767	38,350	922	46,100
7.1–8.0	40	988	39,520	1135	45,400
8.1–9.0	20	1273	25,460	1396	27,920
9.1–10.0	30	1640	49,200	1718	51,540
10.1–11.0	30	2113	63,390	2114	63,420
11.1–12.0	10	2722	27,220	2600	26,000
12.1–13.0	10	3507	35,070	3199	31,990
Total population fruit yield/ha:			1,203,560		1,616,060

of this fruit is produced by individuals in the smaller size classes. Based on an average fruit weight of 7.91 ± 0.23 (mean ± standard error; n = 500), the total fruit yield in 1984 and 1985 is equivalent to 9.5 and 12.7 t/ha/year, respectively.

Using a similar analysis, total fruit production by the *G. peruviana* study population was estimated at 8581 fruits/ha/year (Table III). As a mature fruit weighs 271.1 ± 5.1 g (mean ± standard error; n = 42), this figure represents a total annual fruit yield of 2.3 t/ha/year.

Given the small sample size and the lack of a significant relationship between tree diameter and fruit production, estimating size-specific fecundity and total fruit yield for the *S. mombin* population was less precise (Table IV). The population was divided into seven 10 cm diameter classes. A mean fecundity value (n = 3) was used for the 40–50 cm diam. class, while production data for the 30–40 cm and 60–70 cm classes were based on collections from one tree in each class. None of the sample trees were in the 50–60 cm diam. class, and, therefore, an approximate fecundity value was calculated using data from the 40–50 cm and 60–70 cm size classes. Based on

Table III

Total annual fruit production for a 1.0 ha population of *Grias peruviana* growing along Iricahua river, Loreto, Peru during 1985. N is equal to the number of adult trees/size class; class total represents fruit yield/size class/year

Diameter class (cm DBH)	N	Size-specific fecundity	Class total
8.1–10.0	44	23	1012
10.1–12.0	41	29	1189
12.1–14.0	34	36	1224
14.1–16.0	22	45	990
16.1–18.0	22	56	1232
18.1–20.0	6	69	414
20.1–22.0	9	87	783
22.1–24.0	6	108	648
24.1+	9	121	1089
Total population fruit yield/ha:			8581

Table IV

Total annual fruit production by a 1.0 ha population of *Spondias mombin* growing near Sahua Cocha, Loreto, Peru during 1986–1987 fruiting season. N is equal to the number of adult trees/size class; class totals represent fruit yield/size class/year

Diameter class (cm DBH)	N	Size-specific fecundity	Class total
20.1–30.0	3	624	1872
30.1–40.0	2	624	1248
40.1–50.0	1	15,775	15,775
50.1–60.0	1	15,934	15,934
60.1–70.0	2	16,093	32,186
70.1–80.0	1	2702	2702
80.1–90.0	1	2702	2702
Total population fruit yield/ha:			72,419

these calculations, total fruit yield for *S. mombin* was estimated at 72,419 fruits/ha/year. The weight of a mature *S. mombin* fruit is extremely variable (3.6–13.3 g). Using a mean fruit weight of 8.7 ± 0.98 g (mean ± standard error; n = 527), total annual fruit production by the 11 adult trees on Site 2 equals 630 kg/ha/year.

Discussion

While the objective of this study was to determine fruit yield from natural populations of *M. dubia, G. peruviana,* and *S. mombin,* the data collected on the abundance, spatial distribution and reproductive phenology of each species merit discussion. This type of information is extremely important from a management standpoint, and provides a framework for evaluating the exploitation potential of these forest resources. The data are also of interest given the limited number of studies which have focused on the ecology of Amazonian forest trees.

ABUNDANCE

Floristic studies conducted in Amazonia have shown that most tree species occur at densities of 1–3 individuals per hectare (Cain et al., 1956 and Pires et al., 1953 for trees ≥10 cm DBH; Prance et al., 1976 for trees ≥15 cm DBH). *Myrciaria dubia, G. peruviana,* and *S. mombin,* however, form natural populations which contain 10 to 1000 times more individuals than this. Although dramatic, these high-density populations should not be viewed as ecological curiosities or phenomena restricted to Peruvian Amazonia. Dense stands of *M. dubia* have also been reported in Brazil (Keel & Prance, 1979), and *S. mombin* is an abundant canopy tree in many neotropical forests (Gentry, 1982; Janzen, 1985). Several genera of Lecythidaceae (e.g., *Eschweilera, Gustavia*) occur naturally in high densities (see review in Prance & Mori, 1979), but the present study is apparently the first documentation of this for *Grias.*

A common characteristic of the size-class distributions of many tropical trees is a pronounced absence of saplings and juveniles (Richards, 1952; Sarukhan, 1980; Whitmore, 1975). The canopy may contain numerous adult trees, and the understory may be carpeted with seedlings, yet the population contains very few individuals of intermediate size. This type of size structure indicates that the regeneration of a species is severely limited for some reason, with most seedlings dying before becoming established due to suppression, predation, physical damage or competition. Seedling establishment would seem to be especially problematic for *M. dubia, G. peruviana* and *S. mombin* because these species are exploited every year by local fruit collectors. The fruits of *S. mombin* and *G. peruviana* are collected sporadically from Sites 2 and 3 for consumption or sale in Jenaro Herrera, and a considerable quantity of *M. dubia* fruit is harvested each year from Site 1 and shipped to Iquitos. Yet, in spite of the large number of fruits and seeds which are removed from the forest, the size-class distributions of all three study species display a progressive increase in the number of individuals from the larger to the smaller size classes, and the intermediate size classes are well-represented. Each of the populations appears to maintain a continual input of newly established seedlings.

SPATIAL DISTRIBUTION

The results from the analysis of spatial pattern, together with simple observations made in the field, provide a general picture of the regeneration strategy used by each species. For example, *M. dubia* and *S. mombin* both display an aggregated spatial pattern suggesting that seedling establishment is more successful in some microsites than in others. There is little doubt that the pronounced aggregation exhibited by *S. mombin* results from enhanced seedling establishment in large treefall gaps. Without exception, all of the saplings recorded in the inventory of Site 2 were clumped in areas where the canopy had been opened by the fall of a large tree. The aggregated spatial pattern shown by *M. dubia,* on the other hand, seems to have a different origin. The individuals of this species are segregated spatially in contours or waves which parallel the shoreline, seedlings being confined to the water's edge, juveniles growing behind them, and adults of varying size being dispersed further upslope. Apparently, *M. dubia* seedlings are very shade intolerant and can only become established in fully illuminated, newly deposited beach sediments. The random distribution of *G. peruviana* suggests that seedling establishment may be independent of canopy cover in this species. The extremely large seeds (and abundant seed reserves) possessed by *G. peruviana* lends credibility to this conclusion, as seedlings probably can withstand prolonged periods of suppression under a closed canopy (Foster, 1986).

REPRODUCTIVE PHENOLOGY

Most phenological studies of tropical trees have attributed periodic flowering and fruiting to variation in local rainfall patterns (Frankie et al., 1974; Medway, 1972; Opler et al., 1980). A clear example of this is the flowering peak exhibited by many tree species in semi-evergreen forests during the dry season (Fournier & Salas, 1966; Janzen, 1967), seedfall coinciding with the onset of the rainy season. The salient feature of the reproductive phenologies of *M. dubia, S. mombin* and, to a lesser degree, *G. peruviana,* is that the seasonality of flowering and fruiting appears to be more closely linked to the rise and fall of the Ucayali river than to local precipitation patterns (Fig. 6). Flowering by *M. dubia* is initiated when the crowns of adult trees are completely out of the water and the river is at its low point. Fruit production for the year ceases only after all plants are inundated. Mature fruits drop into the lake where they are eaten and consequently dispersed by *Colossoma macropomum* ("gamitana") and several other species of large fish (unpubl. field data; see Goulding, 1980; Smith, 1981). Seeds that are not dispersed remain dormant on the lake bottom for 6–7 months until the water level falls again. Flower production by *S. mombin* coincides with the initial rise of the Ucayali river in September. The levees on which the tree grows start to flood about the time that mature fruit are being formed. Both fruits and seeds float; dispersal is effected by the current. *Grias peruviana,* in contrast, flowers and fruits almost continually throughout the year, except during the floodpeak and two or three months thereafter. As might be expected given this pattern, prolonged flooding destroys the large fleshy seeds of *G. peruviana,* but has little effect on established seedlings.

FECUNDITY

In spite of the obvious importance of seed production in the regeneration of a plant population, detailed studies of size-specific fecundity in tropical trees are virtually non-existent. The little information that does exist is usually related to cultivated fruit trees (Falcão & Lleras, 1980, 1981; Purseglove, 1975; Valmajor et al., 1965). Janzen (1978) has provided, perhaps, the most comprehensive treatment of seeding patterns in tropical forest trees to date, yet quantitative data are presented for only two species. Collecting fruit production data for forest trees is tedious, time-consuming and fraught with methodological problems, and the existing data base reflects this.

The data collected in this study represent the first estimates of size-specific fecundity and total population fruit yield for *M. dubia, S. mombin,* and *G. peruviana*. All other information related to fruit production by these species is anecdotal. Calzada-Benza (1980) reports that *M. dubia* plants growing near Iquitos produce an average of 119 fruits, but only part of the fruiting season was sampled and the size of the plants is not given. Janzen (1985) estimates that large *S. mombin* trees in Costa Rica "may bear as many as 10,000 fruits." In the absence of comparative data, it is impossible to assess the accuracy of the fecundity values calculated for each species. However, two limitations of the data set should be noted. First, fruit production rates vary from year to year. As *S. mombin* and *G. peruviana* were sampled over only one fruiting season, the magnitude of this variation cannot be determined for these species. Second, fruit production rates undoubtedly vary from site to site. Soil fertility, flooding regimes, pollinator abundance, and proximity of competitors can all affect the reproductive output of a tree; replicate study populations would be required to evaluate these relationships.

PRODUCTIVITY

In terms of total yield, natural populations of *M. dubia, G. peruviana* and, to a lesser extent, *S. mombin* compare favorably with intensively managed plantations of many tropical fruits. For example, avocado (*Persea americana* Mill.) plantations produce from 2 to 10 t/ha/year (Ochse et al., 1961), while mangoes (*Mangifera indica* L.) average 3.5 to 6.5 t/ha/year under cultivation (Purseglove, 1975). The impressive productivity exhibited by wild stands of *M. dubia, G. peruviana* and *S. mombin* is at least partially the result of edaphic factors. Flooded forests contain some of the richest and most productive soils in Amazonia (Sanchez, 1976; Cochrane & Sanchez, 1981), these habitats being naturally "fertilized" each year by the rise and fall of the river.

MARKET VALUE

In 1986, the average price of a 500 g bag of "camu-camu" or "uvos" was approximately 30 cents (U.S.) in the Iquitos market. A single "sacha mangua" fruit sold for about 50 cents (U.S.). These prices, together with the yield data collected in this study, can be used to estimate the current market value of the fruit produced by natural populations of *M. dubia, G. peruviana,* and *S. mombin.* The results from such an analysis reveal that a 1.0 ha population of *M. dubia* produces from $5700 to $7620 (U.S.) worth of fruit each year. The total market value of the fruit produced by *G. peruviana* is estimated at $4242 (U.S.)/ha/year, and *S. mombin,* which averages only 11 adult trees/ha, produces approximately $378 (U.S.) worth of fruit/ha/year. While these calculations are based on several unlikely assumptions, i.e., that all of the fruit produced by each population is collected, and that fruit prices remain stable regardless of the supply, they do provide a general idea of the economic potential of these important forest resources.

Conclusion

In response to the accelerating deforestation in the Amazon and the decline or total loss of many species, much attention has been focused on developing sustainable methods for exploiting the native plant resources of the region (e.g., Fearnside, 1979, 1983; Goodland & Irwin, 1975; Hecht, 1982; IUCN, 1975; National Research Council, 1982). Of particular interest are land-use systems that integrate the utilization and conservation of intact forest. Based on the results of this study, controlled harvests of the fruit produced by high-density populations of *M. dubia, G. peruviana,* and *S. mombin* represent an appropriate example of such a system.

In contrast to most other forms of resource exploitation practiced in the tropics, fruit collection has little effect on the structure and function of natural forests. Canopy cover is maintained, nutrient and hydrological cycles remain essentially undisturbed, and genetic diversity is preserved. Although quantitative data on the nutrient content of Ucayali floodwater are lacking, nutrient losses resulting from intensive fruit harvests are probably replenished by annual sediment deposition. The fact that *M. dubia,* and occasionally *S. mombin,* are harvested from a boat during peak flooding further reduces the potential for damage to the forest. Special care, however, must be taken to ensure that repeated harvests do not alter the size-class distribution of the species being exploited. Given that the long-term sustainability of the system is contingent upon the population maintaining an adequate level of regeneration, this caveat is of critical importance.

The economics of exploiting natural populations of *M. dubia, G. peruviana,* and *S. mombin* also seem favorable. No initial investment is required, energy inputs such as fertilizers, pesticides and cultural practices are minimal, and there is no waiting period between planting and first production as is the case with plantation establishment. The technology required to exploit these forests is available to even the poorest rural populations. Expanded markets and innovative processing technologies, however, are sorely needed for these three fruits. Detailed studies of the current and potential role of each fruit in market and subsistence economies would also be extremely useful.

Acknowledgments

This study was conducted as part of a cooperative agreement between the Instituto de Investigaciones de la Amazonía Peruana (I.I.A.P.) of Iquitos, Peru and the Institute of Economic Botany of The New York Botanical Garden. The support of both these institutions is gratefully acknowledged. We thank Dr. Jose López Parodi and the entire staff at the Jenaro Herrera research station for providing a comfortable place to work. Humberto Pacaya and Jose Tuanama provided invaluable assistance in the field. We also thank Dr. Anthony Anderson and one anonymous reviewer for helpful comments on the manuscript. Funding for the research on *M. dubia* and *G. peruviana* (CMP) was provided by the Exxon Corporation; the *S. mombin* study (EJH) was supported by a grant from the World Wildlife Fund. This is publication number 137 in the series of the Institute of Economic Botany of The New York Botanical Garden.

Literature Cited

Cain, S. A., G. M. D. Castro, J. M. Pires & N. T. de Silva. 1956. Application of some phytosociological techniques to Brazilian rain forest. Amer. J. Bot. **43**: 911–941.

Calzada-Benza, J. 1980. Frutales Nativas. Libreria El Estudiante, Lima, Perú.

Cavalcante, P. 1976, 1978, 1980. Frutas comestíveis da Amazônia. Vols. I, II & III. Museu Paraensis Emilio Goeldi, Belém.

Cochrane, T. T. & P. A. Sanchez. 1981. Land resources, soils and their management in the Amazon region: A state of knowledge report. Pages 137–209 *in* S. B. Hecht (ed.), Amazonia: Agriculture and land use research. Centro Internacional de Agricultura Tropical (CIAT), Cali, Colombia.

Croat, T. B. 1974. A reconsideration of *Spondias mombin* L. Ann. Missouri Bot. Gard. **61**: 483–490.

Encarnación, F. 1985. Introducción a la flora y vegetación de la Amazonia peruana: Estado actual de los estudios, medio natural y ensayo de una clave de determinación de las formaciones vegetales en al llanura amazónica. Candollea **40**: 237–252.

Falcão, M. A. & E. Lleras. 1980. Aspectos fenológicos, ecológicos e de productividade do Umari (*Poraqueiba sericea* Tul.). Acta Amazonica **10**: 445–462.

———— & ————. 1981. Aspectos fenológicos, ecológicos e de productividade da Sorva (*Couma utilis* Muell.). Acta Amazonica **11**: 729–741.

Fearnside, P. M. 1979. The development of the Amazon rainforest: Priority problems for the formulation of guidelines. Interciencia **4**: 338–343.

————. 1983. Development alternatives in the Brazilian Amazon: An ecological evaluation. Interciencia **8**: 65–78.

Ferreyra, R. 1959. Camu-camu, nueva fuenta nacional de vitamina C. Bol. Exp. Agropecu. **7**: 28.

Foster, S. A. 1986. On the adaptive value of large seeds for tropical moist forest trees: A review and synthesis. Bot. Rev. **52**: 260–299.

Fournier, L. A. & S. Salas. 1966. Algunas observaciones sobre la dinámica de la floración en el bosque tropical húmedo de Villa Colon. Rev. Biol. Trop. **14**: 75–85.

Frankie, G. W., H. G. Baker & P. A. Opler. 1974. Comparative phenological studies of trees in tropical wet and dry forests in the lowlands of Costa Rica. J. Ecol. **62**: 881–919.

Gentry, A. H. 1982. Patterns of neotropical plant species diversity. Evol. Biol. **15**: 1–84.

Goodland, R. J. A. & H. S. Irwin. 1975. Amazon jungle: Green hell to red desert? Elsevier, Amsterdam.

Goulding, M. 1980. The fishes and the forest: Explorations in Amazonian natural history. University of California Press, Berkeley.

Greig-Smith, P. 1983. Quantitative plant ecology. Blackwell Scientific Publications, Oxford.

Hecht, S. B. (ed.). 1982. Amazonia: Agriculture and land use research. University of Missouri Press, Columbia.

Holdridge, L. R., W. C. Grenke, W. H. Hatheway, T. Liang & J. S. Tosi. 1971. Forest environments in tropical life zones: A pilot study. Pergamon Press, Oxford.

INDDA. 1984. Unpublished laboratory results. Laboratorio de Bromatología, Instituto Nacional de Desarrollo Agroindustrial. Lima, Peru.

IUCN. 1975. The use of ecological guidelines for development in the American humid tropics. International Union for the Conservation of Nature and Natural Resources, New Series No. 31, Morges, Switzerland.

Janzen, D. H. 1967. Synchronization of sexual reproduction of trees within the dry season in Central America. Evolution **21**: 620–637.

————. 1978. Seeding patterns of tropical trees. Pages 83–128 *in* P. B. Tomlinson & M. H. Zimmerman (eds.), Tropical trees as living systems. Cambridge University Press, Cambridge.

————. 1985. *Spondias mombin* is culturally deprived in megafauna-free forest. J. Trop. Ecol. **1**: 131–155.

Keel, S. H. & G. T. Prance. 1979. Studies on the vegetation of a white-sand blackwater igapó (Rio Negro, Brazil). Acta Amazonica **9**: 645–655.

Leak, W. B. 1965. The J-shaped probability distribution. Forest Sci. **11**: 405–419.

Le Cointe, P. 1934. Arvores e plantas úteis (indígenas e aclimadas): Nomes vulgares, classificação botánica, habitat, principais aplicações e propriedades. Lic. Classica, Belém.

Macbride, J. F. 1951. Flora of Peru. Field Mus. Nat. Hist., Bot. Ser. **29(8)**: 395–632.

McVaugh, R. 1963. Tropical American Myrtaceae, II. Notes on generic concepts and descriptions of previously unrecognized species. Fieldiana: Botany **29(8)**: 393–532.

Medway, L. 1972. Phenology of a tropical rain forest in Malaya. J. Linn. Soc., Biol. **4**: 117–146.

Meyer, H. A. 1952. Structure, growth and drain in balanced, uneven-aged forest. J. Forestry **50**: 85–92.

Morisita, M. 1959. Measuring the dispersion of individuals and analysis of distributional patterns. Mem. Fac. Sci. Kyushu Univ., Ser. E, Biol. **2**: 215–235.

National Research Council. 1982. Ecological aspects of development in the humid tropics. National Academy Press, Washington, D.C.

Ochse, J. J., M. J. Soule, M. J. Dijkman & C. Wehlburg. 1961. Tropical and subtropical agriculture. Macmillan, New York.

Opler, P. A., G. W. Frankie & H. G. Baker. 1980. Comparative phenological studies of shrubs and treelets in wet and dry forests in the lowlands of Costa Rica. J. Ecol. **68**: 167–186.

Pires, J. M., T. Dobzhansky & G. T. Black. 1953. An estimate of the number of species of trees in an Amazonian forest community. Bot. Gaz. **114:** 467–477.

Prance, G. T. & S. Mori. 1979. Lecythidaceae, Part I. The actinomorphic flowered New World Lecythidaceae. Flora Neotrop. Monogr. **21:** 1–270.

——, **W. A. Rodrigues & M. F. de Silva.** 1976. Inventário florestal de um hectare de mata de terra firme km 30 da estrada Manaus-Itacoatiara. Acta Amazonica **6(1):** 9–35.

Purseglove, J. W. 1975. Tropical crops. Dicotyledons. Longman, London.

Richards, P. W. 1952. The tropical rain forest. Cambridge University Press, Cambridge.

Roca, N. 1965. Estudios químico-bromatológico de la *Myrciaria paraensis* Berg. Tesis Química. Univ. Nac. Mayor San Marcos, Lima, Peru.

Romero-Castaneda, R. 1961. Frutas Silvestres de Colombia. Bogotá, Colombia.

Sanchez, P. 1976. Properties and management of soils in the tropics. Wiley, New York.

Sarukhan, J. 1980. Demographic problems in tropical systems. Pages 161–188 *in* O. Solbrig (ed.), Demography and evolution of plant populations. Blackwell Scientific Publications, Oxford.

Smith, N. J. H. 1981. Man, fishes, and the Amazon. Columbia University Press, New York.

Valmajor, R. V., R. E. Coronel & D. A. Ramirez. 1965. Studies on floral biology, fruit set and fruit development in Durian. Philipp. Agric. **47:** 355–365.

Whitmore, T. C. 1975. Tropical rain forests of the Far East. Clarendon Press, Oxford.

The Genus *Caryocar* L. (Caryocaraceae): An Underexploited Tropical Resource

Ghillean T. Prance

Table of Contents

Abstract

PRANCE, G. T. (Director, The Royal Botanic Gardens, Kew, Richmond, Surrey TW9 3AB, United Kingdom). The genus *Caryocar* (Caryocaraceae): An underexploited tropical resource. Advances in Economic Botany **8**: 177–188. 1990. The genus *Caryocar* already has many local uses throughout its natural range. The wood is durable, finishes well, is extremely resistant to insect attack and is much used in boat building. The fruit of most species has an edible mesocarp and a kernel that can be used like a nut. Pulp and kernel contain an oil rich in oleic acid. The fruit is used as a fish poison and the mesocarp to prepare a liqueur. The sixteen species range from Costa Rica to southern Brazil and occupy a wide range of habitats. This ecological amplitude and many possible uses make *Caryocar* a genus of much greater economic potential than realized by its current uses, especially in agroforestry systems where multipurpose crops are required. The genus merits further research and attention, representing an underexploited neotropical resource.

Key words: *Caryocar*; edible oil; fish poison; wood; edible nut; multipurpose crop.

Introduction

The small neotropical plant family Caryocaraceae (order Theales), contains only 25 species in two genera. It is, however, widespread and its species are often important components of the vegetation types in which the family occurs. The various species are adapted to a wide range of vegetation types with the greatest number of species occurring in tropical rainforest. The genus *Anthodiscus* G. F. W. Meyer, with nine species, has not been widely used except as a local source of timber. Most of the 16 species of *Caryocar* L. have been used in some way and certainly have the potential to become much more important economically. The purpose of this paper is to draw attention to the entire genus *Caryocar* as an as yet under-utilized resource and to review the extensive literature that now exists on its uses.

Although *Caryocar* was one of the plants described and listed by the U.S. National Academy of Sciences (NAS, 1975) as one of their selected underexploited tropical plants with promising economic value, relatively little has been done since to increase the use of *Caryocar*. Much progress, however, has been made in the development of some of the other species selected by the NAS panel, for example, babassu (*Orbignya*) and *Leucaena*. This disparity is not because *Caryocar* has been found to be unpromising, but rather because it has not yet received much further research. The genus is of considerable potential because of the many ways in which it has been used, and because of the ecological amplitude covered by the different species. For example, *Caryocar villosum* and *C. glabrum* are rainforest species, *C. microcarpum* occurs in floodplain forest, *C. gracile* in Amazonian white sand caatinga and *C. brasiliense* in cerrado of Central Brazil. *Caryocar* was also included by Schultes (1980) in his review of possible new economic plants from Amazonia.

The Caryocaraceae was monographed by Prance and Silva (1973) who recognized fifteen species in the genus *Caryocar*. Since that time one new species, *C. harlingii* Prance (1987), has been added. *Caryocar* is distributed from Costa Rica and Chocó, Colombia through lowland South America east of the Andes to Paraguay and the State of Paraná, Brazil.

Table I is a synopsis of the species of *Caryocar*, their distribution, ecology and known uses. It shows that the genus is widespread, that species occur in a number of different major habitat types such as non-flooded forest, floodplain forest, cerrado and caatinga, and that most of the species have been used in some way at least on a local basis. The principal uses are the edible fruit, the use of the fruit for the extraction of cooking oil, and the good timber of most species. A genus with a good number of wild species, that are adapted to a wide range of habitats, certainly has considerable potential for the generic improvement and alteration of any of the species with useful properties.

Generic Description

Caryocar Linnaeus, Mantissa plantarum 2: 247. 1771.

Synonyms: *Pekea* Aublet, Pl. Guiane 1: 594. 1775.
Souari Aublet, Pl. Guiane 1: 599. 1775.
Barollaea Necker, Elem. Bot. 2: 322. 1790.
Rhizobolus Gaertner ex Schreber, Linn. Gen. Pl. ed. 8, 1: 369. 1789.
Acantacarix Arruda da Camara ex Koster, Trav. Bras. 491. 1816.

Type species: *Caryocar nuciferum* L.

Large *trees* or rarely (in *C. brasiliense*) shrubs or suffrutices, with opposite, often horizontal branches. *Stipules* soon caducous, or absent. *Leaves* opposite, usually long-petiolate, rarely almost sessile, compound-trifoliolate; leaflets with short petiolules, pinnately nerved, the margins serrate, crenate, dentate or rarely entire. Often with 2–4 stipels at base of leaflets, the stipels persistent or caducous, sometimes with two large and two small stipels. *Inflorescences* of terminal racemes with short rachis, often rather corymbose; pedicels articulate near to flower; bracteoles lateral, alternate, small, subpersistent or caducous. *Flowers* hermaphrodite, large, generally opening at night. Calyx distinctly five- (rarely six-)lobed, imbricate. Petals five, or rarely six, imbricate, fused at base together with filaments and often, after anthesis, caducous together with the stamens in a single unit. Stamens numerous, 57–750, exceeding the petals; filaments bent into an S in bud, those on the interior shorter and sterile or with smaller anthers, often with an inner row of short sterile staminodes, the basal portion of which form a glandular nectar-secret-

Table I

The species of *Caryocar*: distribution, ecology and uses

	Distribution	Habitat	Uses
C. *amygdaliferum* Mutis	Colombia: Magdalena Valley	—	—
C. *amygdaliforme* G. Don	Peru: Andean foothills	Forest on terra firme	Fruit: edible
C. *brasiliense* Camb.	Central Brazil, Paraguay	Cerrado	Fruit: edible pulp, oil, liqueur
			Wood: yellow dye, fence posts, charcoal
C. *coriaceum* Wittm.	N.E. Brazil	Caatinga woodland	Fruit: edible, oil
C. *cuneatum* Wittm.	N.E. and Central Brazil	Cerrado, Cerradão	Fruit: edible
C. *costaricense* J. D. Smith	Costa Rica	Forest on terra firme	Wood
C. *dentatum* Gleason	W. Amazonia: Bolivia and Brazil	Forest on terra firme	—
C. *edule* Casar.	Atlantic coastal Brazil	Coastal forest beside streams	Fruit: edible
C. *glabrum* (Aubl.) Pers	Guianas, widespread in Amazonia	Forest on terra firme	Fruit: edible
			Wood: fish poison, boats
C. *gracile* Wittm.	N.W. Amazonia: Brazil, Colombia, Venezuela	Amazonian caatinga	Fruit: fish poison
			Leaves: poison
C. *harlingii* Prance	Amazonian Peru	Forest on terra firme	Fruit: edible
			Wood
C. *microcarpum* Ducke	Widespread in Amazonia	Várzea and igapó	Fruit: edible
			Leaves: fish poison, soap
C. *montanum* Prance	Guayana Highland	Lower montane forest	—
C. *nuciferum* L.	Guianas, cult. in W.I. Colombia: Chocó	Forest on terra firme	Fruit: edible
C. *pallidum* A. C. Smith	Widespread in Amazonia	Forest on terra firme	Wood
C. *villosum* (Aubl.) Pers.	Guianas, Amazonian Brazil	Forest on terra firme	Fruit: edible, oil
			Wood: construction, boats

ing tissue, the apical portion of filaments tuberculate and entire length of smaller filaments sometimes tuberculate; anthers bilocular, oblong, introrse, basifixed or dorsifixed, longitudinally dehiscent. Ovary 4(–6)-locular, with a single ovule in each loculus. Styles four, long, filamentous. *Fruit* a drupe, 4–6-locular, with from 1–4 loculi developing into seeds, each loculus forming a 1-seeded coccus; mesocarp thick and fatty or fleshy; endocarp woody, muricate, tuberculate or spinous on exterior; seeds reniform or subreniform, the embryo with a straight to arcuate radicle. Germination cryptocotylar, first leaves opposite. *Chromosome* number 2n = 46 (fide Ehrendorfer et al., 1984).

Key to the Species of *Caryocar*

1. Leaflets obtuse to rounded at apex.
    2. Leaflets usually densely tomentose, rarely only sparsely hirsutulous beneath, usually tomentellous above, the terminal lamina 10–18 cm long, the venation prominent beneath; Planalto of Brazil and Paraguay. . . . . .1. *C. brasiliense.*
    2. Leaflets entirely glabrous, the terminal lamina 5–10 cm long, the venation prominulous beneath; northeastern Brazil. . . . . .2. *C. coriaceum.*
1. Leaflets distinctly acuminate at apex.
    3. Stipels absent or small and early caducous.
        4. Leaf underside with conspicuously reticulate and prominent venation, usually villous, rarely glabrous; pedicels bracteolate; Guianas and eastern Amazonia. . . . . 3. *C. villosum.*
        4. Leaf underside with plane or prominulous venation, glabrous or with a barbate junction of midrib and veins only; pedicels usually ebracteolate except *C. cuneatum.*
            5. Calyx ca. 2 cm long; corolla lobes 6–7 cm long; filaments 7–8.5 cm long; stamens over 700; Guianas and Venezuela. . . . . 4. *C. nuciferum.*
            5. Calyx 4–12 mm long; corolla lobes 1–3 cm long; filaments 4–6.5 cm long; stamens not exceeding 520.
                6. Pedicels bracteolate; leaf apices mucronate; northern Planalto of Brazil. . . . .
                . . . . 5. *C. cuneatum.*
                6. Pedicels ebracteolate; leaf apices usually acuminate.
                    7. Inflorescence rachis elongate, 6–14 cm long; calyx 4–6 mm long; western Amazonia. . . . . 6. *C. gracile.*
                    7. Inflorescence rachis 1.5–6 cm long; calyx 8–12 mm long.
                        8. Apex of peticels and exterior of calyx tomentellous; pedicels 3–5 cm long, not crustaceous; leaf margins irregularly dentate, the teeth mostly spaced far apart; Peru. . . . . 7. *C. amygdaliforme.*
                        8. Apex of pedicels and exterior of calyx glabrescent; pedicels 1–2.6 cm long, usually crustaceous, sometimes not so; leaf margins entire or weakly crenulate.
                            9. Leaf margins entire or only slightly undulate.
                                10. Peduncles and pedicels crustaceous lenticellate; center petiolule 5–10 mm long; filaments usually deep red, rarely white; Venezuela, Guianas, Amazonia. . . . . 8. *C. glabrum.*
                                10. Peduncles and pedicels without lenticels or sparingly lenticellate but never crustaceous; center petiolule 8–17 mm long; filaments always white; Venezuela and Brazil, Terr. Roraima. . . . . . . 9. *C. montanum.*
                            9. Leaf margins deeply crenulate; Amazonia, Venezuela. . . . 10. *C. pallidum.*
    3. Stipels present and persistent.
        11. Leaf underside with hirsute mass at junction of primary veins and midrib.
            12. Inflorescence rachis 5–9 cm long.
                13. Terminal leaflet 12–16 cm long; peduncles 12–14 cm long. . . . . 8. *C. glabrum.*
                13. Terminal leaflet 7–11.5 cm long; peduncles 6.5–10 cm long; eastern-central Brazil.
                . . . . 11. *C. edule.*
            12. Inflorescence rachis 2.5–3.5 cm long.
                14. Peduncles 4–7 cm long, conspicuously lenticellate; terminal petiolule 6–13 mm long, deeply canaliculate; leaf margins crenate; eastern Brazil. . . . . 11. *C. edule.*
                14. Peduncles 9–15 cm long, not conspicuously lenticellate; terminal petiolule 2–3 mm long, terete or shallowly canaliculate; leaf margins serrate; Costa Rica. . . . .
                . . . . 12. *C. costaricense.*
        11. Leaf underside entirely glabrous or with few hairs on midrib and primary veins but no hirsute mass at their junction.

15. Leaf margins deeply and conspicuously serrate.
    16. Calyx lobes tomentellous on exterior; petioles usually puberulous; Bolivia, Brazil (Terr. Rondônia). .... 13. *C. dentatum.*
    16. Calyx lobes glabrous; petioles glabrescent.
        17. Leaves narrowly oblong; stipels small and inflated, to 3 mm long; Colombia. . ... 14. *C. amygdaliferum.*
        17. Leaves elliptic; stipels large and recurved, 5–10 mm long; Amazonia and Venezuela. .... 10. *C. pallidum.*
15. Leaf margins entire to slightly crenulate-serrate.
    18. Leaves coriaceous; peduncle and pedicels often crustaceous lenticellate; stamens 250–350.
        19. Peduncles round, lenticellate; stamens red to pink; leaves not articulated; stipules convex, 6–10 mm long. .... 8. *C. glabrum.*
        19. Peduncles flattened quadrangular, not lenticellate; stamens yellow; leaves articulated at junction of petiole and leaflets; stipules concave, 15–25 mm long. ... . 16. *C. harlingii.*
    18. Leaves membraneous-chartaceous; peduncle rachis and pedicels sparsely lenticellate, but never crustaceous; stamens ca. 60; Guianas, Amazonia, Venezuela. .... 15. *C. microcarpum.*

The Uses of *Caryocar*

ETHNOBOTANY

As with many of the plants which they use, the Indians of Brazil have various legends attached to the piqui or *Caryocar brasiliense* tree. It is an important part of the life of the Mehinaku Indians whose legend tells of the following origin for the piqui tree:

> Two beautiful women were bathing in the river when a large alligator came by. A male spirit stepped out of the alligator skin and had sexual relations with the women. This continued for many months until the men of the village found out about the alligator and killed it. They planted the sexual organs of the alligator and up sprang a tree which they called piqui because of the spinous covering of the seeds.

The Mehinaku have an elaborate piqui festival which takes place over a one month period and is essential to please the spirits of the piqui to ensure a good harvest. The ceremony is divided into four stages over about a one month period.

Matapu

The spirit is represented by a carved board which makes a roaring noise when whirled in the air. It must be fed once it has been painted. The matapu is taboo for women to touch, but they may see it.

Upe

Upe is the anteater and, the image is brought into the village to be fed. It is a long series of branches with an anteater-like front. It is borne by many men and taken from house to house to taunt the women who beat its nose with sticks. It is finally beaten to 'death' in the main square of the village.

Alucaca and Yuyoto

The men blacken their bodies with a greasy mixture of charcoal and piqui oil. This makes their bodies supple and slippery. A long line (the alucaca) is formed, led by the chief with the other elders in front and, in order of rank, to the youngest men at the rear. The women break up the alucaca segment by segment and attack the men. Those who end up with the most women are declared the winners. The next day men get their revenge by forming a Yuyoto or bee which is a line that swarms through houses and drives all the women out into the central plaza attacking them by smearing their bodies with the black grease.

Mupulawaja

This final ritual lasts for four days. Two men from each piqui orchard hold piqui leaves up in the air. This combination represents the spirit of the piqui and must be caught to ensure a good crop. The women feed them with a gruel made from manioc and the outer mesocarp of piqui. They then carve wooden birds to embody the spirits. The birds on poles are planted in the main plaza and become spirits after they are painted red with urucu (*Bixa*) paint. These humming bird-like images are then carried back to the piqui groves with much dancing and ritual.

This is one of several rituals involving the piqui that is practised by the Indians of Central Brazil who have used the fruit of this tree as a food and a source of oil and dye for many centuries. Since piqui is such an abundant tree in the Cerrado the Indians used wild trees, rather than cultivating them in their fields. The point of including details about a piqui ritual is to show the importance of the tree to these people. The piqui and other economic plants used by the Indians command a certain sanctity and respect which in turn will lead to protection and wise management of the resource rather than abuse and overexploitation.

There are many other piqui legends and rituals in Central Brazil and as yet they remain poorly studied. Further ethnobotanical work on piqui would certainly make a fruitful and interesting study.

CARYOCAR AS A SOURCE OF OIL

The most significant use of *Caryocar* has been as a source of oil from the fruit, and it is this aspect that probably has the most future potential. Lane (1957) presented a fascinating account of the efforts of Sir Henry Wickham to develop the *Caryocar* oil industry. After Wickham had successfully transferred the *Hevea* rubber from Brazil to England in 1876 so that it could be sent on to tropical Asia, he devoted his efforts in the late nineteenth century to the piquiá tree (*Caryocar villosum*) as a source of oil. As a result of Wickham's work, piquiá was introduced to a plantation in Malaya, but it never caught on in the way that rubber did, in spite of various shortages of fats and oils on the international market.

Both the pulpy mesocarp and the kernel of most species of *Caryocar* contain oil. The species that have been most used are *C. brasiliense, C. cuneatum* and *C. villosum* which are a group of closely related species (Fig. 1). Considerable morphological details of the fruit of *C. brasiliense* were given by Barradas (1973).

There have been many papers on the properties and fatty acid content of *Caryocar* oils. For example, reports on the oil of *C. villosum* were given by Georgi (1929) and by Hilditch and Rigg (1935), on *C. brasiliense* by Ferreira and Motidome (1962) and Handro and Barradas (1971),

and on *C. coriaceum* by Alencar et al. (1983), Lima et al. (1981) and Sales (1973). Table II summarizes the fatty acid content of the mesocarp and kernel oils of the species analyzed. This shows that *Caryocar* oils are rich in palmitic and oleic acids, and that there is little difference between the mesocarp and kernel oils. Since the differences are just in a few minor components the two oils can be mixed, which facilitates their use. There are also no significant differences between the oils of *C. brasiliense* and *C. villosum,* the two principal oil-yielding species. The high content of the unsaturated oleic acid adds to the potential use of *Caryocar* oil as a comestible product.

Caryocar brasiliense is a common small tree or shrub of the cerrado of Central Brazil, *C. villosum* is a large tree of the Guianas and Amazonian terra firme forests, and *C. cuneatum* is a medium sized tree of dry cerrado and caatinga forest of northeastern Brazil. Thus, the three principal oilseed species vary in ecology, habit, and climate conditions, enhancing the possibilities of the genus *Caryocar.*

USES OF MESOCARP

In addition to its use for oil, the mesocarp of various species of *Caryocar* is eaten. This pulpy layer is colored yellow because of the high carotene content, and has an agreeable odor, which makes it popular in Northeast Brazil in rice dishes. The rice becomes a rich yellow color when piqui is added. The pulp is also used to make liqueurs and a yellow juice. Carvalho and Burger (1960) analyzed the pulp and found it to be rich not only in carotene, but also Vitamin A, riboflavin, niacin, iron and phosphorus. It is therefore highly nutritious and is an important minor addition to the diet of many people in northeastern Brazil.

The liqueur is a major product of the states of Mato Grosso and Minas Gerais, Brazil, and is called *licor de piqui.* It is a very sweet syrupy liquid with a twig in the bottle upon which the super-saturated sugar solution has crystallized. It has the distinct taste of mesocarp of *Caryocar,* and is one of the most popular liqueurs in Brazil.

CARYOCAR SEEDS

A glance at the specific names in the genus *Caryocar* beginning with the type species, *C. nu-*

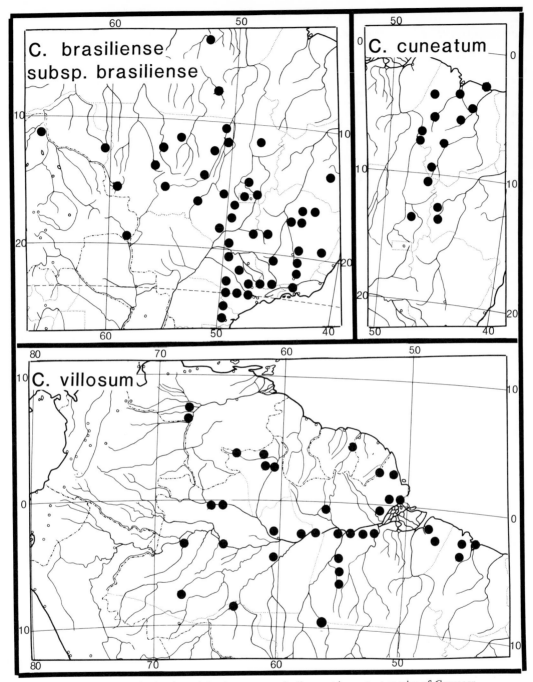

FIG. 1.   The distribution of the three economically most important species of *Caryocar*.

ciferum, is enough to make it obvious that the seeds of *Caryocar* are used extensively as a "nut." These names include *C. amygdaliferum, C. amygdaliforme* and *C. edule*, all referring to the edible seed. The best known product is the souari, sowari, or butternut, a seed that used to be exported commercially to Europe from the Guianas and is used widely by the indigenous pop-

Table II

Fatty acid content of *Caryocar* species, adapted mainly from Handro and Barradas (1971), and Hilditch (1956)

Fatty acid	C. brasiliense mesocarp	C. brasiliense kernel	C. villosum mesocarp	C. villosum kernel	C. nuciferum kernel	C. edule mesocarp
Caproic	0.8	—	—	—		
Caprylic	0.1	—	—	—		
Capric	0.1	—	—	—		
Lauric	0.1	—	—	—		
Myristic	0.3	0.4	1.5	1.4–1.6		
Palmitic	39.0	32–33	41–45	48–50.6		
Palmitoleic	1.6–1.7	0.4–0.7	—	—		
?	0.2	0.2–0.3	—	—		
?	—	0.2	—	—		
Stearic	0.7–1.2	2.6–3.0	0.9	1.0–1.8		
Oleic	51.7–54	44–47	46–54	43.7–49.6		
Linoleic	2.0–3.5	15–19	2.6–3.3	2.0–3.3		
Linolenic	1.0–1.2	0.3–0.6	—	2.6–3.2		
Arachidic	—	0.5–0.8	—	—		
Saponification value	195–204	192–202	199–205	198–203	197.6	197
Oil yield, % fresh wt.	5–6%	36.5%	47%	45%	70%	63%
Hanus iodine value	56–58	69–71	46–48	48–52	41.9	77

ulations as a source of food. These seeds are usually called nuts in the literature, and their use is the same as for most true nuts, and they are eaten raw. However, they are botanically seeds rather than nuts. Souari has also been grown in plantations in the West Indies. The *Caryocar* seeds were one of the tropical fruits described and extolled by Fouqué (1973). *Caryocar nuciferum* is a fast growing and easily cultivated tree, and it can also be used as a source of *Caryocar* oil. It is widely used in local soap manufacture in the Guianas. The species occurs naturally in the Guianas and in Chocó, Colombia. The disjunct distribution is either a natural one or one caused by transport of a useful plant by pre-Colombian populations of Indians.

THE WOOD OF *CARYOCAR*

Most species of *Caryocar* yield a usable wood, and several have entered into the commercial timber industry throughout the range of the genus. It is a hard and durable wood that is employed mainly for uses where resistance is an important property. For example, *Caryocar* wood is much used in fence posts, pilings and in boat building rather than as a finished lumber. It does, however, finish into an attractive dark-grained wood. Notes on the uses of a few species are given below. The wood of *C. coccineum* was found to

be the most resistant to attack on exposure to five species of wood rotting fungi (González & Icochea, 1981).

An anatomical study of the wood of five species of *Caryocar* was made by Mello (1970) and of twelve species of Caryocaraceae by Araujo and Mattos Filho (1973). Both are detailed descriptive works at the level of light microscopy.

Caryocar costaricense is confined to Costa Rica and is becoming rare through overexploitation. It is known locally as *cagui* or as *ajo* or *ajillo* (=garlic) because of the garlic-like scent of the flowers. The wood is valued locally because it is hard and durable. It is used mainly for fence posts and pilings because of its resistance to termite damage. In 1980–1983 only 500 cubic feet of *ajo* were available to the lumber yards of San José because the tree has now become rare.

This is certainly a species that is both in need of conservation and with potential in reforestation projects. The wood of this species was described by Barghoorn and Renteira (1967).

Caryocar brasiliense, the true *piqui,* is the smallest tree in the genus and therefore does not produce commercial lumber. However, the wood is much used locally in Central Brazil because of its durability and mechanical strength. It is used mainly for fence posts and in primitive machinery where wood parts are required (for example, cart wheels and tool handles). It is also the preferred wood for the construction of corrals for cattle because it is one of the few woods that is resistant to the rotting effect of the urine-polluted environment. It is also used as a flooring material and has been used to extract a vivid yellow dye. The overuse of the wood of this species is making it become rare in some areas, and is threatening its availability as an oilseed crop. It is certainly time to protect the piqui and collect germ plasm of it from the areas of the cerrado from which it is being eliminated through overuse and agricultural development projects. Heringer (1959) called the wood of piqui "first place among the better Brazilian woods" because of its durability and mechanical properties. The main hindrance to more extensive use of the wood of piqui is its tortuous growth form caused by the effects of cerrado fires during its growth.

Caryocar villosum or *piquiá* is one of the largest species of the genus and often reaches thirty meters in height and a trunk diameter of one meter. It is a quick-growing, light-demanding species (Vastano & Barbosa, 1983) that is widespread in the Amazonian terra firme forest, and can be harvested for timber after 20–25 years. The wood is quite heavy (specific gravity 0.80–0.91) and is easily worked into an attractive finish. It has been used for railroad ties, the interior framework of ships, civil and naval construction, cabinet work, plywood and posts. The wood of *C. villosum* was recommended by Senft and Lucia (1979).

Caryocar glabrum or *piquiarana* is the most widespread species of *Caryocar* in the Guianas and Amazon terra firme forest. Throughout the region the wood has been used. In the Peruvian Amazon, where it is called *almendro,* it is used in rural construction, for fence posts, in boat building and flooring. It is used mainly for boat building in Brazil, but has also entered the lumber trade from time to time.

Overall, the wood of Caryocaraceae is of good quality because of its hardness, resistance to humid conditions and to attack by wood boring insects. It is of medium texture and easily workable, as was noted by Record and Hess (1943) and Araújo and Mattos Filho (1973). Although it has been used largely for boat building, posts, and handles for tools, it has a much wider potential for other uses. Combined with the other uses such as the edible fruits and the oil, the species of *Caryocar* are prime candidates for use in managed agroforestry systems where multipurpose trees are needed.

FISH POISONS FROM *CARYOCAR*

A review of the use of Caryocaraceae as a source of fish poisons was recently published by Kawanishi et al. (1986). The use of the fruit pulp of three species of *Caryocar, C. glabrum, C. gracile* and *C. microcarpum,* as fish poisons has been reported. The use of *Caryocar* as a fish poison appears to be confined to the northwest of Amazonia in Colombia, Venezuela and Brazil and is practised mainly by the Tukano and the Maku Indians.

The Maku use the fruit of *Caryocar glabrum* (see Prance, 1973) in this way. The mesocarp is ground up into a pulp in a muddy hole in the ground. The soapy mixture of mud and pulp is then thrown into a small stream and is a most effective fish poison. The stunned fish soon begin to float to the surface of the stream and are collected from downstream of the place where the poison was introduced to the water. The use of this fish poison produces abundant foam on the water indicating a high content of saponins. In some places the leaves and fruit of *C. microcarpum* are used as a soap substitute for washing clothes.

ANTI-TUMOR ACTIVITY

Oliveira et al. (1968, 1970) mention that an alcohol extraction from the leaves of *Caryocar brasiliense* was found to be active against Sarcoma 180 by the U.S. National Cancer Institute. Oliveira and his colleagues carried out a chemical analysis and tests on mice with Sarcoma 180 and found that oleanolic acid in *Caryocar* was responsible for this anti-tumor activity.

THE GROWTH OF *CARYOCAR BRASILIENSE*

Since *Caryocar brasiliense* is the most important source of *Caryocar* oil it has been studied more than the other species. Its abundance in the cerrado vegetation makes it a particularly important species for the future development of that region. Some details about the growth of this species were provided to me by the late Dr. Ezechias P. Heringer (pers. comm., 1984) and the data given below are based mainly on his work.

Table III

Flower production in October 1983 from a single tree of *Caryocar brasiliense* (from data collected by E. P. Heringer)

Date	Number of flowers on ground	Weight of young fruits fallen on ground in g	Average no. of stamens/ flower
Oct 5	9	—	505
6	30	—	228
7	39	—	492
9	34	—	573
10	46	—	428
11	142	—	492
12	226	1.3	505
13	267	1.7	495
14	576	2.4	423
15	134	—	351
16	105	—	444
17–19	592	2.7	540
20	615	3.1	439
21–23	209	—	557

Table IV

Germination experiment with seeds of *Caryocar brasiliense* from a total of 5450 scarified seeds and 300 untreated seeds planted on 23 January 1959 by E. P. Heringer

Date	Scarified (n = 5450) Germination	Scarified (n = 5450) Seedling fatalities	Untreated (n = 300) Germination	Untreated (n = 300) Seedling fatalities
28 Mar 59	89	—	1	—
30 Mar 59	210			
6 Apr 59	240			
13 Apr 59	300			
20 Apr 59	350			
27 Apr 59	354			
4 May 59	357			
2 Jun 60	1101	327	137	10

Phenology

Through the entire cerrado region flowering of *Caryocar brasiliense* occurs between September and January. The actual flowering of any local subregion lasts about six weeks and depends on local temperature and light so that there is generally about a one month difference in flowering period between the northern and southern part of the cerrado. Fruits develop quickly after flowering and so there is the same difference in fruit production. The fruits are produced between the months of January and March. A summary of the flower production of a single tree is given in Table III.

Propagation and Growth

The general method of propagation is by seeds, which are planted directly in the place where the tree is to be grown. Germination is much faster when the seeds are scarified (see Table IV). This shows that scarification of the seeds, including removal of the pulpy mesocarp, permits much more rapid germination. The germinated seedlings begin to die once the rainy season ends, but seedling mortality can be avoided by watering three times a week. The use of water is much more critical than soil type. Piqui grows well in the normal cerrado soil without any fertilizer or insecticide. In a nursery of piqui, plants that are irrigated twice a week grow twenty times faster than those that were left to experience the normal cerrado rainfall. Plants that were irrigated grew to 2.5 m in the first year and those without irrigation averaged only 40 cm in height. For piqui to reach maturity in a reasonable time in cultivation it seems that irrigation is necessary during the dry season. With proper irrigation, the trees reach mature size and an economic level of production within ten years.

Piqui can be propagated by cuttings. In mature trees shoots also arise from the roots to form new trunks, a phenomenon that is common in many fire-adapted trees of the cerrado. These basal sprouts can also be used as propagules. The presence of mycorrhizae in the roots of *Caryocar brasiliense* was established by Casagrande (1981), but she did not identify the fungal species involved. I am not aware of any attempts to hybridize this or any other species of *Caryocar*.

Predators

The main predators of piqui are rodents which quickly take fruit that have fallen to the ground and eat the kernels of the fruit. They are presumably natural dispersal agents as well. The other predators on fallen fruit are termites, which are abundant and ubiquitous in the cerrado. Termites eat both the mesocarp and kernels of fruit left by the rodents. Both these predators attack only fruits which have fallen to the ground and so rapid harvest avoids loss. A species of thrips

(*Holopothrips anacardii*) was listed as a pest of *Caryocar villosum* by Adis and Kerr (1979).

Conclusions

The review of the various uses of the genus *Caryocar* demonstrates that it has broader application and further potential as a multipurpose crop than its current use would indicate. The numerous recent papers about the growth, propagation and management of various species of *Caryocar* will facilitate future use (e.g., Barradas, 1972; Labouriau et al., 1963a, 1963b; Vastano & Barbosa, 1983). *Caryocar* is valuable both in diversity of uses and ecology. With species adapted to each of the major lowland neotropical habitats the genus can be used in many different types of soil and climate (e.g., *C. gracile* on white sands, *C. brasiliense* in savanna and cerrado, *C. coriaceum* in arid areas and *C. microcarpum* in floodplains).

Caryocar brasiliense of the cerrado of Central Brazil has the best potential for oil. As a typical cerrado tree it is low in height with twisted tortuous branches and is therefore easy to harvest. It is also one of the commonest trees of the cerrado and is widely distributed throughout the cerrado region. Indians encourage orchards of this species to form, and its abundance and adaptation to the cerrado make it an ideal crop for central Brazil. *Caryocar nuciferum* is another possible source of oil because it also grows well in plantations and has already been shown to be productive. *Caryocar cuneatum* is a good oilseed for the drier regions. *Caryocar villosum* has potential both as an excellent timber and as an oilseed. It is fast growing and produces fruit after a few years of cultivation while growing to the size of commercial timber.

All of the above species are easy to propagate from seeds and at least two (*C. brasiliense* and *C. villosum*) are known to reproduce from cuttings (Heringer, pers. comm., 1984; Lane, 1957). There are no serious problems apparent with pollination.

The emphasis on these four species does not exclude the use of any of the other 12 species of the genus, all of which have oily cotyledons and hard durable wood. Another possible use for some species of *Caryocar* is as a street tree, although they have not yet been so used. The production of quantities of large flowers and heavy fruit that fall to the ground might be a disadvantage to such use, however. Perhaps *Caryocar gracile,* with its small fruits and beautiful pink flowers, would make the most attractive ornamental.

The entire genus is primarily pollinated by bats, and at least *Caryocar brasiliense* is self compatible (Barradas, 1972).

Caryocar is one of the many examples of economically important tropical plants that is bat pollinated. It provides a strong argument for the importance of bat conservation since the fruit crop will not thrive without the bats.

Acknowledgments

I thank the various people who supplied information about the uses of *Caryocar,* especially Filomeno Encarnación, Luis D. Gómez, the late Ezechias P. Heringer, Walter B. Mors and Marlene F. da Silva. Fieldwork on the Caryocaraceae was funded by support from the National Science Foundation. I am grateful to Michael J. Balick, H. David Hammond, David Johnson and Scott A. Mori for critical readings of an earlier draft of this manuscript. This is publication number 77 in the series of the Institute of Economic Botany of The New York Botanical Garden.

Literature Cited

Adis, J. & W. E. Kerr. 1979. Um trips como praga do piquiá. Acta Amazonica **9:** 790.

Alencar, J. W., P. B. Alves & A. A. Craveiro. 1983. Pyrolysis of tropical vegetable oils. J. Agric. Food Chem. **31:** 1268–1270.

Araújo, P. A. de M. & A. de Mattos Filho. 1973. Estrutura das madeiras de Caryocaraceae. Arch. Jard. Bot. Rio de Janeiro **19:** 5–47.

Barghoorn, A. W. & M. Renteira R. 1967. Estudio anatomico y fisico-mecanico del cagui (*Caryocar costaricense* Donn. Sm.). Bol. Inst. Forest. Latinoamer. Invest. **24:** 35–57.

Barradas, M. M. 1972. Informações sôbre floração frutificação e dispersão do piqui *Caryocar brasiliense* Camb. (Caryocaraceae). Ciência e Cultura **24:** 1063–1068.

———. 1973. Morfologia do fruto e da semente de *Caryocar brasiliense* (piqui), em várias fases de desenvolvimento. Revista de Biol. (Lisboa) **9:** 69–84.

Carvalho, M. C. & O. N. Burger. 1960. Contribuição ao estudo piqui de Brasília. Coleção Estudo e Pesquisa Alimentar, Serviço de Assistência e Previdência Social (SAPS) **50:** 7–15.

Casagrande, L. I. T. 1981. Estudo sobre a associação

micorrhizal em *Caryocar brasiliensis* Camb. Revista Agric. **56**: 5–8.

Ehrendorfer, F., W. Morawetz & J. Dawe. 1984. The neotropical angiosperm families Brunelliaceae and Caryocaraceae: First karyosystematical data and affinities. Plant Syst. & Evol. **145**: 183–191.

Ferreira, P. C. & M. Motidome. 1962. Estudo químico do óleo de piquí. An Fac. Farm. Odont. Univ. S. Paulo **19(1)**: 25–30.

Fouqué, A. 1973. Espèces frutières d'Amerique tropicale: Bombacacées, Caryocaracées et Sterculiacées. Fruits **28(4)**: 290–299.

Georgi, C. D. V. 1929. Piqui-a fruit oils. Malayan Agric. J. **17**: 166–170.

González, V. R. & T. Ames de Icochea. 1981. Pudrición de la madera de diez especies forestales por acción de cinco hongos xilofagos. Revista For. Perú **10**: 102–137.

Handro, W. & M. M. Barradas. 1971. Sôbre os óleos do fruto e da semente do pique—*Caryocar brasiliense* Camb. (Caryocaraceae). Pages 110–113 *in* III Simpósio sôbre o cerrado. Ed. E. Blücher Ltd., Univ. São Paulo.

Heringer, E. P. 1959. Trés árvores úteis do cerrado mineiro. Bol. Agric. Minas Gerais **7**: 59–62.

Hilditch, T. P. 1956. The chemical constitution of natural fats, ed. 3. Wiley, New York. Pages 160, 358.

——— **& J. G. Rigg.** 1935. The component glycerides of piqui-a fats. J. Soc. Chem. Industu. **54**: 109.

Kawanishi, K., R. F. Raffauf & R. E. Schultes. 1986. The Caryocaraceae as a source of fish poisons in the northwest Amazon. Bot. Mus. Leafl. **30**: 247–253.

Labouriau, L. G., I. F. Marques Válio & E. P. Heringer. 1963a. Sôbre o sistema reproductivo de plantas de cerrados—1. Anais. Acad. Brasil. Cienc. **36**: 449–464.

———, ———, **M. L. Salgado Labouriau & W. Handro.** 1963b. Nota sôbre a germinação de sementes de plantas de cerrados em condições naturais. Revista Brasil. Biol. **23**: 227–237.

Lane, E. V. 1957. Piqui-á—Potential source of vegetable oil for an oil-starving world. Econ. Bot. **11**: 187–207.

Lima, M. T., G. A. Maia, Z. B. L. Guedes & H. F. Oria. 1981. Composição de acidos graxos da fração lipidica do piqui (*Caryocar coriaceum* Wittm.). Ciencia Agron. **12**: 93–96.

Mello, E. C. 1970. Estudo anatômico das madeiras do gênero *Caryocar* Linn. Brasil Flor. **1(2)**: 54–62.

National Academy of Sciences. 1975. Underexploited tropical plants with promising economic value. Washington, D.C.

Oliveira, M. M. de, B. Gilbert & W. B. Mors. 1968. Triterpenes in *Caryocar brasiliense*. Anais. Acad. Brasil. Cienc. **40**: 451–452.

———, **R. P. Sampaio, W. Giorgi, B. Gilbert & W. Mors.** 1970. *Caryocar brasiliense*—Isolamento e identificação de algumas substâncias: Atividade biologica sôbre o Sarcoma 180. Arquiv. Inst. Biol. **37(Supl. 1)**: 25–27.

Prance, G. T. 1973. Ethnobotanical notes from Brazil. Econ. Bot. **26**: 221–237.

———. 1987. An update on the taxonomy and distribution of the Caryocaraceae. Opera Bot. **92**: 179–183.

——— **& M. F. da Silva.** 1973. Caryocaraceae. Fl. Neotrop. Monogr. **12**: 1–75. Hafner, New York.

Record, S. J. & R. W. Hess. 1943. Timbers of the New World. Yale Univ. Press, New Haven. Pages 118–119.

Sales, F. J. M. 1973. O oleo no fruto de piquizeiro, *Caryocar coriaceum* Wittm. Turrialba **23**: 108–109.

Schultes, R. E. 1980. The Amazonia as a source of new economic plants. Econ. Bot. **33**: 259–266.

Senft, J. F. & R. M. D. Lucia. 1979. Increased utilization of (Brazilian) tropical hardwoods through species—Independent structural grading (*Virola* spp., *Tachigalia paniculatum, Caryocar villosum*). For. Prod. J. **29**: 22–28.

Vastano, J. W. & A. P. Barbosa. 1983. Propagação vegetativa do piquiá (*Caryocar villosum* Pers.) por estaquia. Acta Amazonica **13**: 143–148.

Three Sub-tropical Secondary Forests in the U.S. Virgin Islands: A Comparative Quantitative Ecological Inventory

Anne E. Reilly, John E. Earhart and Ghillean T. Prance

Table of Contents

Abstract

REILLY, A. E. (Institute of Economic Botany, The New York Botanical Garden, Bronx, New York 10458-5126, U.S.A.), J. E. EARHART (World Wildlife Fund and the Conservation Foundation, 1250 24th Street, N.W., Washington, D.C., U.S.A.) and G. T. PRANCE (The Royal Botanic Gardens, Kew, Richmond, Surrey TW9 3AB, United Kingdom). Three sub-tropical secondary forests in the U.S. Virgin Islands: A comparative quantitative ecological inventory. Advances in Economic Botany 8: 189–198. 1990. Permanent plots totalling 2.0 hectares were established in 1986 in secondary forest on the island of St. John, U.S. Virgin Islands. These three plots are representative of ecological zones present on the island and include upland moist forest, gallery moist forest and dry evergreen woodland. All stems measuring ≥5 cm diameter breast height were tagged, mapped and measured. Species and family importance values have been calculated for each site. One exotic species, *Melicoccus bijugatus* Jacq., makes up a significant percentage of the importance value (15.5%) at the most disturbed site while in the older stands native species dominate.

Key words: Caribbean; Virgin Islands; forest inventory; land use; secondary vegetation.

Introduction

There have been a number of studies focusing on the vegetation of St. John (Robertson, 1957; Little and Wadsworth, 1964; Ewel and Whit- more, 1973; Little et al., 1974; Forman and Hahn, 1980; Teytaud, 1983; Woodbury and Weaver, 1987). However, none of these have looked at the long term changes in the forest or the effects of historical land use practices on stand development. The St. John Forest Dynamics Project is examining the mechanisms of forest recovery on a small, sub-tropical island in the U.S. Virgin Islands. The primary objective of this research is to monitor regeneration of the forest in permanent study plots over a thirty year period. Since many of the species found on St. John are distributed throughout the Greater and Lesser Antilles, the information gathered on vegetation dynamics and stand development can be used in forest management decisions throughout the Caribbean. This paper reports on the first phase of the project; identification of the floristic composition of the forests and determination of species and family importance values.

The history of resource use on St. John is typical of many Caribbean islands. Originally the island was almost completely forested, but approximately 90% of those forests were removed with the advent of agricultural production, during the plantation era in the 1700's and 1800's. As slavery was abolished, resource utilization shifted from large plantations to small scale intensive farming and cattle ranching. This second agricultural phase lasted approximately fifty years and eventually deteriorated due to problems associated with declining soil fertility, increasing erosion and a reduced labor force. As the agricultural phase came to a close the natural vegetation began to recover. The resultant secondary forests are representative of an ecosystem in which human intervention has been significant.

St. John is unique among Caribbean islands in that nearly 70% of its land surface is now protected as a National Park and has been designated a Man and the Biosphere Reserve. This provides a rare ecological research opportunity because forests on most Caribbean islands continue to be exploited, whereas the forests in the park on St. John are returning to a semi-natural state.

Study Sites

The island of St. John, located in the Caribbean Sea, is situated in an archipelago between 18°18′ and 18°22′N latitude and 64°40′ and 64°48′W longitude (Fig. 1). Occupying 5180 ha, St. John is 11 km long and five km across at its widest point. It is underlain by mildly deformed Cretaceous volcanic and limestone sediments (Rivera et al., 1966) and is at the eastern end of the Greater Antilles chain. The island lies in an east–west direction 4.8 km east of St. Thomas and 88 km east of Puerto Rico.

The topography of St. John is mountainous with slopes greater than 30% over much of the island. The highest point on the island, Bordeaux Mountain, reaches an elevation of 387 m above sea level. There are no permanent streams or rivers on the island.

The dominant soil series on the island is the Cramer clay loam. These soils are shallow, well-drained and found in areas which are moderately to steeply sloped. They are subject to excessive run-off and have a low water holding capacity. Depth to bedrock ranges from 25 to 50 cm (Rivera et al., 1966, 1970); although on-site observations revealed variation in depth as well as the presence of large stones and boulders.

The precipitation on St. John is heaviest from May through November with a range of 890 to 1400 mm/yr. Seasonal hurricanes, easterly waves and polar fronts contribute significantly to annual rainfall quotas but are not continuous sources throughout the year. The most common source of rainfall is orographic.

The climate of the area is relatively dry due to long periods of daylight, high temperatures and continuous trade winds which cause high evapotranspiration rates. Water is limited in supply because of high evaporation rates and rapid run-off from steep slopes. The average temperature is 27°C with 75 percent relative humidity.

Three study sites have been established within the boundaries of the park (Fig. 2). These sites are representative of ecological zones present on the island and include upland moist forest (Bordeaux), gallery moist forest (L'Esperance) and dry evergreen woodland (Hawksnest); which are 1.0, 0.5, and 0.5 ha in size, respectively. Each plot is part of a separate watershed system and contains at least one intermittent stream.

All three study plots are on soils that are volcanic in origin with pH values in the range of 6.1–7.3 (Rivera et al., 1970). The soil types in the plots on Bordeaux Moutain and L'Esperance are of the Cramer (CrF) series. The Cramer gravelly clay loam occurs in very steep mountainous sites with slopes of 40–60 percent. The Hawksnest site is underlain by a gravelly clay loam as well but the topography is somewhat less sloped and is considered to be of the Cramer (CrE) series.

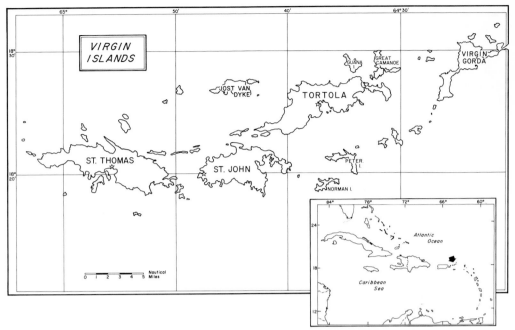

FIG. 1.   Map of St. John, U.S. Virgin Islands.

The Bordeaux Mountain plot is located in the central part of the island approximately 8.0 km east of Cruz Bay Harbor with an elevation of 280 m. It is located on very steep slopes across an intermittent stream allowing for both north and south facing aspects. The vegetation present at this site has been classified as upland moist forest (Woodbury & Weaver, 1987). The canopy is con-

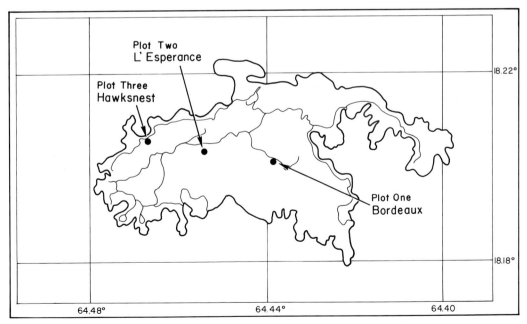

FIG. 2.   Locations of three permanent study sites on St. John.

tinuous at 15–20 m with occasional emergents ranging in height to 25 m. A suppressed stratum is attained at a height of 5–10 m.

The area was abandoned approximately one hundred years ago and has remained relatively undisturbed since the decline of the sugar plantation economy (Tyson, 1986). The presence of small bench terraces, exotic fruit trees and farm tool artifacts provide evidence of past farming activity.

L'Esperance, the second study site, has an elevation of 195 m and is oriented in a north–south direction. It is approximately 5.0 km east of Cruz Bay Harbor. The secondary forest vegetation present on this site has been classified as gallery moist forest (Woodbury & Weaver, 1987). There are three distinct strata present but tree heights are smaller than on Bordeaux. The physical presence of terracing as well as historical information indicates that this area was also the site of a sugar plantation. We estimate that the stand has remained relatively undisturbed for sixty-five years.

The Hawksnest plot is located on the north side of the island and is approximately 5.0 km east of Cruz Bay Harbor. The plot is laid out in an east–west orientation across Hawksnest stream. The east side of the plot is moderately sloped and the west side is located on very steep terrain. The soil is shallow with the presence of numerous rock outcrops. The elevation of the area is 12 m above sea level.

The secondary forest vegetation present on this site has been classified as dry evergreen woodland (Woodbury & Weaver, 1987). The area has two strata; one that is five to ten meters in height and another that is ten to twelve meters in height. There are also occasional emergents in the fifteen to twenty meter range. The trees present in this plot are generally shorter than in the Bordeaux and L'Esperance plots.

This area is the most recently abandoned of the three plots. It was subject to a sugar plantation economy for nearly two hundred years (Tyson, 1986). Erosion occurring on this site has been intense due to the steep slopes and the continuous use of the land. Consequently, there is very little soil remaining on the site.

Methods

There has been little consistency in the layout of quantitative forest inventory studies (Camp-

bell et al., 1986). However, with the exception of Gentry (1982), who uses 1000 m² plots, most of the available floristic data for tropical forests have been taken on single plots 0.5 to 1.0 ha in size (Lang, 1969). During the initial reconnaissance of the area we completed a study to determine the appropriate size of the research plots. The calculation of a species area curve indicated that a 1.0 ha size plot would be appropriate for both Bordeaux Mountain and L'Esperance and a 0.5 ha size plot for the Hawksnest study area. However, time constraints led to the establishment of a 0.5 ha plot for L'Esperance with the intention of expansion in the near future.

In these study plots a thorough botanical inventory was conducted of all stems ≥5 cm at diameter breast height (dbh). These individuals were identified to the species level using the nomenclature of Little and Wadsworth (1964), Cronquist (1968), Little et al. (1974), and Acevedo-Rodriguez (1985). The small trees were considered because of the potential to detect patterns in vegetation development. Dbh was measured at 1.3 m and marked with spray paint to allow for the accuracy of future measurements. Each stem was tagged with a numbered aluminum marker and recorded on a species list to ensure prompt identification at a later date. The nail was placed approximately 20 cm above dbh. If the tree branched below the level of the dbh measurement then two or more trees were recorded (i.e., multiple stem). However, if the tree branched above dbh then only one tree was recorded (i.e., single stem). In addition, each marked tree was measured to the nearest meter and given x and y coordinates that corresponds to a precise location relative to each subplot and the entire plot.

Information was also collected on tree heights. We determined the number of species found within each site and then sampled a variety of size classes based on the dbh measurements. At each site if a species population consisted of twenty trees or less the height of each stem was measured. However, if a species had a population of twenty or more trees then the individuals measured were representative of all the size classes present on the site. Heights were determined using one of two methods, depending on site conditions, either a range finder or a clinometer with trigonometric triangulation.

Species (SIV) and family (FIV) importance values were calculated following the standard

Table I

Characteristics of three permanent plots[a] in secondary forest, St. John, United States Virgin Islands. Data refer to stems ≥ 5 cm dbh

	Plot B	Plot L	Plot H	Total
Area (ha)	1.0	0.5	0.5	2.0
Number of stems	2350	1198	1326	4874
Number of species	62	39	51	87
Number of genera	49	33	37	63
Number of families	33	26	26	39
Basal area (m² ha⁻¹)	30.9072	15.6105	13.2645	59.7822
Age of stand (yr)	100	80	50	—
Type of land-use	sugar	sugar	sugar	—
Length of impact (yr)	100	150	200	—

[a] B = Bordeaux; L = L'Esperance; H = Hawksnest.

practices as discussed in Cain et al. (1956), Mori et al. (1983) and Boom (1986). By summing the relative frequency, relative density, and relative dominance, one obtains the importance value (IV) for each species. Relative frequency is the number of times a particular species occurs divided by the total number of occurrences of all species, relative density is the number of trees of a species divided by the total number of trees recorded, and relative dominance is the basal area of a species divided by the total basal area for all trees recorded. The FIV is simply a summation of three percentages (relative diversity, relative density, and relative dominance) multiplied by 100.

Voucher specimens have been collected for most of the tagged stems present in the plots and deposited at the herbaria of The New York Botanical Garden (NY), the University of the Virgin Islands (UVI) and the Virgin Islands National Park Service (VIIS).

Results

A total of 4874 live trees and lianas ≥ 5 cm dbh were measured in 1986. These stems represent 87 species, 63 genera and 39 families. Basal area, number of stems, species richness and length of disturbance differed among the three plots (Table I). Plot B, the oldest forest, had the highest density (2350), and the highest basal area (30.9072 m² ha⁻¹). Plot L, intermediate in age, had the lowest density (1198), and was intermediate in basal area (15.6105 m² ha⁻¹). Plot H, the most disturbed site, was intermediate in density (1326), and lowest in basal area (13.2645 m² ha⁻¹).

Tables II, III, and IV provide a list of the ten most important species in each plot as ranked according to the importance value calculation. At Bordeaux, these ten species correspond to 71.6 percent of the stand. The ten most important species at L'Esperance contribute 76.9 percent to the total stand IV, while at Hawksnest these ten species make up 67.7 percent of the total.[1]

Discussion

As expected due to the insular nature of the forest and the depauperate nature of the vegetation, the species diversity is much lower than other neotropical forests on the South American mainland (Boom, 1986; Campbell et al., 1986; Gentry, 1982; Hubbell, 1979; Prance et al., 1976).

At the Bordeaux Mountain study plot the ten most important species ranked according to the IV are: *Guapira fragrans* (DC.) Little, *Pimenta racemosa* (Mill.) Moore, *Inga fagifolia* (L.) Willd., *Byrsonima coriacea* (Sw.) DC., *Acacia muricata* (L.) Willd., *Ocotea coriacea* (Sw.) Britt., *Tabebuia heterophylla* (DC.) Britt., *Faramea occidentalis* (L.) A. Rich., *Linociera caribaea* (Jacq.) Knobl., and *Guazuma ulmifolia* Lam. These species have a combined total of 71.6% of the importance value (IV) for the stand.

The stems at L'Esperance were also ranked according to their importance value. The ten most important species share 76.9% of the total IV. These species are: *Ardisia obovata* Desv., *Guapira fragrans*, *Andira inermis* (Wright) HBK, *Inga fagifolia*, *Ocotea coriacea*, *Chrysophyllum pauciflorum* Lam., *Guettarda scabra* (L.) Lam.,

---

[1] The raw data can be obtained from A. Reilly upon request.

Table II

Representation of the 10 most important species at Bordeaux ranked according to importance value, in 1.0 ha of upland moist forest

Species	Density of stems	Dominance (m^2 ha^{-1})	Frequency in plots	IV
Guapira fragrans	293	3.7383	81	35.1878
Pimenta racemosa	316	3.1164	77	33.5100
Inga fagifolia	257	3.6157	77	33.2116
Byrsonima coriacea	157	4.6225	48	23.2123
Acacia muricata	247	4.6170	32	19.1496
Ocotea coriacea	108	0.7201	49	18.6419
Tabebuia heterophylla	64	2.6701	38	16.4541
Faramea occidentalis	65	0.1689	37	13.4375
Linociera caribaea	91	0.4309	31	12.0888
Guazuma ulmifolia	30	1.4045	24	9.9403

Guettarda parviflora Vahl, *Tabebuia heterophylla* and *Hymenaea courbaril* L. Four of these species, *Guapira fragrans, Inga fagifolia, Ocotea coriacea,* and *Tabebuia heterophylla* are also found in the top ten importance values at the Bordeaux plot.

Results from the Hawksnest study plot indicate that the ten most important species as ranked by the importance value are: *Melicoccus bijugatus* Jacq., *Guapira fragrans, Ocotea coriacea, Bursera simaruba* (L.) Sarg., *Eugenia monticola* (Sw.) DC., *Krugiodendron ferreum* (Vahl) Urban, *Guettarda parviflora* Vahl, *Bourreria succulenta* Jacq., *Eugenia biflora* (L.) DC., and *Citharexylum fruticosum* L. Together these make up a total of 67.7% of the IV of the stand. The main difference in this plot is the abundant presence of one exotic species, *Melicoccus bijugatus,* which makes up 15.46% of the IV of the stand. A pos-

sible explanation for this dominance is due to the value placed on fruit, leading to, if not a cultivated situation certainly one where local populations would have encouraged its growth.

Previous studies have classified this area as being drier than the other two sites. Based on our inventory results this would appear to be true, as many of the important species in the plot are indicators of dry or seasonally dry forest. Of the ten most important species in the Hawksnest study area, only two, *Guapira fragrans* and *Ocotea coriacea,* are both ranked in the top ten IV in the other two plots.

In each of the three plots several species have been identified as uncommon because they were represented by four stems or less. The forest stand at Bordeaux has 2.4% uncommon species, L'Esperance has 2.4% and Hawksnest has 2.6%. Thus there is a remarkable consistency between the

Table III

Representation of the 10 most important species at L'Esperance, ranked according to importance value, in 0.5 ha of gallery moist forest

Species	Density of stems	Dominance (m^2 ha^{-1})	Frequency in plots	IV
Ardisia obovata	491	2.3119	50	51.9316
Guapira fragrans	66	3.9012	33	32.1667
Andira inermis	72	0.7859	34	26.3482
Inga fagifolia	59	1.3497	32	25.8569
Ocotea coriacea	48	0.4578	29	21.6464
Chrysophyllum pauciflorum	46	0.4444	28	20.8954
Guettarda scabra	51	0.2911	18	14.0406
Guettarda parviflora	42	0.2915	17	13.1244
Tabebuia heterophylla	43	0.5471	16	13.0313
Hymenaea courbaril	19	1.4628	12	11.6521

Table IV

Representation of the 10 most important species at Hawksnest, ranked according to importance value, in 0.5 ha of dry evergreen woodland

Species	Density of stems	Dominance (m² ha⁻¹)	Frequency in plots	IV
Melicoccus bijugatus	483	3.5413	38	46.3743
Guapira fragrans	76	2.1385	30	27.2844
Ocotea coriacea	79	0.3986	34	25.6542
Bursera simaruba	40	1.7616	19	18.0990
Eugenia monticola	40	0.1907	24	17.4847
Krugiodendron ferreum	74	0.3097	22	17.3052
Guettarda parviflora	72	0.3750	19	15.4191
Bourreria succulenta	41	0.4966	16	12.9453
Eugenia biflora	26	0.0940	16	11.5564
Citharexylum fruticosum	21	0.1958	15	11.0200

three plots in the IV for the ten most important stems as well as in the percentage of uncommon species.

Several hypotheses can be suggested to account for the presence and rank of these species. Many of these trees were cultivated as shade plants to protect cash crops such as coffee (Little & Wadsworth, 1964), which may have been an important consideration when the island was dominated by plantation settlements. Additionally, some of these trees are grown as ornamentals and for honey production which would encourage cultivation by plantation owners. Another possible explanation is that the two older forest sites are returning to a mature forest community which is dominated by native species while in the most disturbed site an exotic, pioneering species is playing a major role in forest development.

Family Importance Values were also calculated for each plot in order to rank the most dominant families. The five most important families found in the Bordeaux plot are Mimosaceae, Myrtaceae, Nyctaginaceae, Malpighiaceae, and Lauraceae. These families share 55.40% of the total FIV for the plot. The ten most important families comprise a total of 83.46% of the total.

The L'Esperance plot is dominated by the Myrsinaceae, Nyctaginaceae, Fabaceae, Mimosaceae, and Rubiaceae. These groups have a combined total of 54.29% of the FIV. The ten most important families of this forest comprise 84.30% of the total FIV.

Hawksnest is dominated to a great degree by one family, the Sapindaceae, which makes up 15.43% of the FIV. This is due to the dominance of *Melicoccus bijugatus,* the only member of the

Sapindaceae in this stand. In total, the top five families make up 48.84% of the FIV and in addition to Sapindaceae include Nyctaginaceae, Lauraceae, Boraginaceae, and Fabaceae. The ten most important families make up 77.45% of the FIV.

The diameter distribution of all three study areas approaches the typical reverse J-shape (Fig. 3). It is interesting to note that while it is assumed that the trees in the Bordeaux plot are older, Hawksnest the most disturbed site has the same percentage of large trees (2%). In addition, when compared to L'Esperance there is actually a smaller percentage of large trees (≥ 30 cm). In the 1 ha plot at Bordeaux, 48 stems fall into this size class (2% of total stems) while in the 0.5 ha plots, 49 trees in L'Esperance (4% of total stems) and 27 in Hawksnest (2% of total stems) fall into this large size class.

Bordeaux has a very high number (39%) of recorded trees in the mid-size dbh class (≥ 10 and ≤ 29.9 cm), while Hawksnest has 28% of its trees in this size class and L'Esperance has only 26%. All three plots demonstrate high percentages of small understory trees (≥ 5 and ≤ 9.9 cm), but Bordeaux at 59% has substantially fewer than the other two plots, both at 70%.

Briscoe and Wadsworth (1970), working in similar forest types in Puerto Rico, found a slightly negative sloping relationship between basal area and diameter class. Corresponding results were obtained in all of our study plots (Fig. 4). In total, the smaller size classes had the greatest basal area, while the smallest basal area was found in the largest size classes. This can be attributed to the large number of stems in the

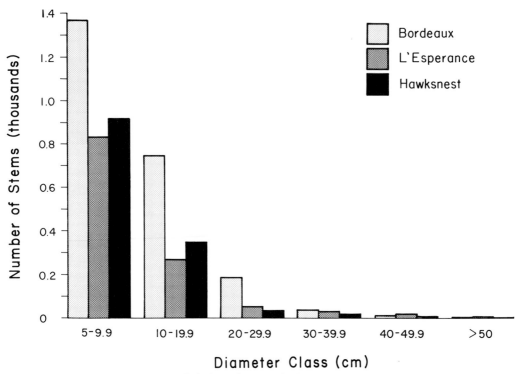

FIG. 3.   Diameter distribution.

smaller size classes and the decreasing number of stems in the other size classes.

Canopy heights on St. John are lower than mainland tropical forests and have fewer layers. Rarely do the tallest trees surpass 30 meters and these are found only in deep protected valleys. Richards (1983) pointed out that it is difficult to accurately identify specific boundaries between canopy classes in mixed forest types. However by delineating a range of heights based upon the mean heights, a canopy classification system can be approximated which includes all stems. The results of the analysis appear to agree with empirical observations.

The lower, over-topped canopy layer (<10 m) in the Bordeaux plot is dominated by a few species with numerous individuals. These include *Faramea occidentalis, Ardisia obovata, Morisonia americana* L., *Bourreria succulenta* and *Capparis* spp. Above this group is a mid-height layer which ranges in height from 10–15 m and is again made up of a few species with several stems, including *Pimenta racemosa, Ocotea coriacea, Guapira fragrans, Casearia* spp., *Cordia* spp., *Guettarda* spp., and *Inga fagifolia*. A third synusium, the continuous canopy (15–20 m) is dominated by *Acacia muricata, Guapira fragrans, Tabebuia heterophylla, Pouteria multiflora* (DC.) Eyma, *Zanthoxylum martinicense* (Lam.) DC., and *Byrsonima coriacea*. Finally, there is an emergent layer made up of a few very tall trees (≥28 m) of *Hymenaea courbaril*.

The trees found in L'Esperance are shorter than those in the Bordeaux plot and there are three distinct canopy classes present. The lower layer (<10 m) is dominated by *Ardisia obovata, Myrciaria floribunda* (West. ex Willd.) Berg., *Guettarda scabra* (L.) Lam., and *Guettarda parviflora* Vahl. The continuous canopy is represented by *Chrysophyllum pauciflorum, Tabebuia heterophylla, Ocotea coriacea, Guapira fragrans, Inga fagifolia, Daphnopsis americana* (Gris.) Nevil, *Margaritaria nobilis* (L.f.) Muell. Arg., and *Andira inermis*. The emergent layer is made up of relatively few individuals which include *Chrysophyllum eggersii* Pierre, *Hymenaea courbaril*, and *Ficus trigonata* L.

The height measurements obtained from the Hawksnest plot indicate that there are two distinct synusia with an occasional emergent. The

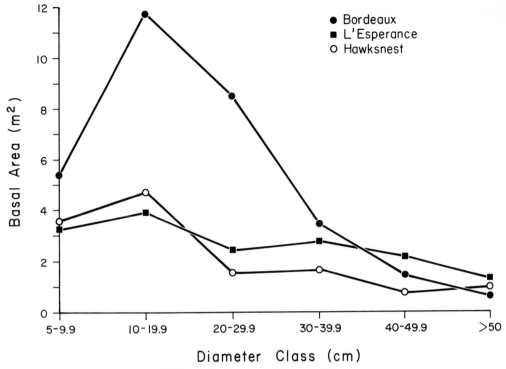

FIG. 4. Basal area vs. diameter class.

lower stratum is made up of *Eugenia biflora, Krugiodendron ferreum, Guettarda parviflora, Eugenia monticola, Maytenus elliptica* (Lam.) Krug. & Urb., and *Capparis* spp. The continuous canopy is dominated by *Meliococcus bijugatus, Bourreria succulenta, Bursera simaruba, Tabebuia heterophylla, Andira inermis, Guapira fragrans,* and *Ocotea coriacea.* There are two emergent species in this plot: *Zanthoxylum martinicense* and *Spondias mombin* L.

Conclusion

The St. John Forest Dynamics Project has the potential to fill crucial information gaps that will aid conservation efforts in the Caribbean. Quantitative data concerning vegetation dynamics and stand development will promote wise and ecologically sustainable use of secondary forests by addressing growth and mortality rates, recruitment, and natural disturbance. Through the establishment of scientifically planned and pragmatically oriented research, a clearer understanding of the impact of human intervention on forest recovery can be developed. Long term monitoring of forest dynamics will provide baseline data to produce sound recommendations for the utilization and protection of this important and fragile ecosystem.

Acknowledgments

This work was supported in part by the Virgin Islands National Park Service, the University of the Virgin Islands Cooperative Extension Service, The New York Botanical Garden and the Yale School of Forestry and Environmental Studies Tropical Resources Institute. This is publication number 96 in the series of the Institute of Economic Botany of The New York Botanical Garden.

Literature Cited

Acevedo-Rodriguez, P. 1985. Los Bejucos de Puerto Rico. Vol. 1. U.S.D.A. Forest Service, GTA-50-58. 331 pp.
Boom, B. M. 1986. A forest inventory in Amazonian Bolivia. Biotropica 18(4): 287–294.

Briscoe, C. B. & F. H. Wadsworth. 1970. Stand structure and yield in the Tabonuco Forest of Puerto Rico. Pages B-79–B-89 *in* H. T. Odum & R. F. Pigeon (eds.), A tropical rainforest. A study of irradiation and ecology at El Verde, Puerto Rico. Division of Technical Information. U.S. Atomic Energy Commission.

Cain, S. A., G. M. de Oliveira Castro, J. M. Pires & N. T. da Silva. 1956. Application of some phytosociological techniques to Brazilian rain forest. Amer. J. Bot. 43: 911–941.

Campbell, D. G., D. C. Daly, G. T. Prance & U. W. Maciel. 1986. Quantitative ecological inventory of terra firme and varzea tropical forest on the Rio Xingu. Brittonia 38(4): 369–393.

Cronquist, A. C. 1968. The evolution and classification of flowering plants. Houghton Mifflin Co., Boston, Massachusetts. 396 pp.

Ewel, J. J. & J. L. Whitmore. 1973. The ecological life zones of Puerto Rico and the U.S. Virgin Islands. U.S.D.A. Forest Service Res. Pap. ITF-18. 72 pp.

Forman, R. T. T. & D. C. Hahn. 1980. Spatial patterns of trees in a Caribbean semi-evergreen forest. Ecology 61(6): 1267–1274.

Gentry, A. H. 1982. Patterns of neotropical plant species diversity. Pages 1–84 *in* M. K. Hecht, B. Wallace & G. T. Prance (eds.), Evolutionary biology. Vol. 15. Plenum Press, New York.

Hubbell, S. P. 1979. Tree dispersion, abundance, and diversity in a tropical dry forest. Science 203: 1299–1309.

Lang, G. E. 1969. Sampling tree density with quadrats in a species-rich tropical forest. M.S. Thesis. University of Wyoming, Laramie. 54 pp.

Little, E. L. & F. H. Wadsworth. 1964. Common trees of Puerto Rico and the Virgin Islands. Vol. 2. Agric. Handbook No. 40, U.S.D.A. Forest Service, Washington, D.C. 548 pp.

———, R. O. Woodbury & F. H. Wadsworth. 1974. Trees of Puerto Rico and the Virgin Islands. Vol. 2. Agric. Handbook No. 449, U.S.D.A. Forest Service, Washington, D.C. 1024 pp.

Mori, S. A., B. M. Boom, A. M. de Carvalho & T. S. dos Santos. 1983. Southern Bahian moist forests. Bot. Rev. 49: 155–232.

Prance, G. T., W. A. Rodrigues & M. F. da Silva. 1976. Inventario florestal de um hectare de mata de terra firme km 30 da estrada Manaus–Itacoatiara. Acta Amazonica 6: 9–35.

Richards, P. W. 1983. The three-dimensional structure of tropical rain forest. Pages 3–10 *in* S. L. Sutton, T. C. Whitmore & A. C. Chadwick (eds.), Tropical rainforest: Ecology and management. Special Publication No. 2 of the British Ecological Society. Blackwell Scientific Publications, Oxford.

Rivera, L. H., W. D. Frederick, C. Ferris, E. H. Jensen, L. Davis, C. D. Palmer, L. F. Jackson & W. E. McKenzie. 1970. Soil survey—Virgin Islands of the United States. USDA Soil Conservation Service, San Juan, Puerto Rico. 78 pp.

———, W. E. McKenzie & H. H. Williamson. 1966. Soils and their interpretations for various uses: St. Thomas, St. John, Virgin Islands. USDA Soil Conservation Service, San Juan, Puerto Rico. 52 pp.

Robertson, W. B. 1957. Biology report: Initial study and development survey. V.I. National Park, U.S.N.P.S., Washington, D.C. 54 pp.

Teytaud, R. 1983. Mapping of climatic, topographic, and edaphic factors, existing vegetation, and potential natural vegetation in St. John, with recommendations for research on the economic impacts of vegetation management. Second Draft, Island Resources Foundation, St. Thomas, U.S.V.I. 70 pp.

Tyson, G. F., Jr. 1986. Historic land use in the Reef Bay, Fish Bay and Hawksnest Bay watersheds. St. John, U.S. Virgin Islands. 1718–1950. VIRMAC II Research Series, Island Resources Foundation, St. Thomas, U.S. Virgin Islands. 49 pp.

Woodbury, R. O. & P. L. Weaver. 1987. The vegetation of St. John and Hassel Island, U.S.V.I. A report of the U.S. National Park. 98 pp.

Variation and Change in Amuesha Agriculture, Peruvian Upper Amazon

Jan Salick and Mats Lundberg

Table of Contents

Abstract

SALICK, J. (Institute of Economic Botany, New York Botanical Garden, Bronx, New York 10458-5126, U.S.A.) and M. LUNDBERG (Department of Cultural Anthropology, University of Uppsala, Sweden). Variation and change in Amuesha agriculture, Peruvian Upper Amazon. Advances in Economic Botany 8: 199–223. 1990. In addition to ecological patterns in Amuesha agriculture (Salick, 1989), the indigenous slash-and-burn system had much quantified variation within and among fields. The causes of this agricultural variation were traced to differences in agricultural practices and socioeconomic conditions, and, further, to inherent limitations, flexibility and buffers in the agricultural system. Change was a major cause of variation, brought about recently by the introduction of a road into the Palcazu Valley, which previously had been only weakly connected to the Peruvian and world economy. The choices faced by the Amuesha in that time of change and their employment of traditional responses are summarized; the question remains whether traditional responses to change can accommodate the changes of scale in economic and social influences which were taking place.

Key words: agricultural ecology, development anthropology, agricultural systems, agroforestry, backyard gardens, indigenous agriculture, intercropping, Amazonian development.

Resumen

Entre los varios principios organizativos de los sistemas de cultivo Amuesha (Salick, 1989) se encuentra un rango amplio de variabilidad dentro y entre los campos de cultivo. Esta variabilidad se debe a diversos factores: diferencias familiares en preferencias y gustos; diferentes estrategias para organizar la economía familiar; otros factores socioeconómicos y limitaciones inherentes; flexibilidad; y mecanismos contra riesgo del mismo sistema de agricultura. Una nueva carretera—que estrecha los nexos entre los Amueshas y la economía regional, nacional, y mundial—tambien introdujo cambios contribuyentes a la variabilidad. Las alternativas enfrentada por los Amueshas en el momento de fuertes cambios y su uso de respuestas tradicionales son resumidos; el futuro va a determinar si estas repuestas tradicionales son suficientes para acomodar los recientes cambios socioeconómicos de una escala mayor a los anteriores.

Introduction

Indigenous slash-and-burn agriculture is not a uniform production system. There is tremendous variation within agricultural fields and among them. Authors describing indigenous agriculture have concentrated on distilling trends or patterns from the confusion; and variation is consequently ignored. In analyzing indigenous Amuesha slash-and-burn agriculture of the Peruvian upper Amazon (Fig. 1; Salick, 1989), it became apparent that in the variation there was method worth exploring. To do so, cross-disciplinary skills were needed to assess both agroecological and anthropological motives. Completion of the study has made it obvious that employing many other disciplines might further broaden our appreciation of variation in indigenous agricul-

FIG. 1. Amuesha indigenous slash-and-burn agricultural plot on alluvial lowlands with a diversity of crops including plantain, cassava, taro, sugar cane, yam and sweet potato.

ture. Many biological and social causes for variation will be discussed at length; one of these causes needs particular introduction, however.

Among the Amuesha Amerindians of east-central Peru (Fig. 2), an important cause of variation in agricultural practices was change. A recent major change in the Amuesha homeland has political roots. The previous president of Peru, Fernando Belaúnde Terry envisioned the future of his nation to lie east of the Andes. To further this vision he increased Peru's national debt considerably and supported construction of the "Carretera Marginal" (Perimeter Road) along the eastern edge of the Andes connecting the major upper Amazon River basins and which, he envisioned, would eventually run from Argentina to Venezuela (Belaúnde, pers. comm.). The road construction and development schemes for the Palcazu Valley, the heart of the Amuesha homeland, and surrounding areas (Projecto Especial Pichis-Palcazu) came under severe criticism from those with environmental and anthropological

concerns as being "A Road to Destruction" (e.g., Santos, 1980b; Smith, 1983; Survival International USA, 1981; Swenson & Narby, 1986). As a result, the development project in the Palcazu—the Central Selva Natural Resource Management Project of the United States Agency for International Development (JRB Assoc., 1981; USAID/Peru, 1981)—was reoriented from Belaúnde's original goals of colonization and maximization of agricultural production.

Instead, many idealistic programs for conservation, resource management, and human rights were incorporated into the development project in the Palcazu. Colonization was strictly limited and there was land titling for native communities. Land use was determined by capability. There was an orientation toward sustainable agriculture and maintaining the subsistence production base. Tropical forest management included natural forest regeneration, communal extraction reserves, protection forests, and the formation of a national park. Indigenous and

FIG. 2. Location of the Palcazu Valley, the principal homeland of the Amuesha people. The Palcazu River in the central lowlands of Peru is a headwater of the Amazon.

national health and education services for the Amuesha were supported. Amuesha ethnobotany and indigenous natural resource management were to be used as both baseline data and development guidance, including the present study. All of the above programs were to support Amuesha rights and concerns.

Nonetheless, the immediate reality was that a road was constructed into the Palcazu Valley. It runs through Amuesha native communities which previously were connected to the outside world only by river, footpath and bushplane. The Amuesha, consequently, were at a pivotal point of change.

HISTORY OF THE AMUESHA

There is always the temptation to speak of "traditional" patterns and "change," but we will attempt an historical perspective of change (e.g., Santos, 1980a; Smith, 1977; Varese, 1973). Although the Amuesha have maintained their cul-

tural identity and unique language, they have faced change throughout their history and persist as a native people through their flexibility. Preliminary archeological excavations in the Pichis (Allen, 1968) and Palcazu (Jimenez, 1986) Valleys date extensive Amuesha habitation in these areas to perhaps 4000 B.P. Oral (Smith, 1977; Stocks, pers. comm.) and Spanish (e.g., Izaguirre, 1923–1929; Larrabure y Correa, 1905–1909; Ortiz, 1967) histories record the Amuesha territory both in the lowland valleys and in the Andean premontane of the Oxapampa and Pozuzo regions. Within the Pozuzo region is located the famous "Cerro de Sal" or Salt Mountain (Varese, 1973) from which trade of this scarce commodity radiated throughout the Amazon Basin. This brought the Amuesha into contact with many different native groups and indigenous commerce.

In 1635, the Amuesha first came into contact with Europeans when a mission post was established at the "Cerro de Sal" (Izaguirre, 1926; Varese, 1973). The Amuesha joined under Juan Santos Atahualpa in 1742 to fend off European invasions of their territory and were successful until 1859 when German immigrants established themselves in the Pozuzo Valley (Ortiz, 1967). Finally, the Amuesha resistance to invasion was broken by epidemics of yellow fever and smallpox in 1879 and 1880 (Reiser, 1943; Smith, 1977). As they dispersed into the forests to isolate themselves from disease, the Franciscans rapidly entered the Oxapampa-Pozuzo region (Izaguirre, 1926; Ortiz, 1967; Smith, 1977).

In the Palcazu Valley, European settlers of German, Swiss and Austrian extraction moved into the area after the Rubber Boom from the lower reaches of the Amazon and from Pozuzo after the German settlement of that region. Until today, descendants of these settlers have cattle ranches on the best lands of the Palcazu which the Amuesha apparently once inhabited in extensive groups (Jimenez, 1986). The Amuesha now live on comparatively marginal lands and work for the colonists for low wages or to repay debts. Often one ranch is associated with one native community, functioning along traditional patron-peon labor patterns. Until recently, the European colonists owned an airline which was the sole buyer and transporter of beef in the valley, effectively monopolizing commerce. Amuesha commercial options were limited to wage

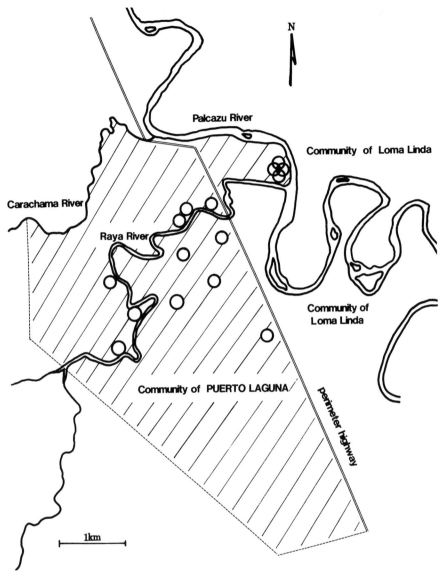

FIG. 3. The Amuesha native community of Puerto Laguna is along the Palcazu River and dissected by the Raya River. Hatching demarcates community lands with a double line approximating the path of the newly constructed dirt road ("Carretera Marginal"). Open circles indicate agricultural areas included in the study.

labor migration mainly to pick coffee or work as domestic labor, and even these menial positions were effectively controlled by European colonists. With the new road the colonists are interested in monopolizing land transportation. Through these colonists the Amuesha have dealt with non-indigenous commerce for a hundred years. Neither change nor the market economy is new to the Amuesha; however, the scale of recent change may overshadow history. Modes of change among the Amuesha have been established over time, but are these sufficient to meet the present situation?

HISTORY OF PUERTO LAGUNA

The Native Community of Puerto Laguna in the Palcazu Valley (Fig. 3) was the site of this study. Choosing one location rather than repeating a broad survey (Salick, 1989) allowed us to detail our study and personally acquaint ourselves with individuals, their lives and the changes they were facing. Puerto Laguna was the dissertation field site of the second author from 1981–1986.

Puerto Laguna is a fairly new native community both as a site of Amuesha habitation and as an officially designated tract of communally owned land. In 1970 the sector of Puerto Laguna broke from the encompassing community of Loma Linda and established independence at the present site. Subsequently, of the twenty-two adults interviewed, sixteen came from outside the surrounding Loma Linda-Laguna area. Both division and migration are characteristic of the Amuesha settlement (Smith, 1977); and most native communities in the Palcazu have been formed from two generations of immigration. In the last few years immigration into native communities has been regulated because of pressure for land from an internally growing population. Native community land titling with concomitant designation of fixed boundaries (Peru, 1974, 1978) has had the negative effect of limiting the amount of land available to the Amuesha.

Several external forces have affected the history of Puerto Laguna. The Summer Institute of Linguistics and Seventh Day Adventists contacted the community of Puerto Laguna over 30 years ago (Stoll, 1985) and the inhabitants continue as fundamentalists today. A farm bordering on the community was claimed about twenty-five years ago by a colonist of mixed German and Italian origin. He used Puerto Laguna residents as his private labor force; goods were traded or minimal wages were paid to his laborers. In 1978, Puerto Laguna founded its cattle cooperative as a conscious entry into the external economic sector.

This short outline of Puerto Laguna history reemphasizes the fact that change cannot be associated solely with the road; it is a continuing historical process. Nonetheless, the last few years have brought accelerated change. The Palcazu Development Project began in 1980 (PEPP, 1981) with USAID providing funds and expertise from 1983 (JRB Assoc., 1981; USAID/Peru, 1981). Two important events transpired in 1984: the community received legal title to its lands and the road arrived, along with a road construction camp, road workers and menial jobs for some Amuesha. This study was conducted in 1984 and 1985 with return trips in 1986. These were unquestionably focal years for incipient change. The Native Community of Puerto Laguna began intensified contact with the greater Peruvian economy. One of our major objectives was to carefully follow the effects of this intensified contact. To do so, we integrated an understanding of the indigenous agricultural systems (Salick, 1989) and knowledge of the introduced factors. We envisioned this study as a basis for continued observation of Amuesha contact with the Peruvian social, political and economic environment.

Methods

In an attempt to be objective in our interpretations of recent agricultural history, we relied initially on quantitative vegetation sampling of agricultural fields and yard gardens which we compared to a valley-wide survey of Amuesha indigenous agriculture (Salick, 1989). We looked for quantitative variation from the normal patterns of Amuesha agriculture and then interviewed agriculturalists and households to interpret the variation. In traditional vegetation ecology, variation is most often viewed as normal and extremes may be ignored as unexplained perturbation. With human manipulation of vegetational communities it is convenient and instructive to ask for explanations. We often found variation to be an indicator of basic agricultural limitations—such as environment, economics, and time—or of change from indigenous agricultural practices.

Amuesha slash-and-burn agricultural plots were sampled and described using ecological vegetation sampling methods and analyses (see Mueller-Dombois & Ellenberg, 1974). Within each field we laid stratified random quadrats 2 m × 5 m (10 m^2) totalling 5–10% of the total plot area. All crops and edible volunteer plants were inventoried in each quadrat, measuring height, estimated percent cover including plants rooted outside the quadrat, distance to another plant of the same species, and the nearest neighboring species of each plant. From these data we calculated density, cover, and diversities (Shannon-Weiner $H' = -\Sigma(p_i)(\log_2 p_i)$) of α, β, and height. Communities were described using polar

Table I

Summary of fields sampled and people interviewed in the Amuesha community of Laguna. Field data are divided by three sites: alluvial lowland fields along the rivers (a) Palcazu and (b) Raya, and (c) acid, infertile, upland fields

	Sites		
	Palcazu	Raya	Upland
Number of fields sampled	14	11	10
Total number of fields	25	27	38
% of fields sampled	56%	41%	26%
Hectares sampled	1.8	1.4	5.1
Total hectares cultivated	2.9	3.8	16.9
% of cultivated hectares sampled	60%	38%	30%
Average field size (ha)	0.12	0.4	0.5
Number of households[a] interviewed: 11			
Total number of households: 27			
% of households interviewed: 41%			
Number of families[b] interviewed: 12			
Total number of families: 33			
% of families interviewed: 36%			

[a] A "household" is considered all people living under the same roof and sharing a joint economy.

[b] A "family" is considered a married couple or single parent plus children.

ordination with presence weighted by density, and gradient analyses using a time sequence.

We interviewed 41% of the households[1] in Puerto Laguna (11 of 27) and sampled all their agricultural fields and yard gardens (Table I). Detailed informal interviews included questions on agricultural practices, problems and expectations, field and homestead histories, household membership, extended family membership, ages, birthplaces, subsequent movements, work, commercial enterprises, and community roles. The households were chosen for occupational and social diversity including: subsistence farmers, cattle raisers, laborers, coffee pickers, the community leader, the health worker and a widow.

Results

VARIATION IN AMUESHA CROPPING SYSTEMS

The basic patterns of Amuesha indigenous agriculture are analyzed in detail elsewhere (Salick, 1989). Here we have directed our attention to those data points from the Native Community

---

[1] Households are considered as persons living under the same roof and sharing joint economies.

of Puerto Laguna which vary from the previous analysis. Since our sample frame was selected to maximize occupational and social diversity, many exceptions were found to the otherwise clear patterns in Amuesha agriculture. First, the variation in the data will be presented and then the associated factors that were attributed by the farming households themselves as causing the variation (see Table II).

Field and Cropping Statistics

There was great range among Amuesha swiddens in the statistical descriptions of both crop diversity and field architecture (Table III). To make this variation meaningful, fields were differentiated by land types: those on sandy alluvial beaches, those on less frequently flooded alluvial lowlands, and those on acid infertile uplands, respectively. Within these land types, there were successional differences in dominant crops among field rotations: bean fields with flooded fallows on beaches, maize-cassava-plantain rotations with bush fallows on alluvial lowlands, and rice-cassava rotations with tree fallows on acid uplands. Salick (1989) presented statistical means and standard deviations characterizing these fields, while here we have concentrated on the extremes of the ranges. For example, while the

Table II

Reference list of causative factors of variation in Amuesha agriculture. This outline corresponds to the results reported in the text and to the numbering in the following Table III and Figures 2 and 3

1. Time of planting and harvesting
2. Cropping cycle
3. Intercropping
4. Microsite utilization
5. Seed storage
6. Experimentation
7. Farming systems
8. Land tenure
9. Family establishment
10. Family stability
11. Labor and time constraints
12. Commercialization
13. Opportunism
14. External introduction

average maize field initiating the rotational sequence on alluvial lowlands is 1500 m², we found a great range of field sizes from 200 m² to a half hectare. Subsequently, we returned to these extreme fields and asked, "Why did you plant so little/much maize; most people we have visited have planted more/less?" In addition to field size in this example, statistical measures of crop and field architecture included planting distance, crop diversity, height diversity, percent cover, and density. Planting distances varied with crops, of course—from an average of a half meter for beans to 4½ meters for plantains—but also within beans from 10 cm in monoculture to 1 m intercropped. The greatest crop diversity (mean H' = 3.19) was in lowland cassava fields although diversity was less if households were under time constraints during the planting season. Height diversity of rice was minimal except, for example, when a particular microsite invited the planting of nutrient demanding papaya. Percent cover ranged completely from 0–100% depending partially on the age of the field; new fields of course are scantily covered. Stem density varies obviously with crops—beans and rice being densely planted and plantains much less so—but it also varies less obviously with sparsely planted commercial fields, to be discussed later.

Table III

Comparative statistics for crops diversity and field architecture in Amuesha agricultural systems. The emphasis in this analysis was placed on extremes of variation and so means and *ranges* are given which can be compared to means and standard deviations in Salick (1989). The agricultural practices and social factors with which extremes were associated are indicated by the outline numeration found in the text and in Table II

| Land Type: | Beach | Lowland | | | Upland | |
Field Type:	Bean	Maize	Cassava transition	Plantain	Rice	Cassava
Successional stage	only	early	middle	late	early	late
Field size × 100 m²	6	15	11	13	38	42
Range	1–10	2–49	6–20	3–28	14–70	25–100
Associated factors	5	3, 5, 6	9, 10, 11	11	5, 8, 11	7, 12
Planting distance (m)	0.56	1.13	1.12	4.39	0.56	1.17
Range	0.1–1.0	0.5–2.25	0.7–4.0	3.0–5.5	0.3–0.7	0.86–1.37
Associated factors	3	3	3	4, 6	4	0
Crop diversity (H')	1.88	2.26	3.19	2.83	0.55	1.70
Range	0–3.41	0–3.31	2.54–3.85	1.16–3.63	0–1.64	0–3.5
Associated factors	3, 4	2, 3, 5, 13	3, 9, 10, 11	1, 2, 12	3, 4	4, 6, 13
Height diversity (H')	1.13	1.52	2.21	2.39	0.85	1.26
Range	0–2.25	0–2.48	1.23–2.86	1.16–2.93	0–1.48	0–2.33
Associated factors	3	3, 4	9, 11	2, 12	3, 4	3, 4, 13
% cover	92	61	42	52	60	31
Range	4–147	0–100	19–79	10–109	30–97	4–72
Associated factors	1	1, 3, 13	1, 2, 3	1, 2, 3, 12	4	1, 13
Density (stems/m²)	3.37	0.95	0.63	0.49	3.04	0.55
Range	1.50–4.69	0.34–1.96	0.33–0.94	0.07–0.90	2.20–1.91	0.27–0.93
Associated factors	3	1, 3, 13	1, 2, 3	1, 2, 3, 12	4	1, 13

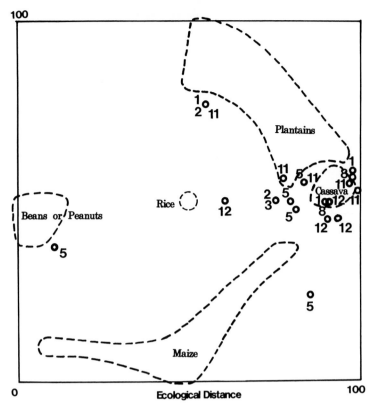

FIG. 4. Outliers in an ordination of Amuesha agricultural fields (ordination from Salick, 1989) are marked with open circles coded by the causes of their variation outlined in Table II. Field clusters conforming to the general patterns in Amuesha agriculture are encompassed by dashed lines and labeled by their dominant crop species.

Field Ordinations

Ecological ordination of Amuesha fields using the dominant crops of the region as poles differentiated field types and defined a time sequence (Salick, 1989). However, not all fields fit neatly within the scheme: a farmer under labor and time constraints cleared a lowland field late in the season and planted a near monocrop of cassava which ordered among the upland cassava monocrops. These outliers, including sparse bean and maize plantings and cassava and plantain monocrops, are graphed in Figure 4 for later discussion of their aberrance.

Field Time Series

A time series analysis of Amuesha agricultural fields on beaches, lowlands and uplands was run on three crop rotational systems (Salick, 1989).

Beans or peanuts are planted each year on beaches during the dry season with flooded fallows in the rainy season. Lowland fields are relay cropped from maize to cassava to plantains; the rotations are typically intercropped with beans, vegetables, and fruit trees respectively and are gradually replaced by a bush fallow. Upland fields have a crop of rice followed by cassava with a tree fallow usually established during the second year. Variants from the time series are graphed in Figure 5; for example, because their beans rotted during the rainy season one year, a family planted a very sparse bean field, more to replace their seed supply than to harvest a crop.

Spatial Distributions

Salick (1989) found a very close correlation between crop diversity and the percentage of

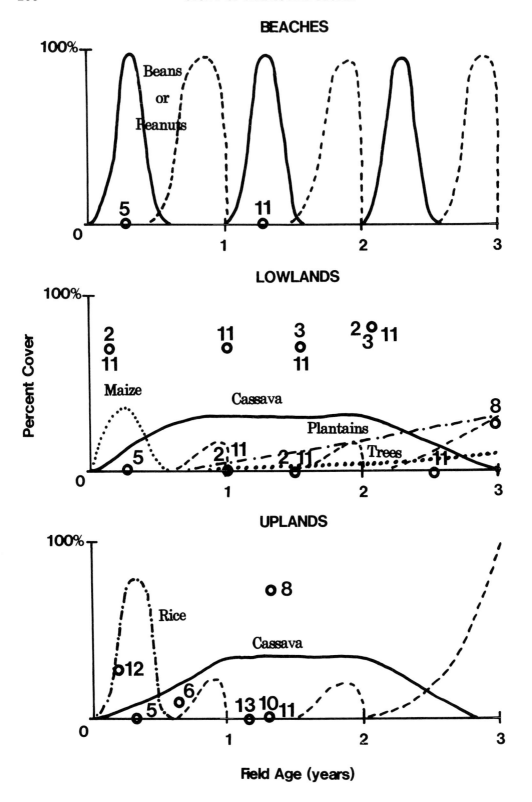

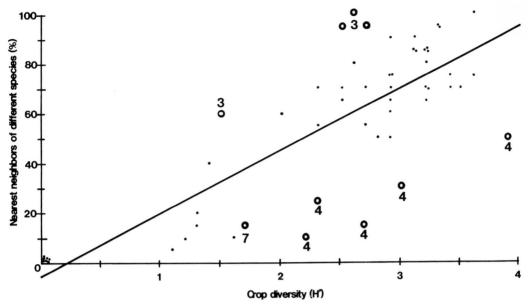

FIG. 6. Exceptions to random spatial distributions were due largely to uniform intercropping and microsite utilization causing uniform and clumped distributions, respectively. Outliers are graphed to a correlation between crop diversity (H') in agricultural fields and the percent of nearest crop neighbors being different species (y = $-5 + 25x$, $r^2 = 0.67$). Outliers are marked with open circles coded by the causes of their variation outlined in Table II.

nearest neighbors being of a different species: a rough indication that crops are randomly distributed within fields. However, there are definite exceptions to this trend (circles on Fig. 6). Some farmers use patchy microenvironments for certain crops forming clumped distributions, and some fields were extreme in their uniformity, such as where beans were uniformly poled on maize.

Yard Gardens

There were three other major types of agricultural variation found among the Amuesha that were not considered at all in the basic trends (Salick, 1989) because their range is more dom-inant than any single trend. Salick (1989) iden-tified 70 species grown in yard gardens including many minor, specialty, herb and fruit crops. The number of species in single yard gardens (Table IV) ranged from none to 26 with strongly pro-fessed attitudes toward yard gardens. The house-holds with the most crops in their yard gardens depended on the gardens for a certain percentage of their subsistence diet; their fields had low crop diversity and they relied on their yard produc-tion. The next group with slightly less yard gar-den diversity produced a greater diversity in their fields and used the yards only for more special-ized crops. A household which had no tenure rights to their land saw no reason to plant fruit trees if some other family would take their land before the fruit was borne. Families dedicated to

←

FIG. 5. Variants from the time series of Amuesha agricultural fields (time series from Salick, 1989) are marked with open circles coded by the causes of their variation in Table II. General patterns in Amuesha agriculture were analyzed in the time series for three land types: beaches, lowlands and uplands. Over a three year period, beans or peanuts were planted each year on beaches during the dry season, and flooded fallows inundated the fields in the rainy season. Over the same period on lowland fields there was a distinct crop succession from maize to cassava to a gradual dominance of plantains, bananas, and fruit trees; fallow species were periodically weeded until they were allowed to gain dominance approximately during the third year. Upland fields have a crop of rice followed by cassava with the fallow well established during the second year. For each series, fallows are marked by dashed lines.

Table IV

Yard garden census ordered by number of edible species (excluding medicinals, herbs, and ornamentals) including households' stated attitudes toward this type of cultivation

Number of species present in yard gardens	Stated attitudes toward yard garden cultivation
26 25 18	Crop diversity low in fields cultivated by these households so they use yard gardens to supplement their diets. Allows easy tending and experimentation near house.
16 15	Crop diversity high in fields cultivated by these households and they depend on yard gardens only for production of specialty crops.
12	Land tenure problems discourage home improvements.
11 11	Head of households work at wage labor and raise cattle so there is little time for home improvements.
9 0 0	Young families not yet established.

Table VI

Area of land cultivated by a household, area of land per household member, land types, and sociological notes. L = lowland, alluvial soils; U = upland, acid infertile soils

Total	Per household member	Land type(s)	Sociological factors
0.19	0.027	L	Widow.
0.21	0.031	L	Family poorly established.
0.42	0.084	L	Land tenure problems.
0.86	0.096	L+U	Professional
0.59	0.12	L	cash income.
0.95	0.16	L+U	
2.25	0.20	L+U	Large family.
1.02	0.20	U	
0.80	0.27	U	Newly
0.70	0.35	U	established
1.18	0.39	U	families.

(Area of land cultivated (ha); Total / Per household member)

cattle and wage labor professed to have no interest or time to grow fruit. Young families were delaying tree planting until the basics were provided.

Table V

Among field (β) diversity numerically ordered with the household economic basis, land tenure status and family establishment noted

Among field (β) diversity	Household economic basis, family establishment and land tenure basis
3.87 2.98 2.37	Swidden agriculturalists with established families and little land tenure limitation.
2.12	Professional with cash income supplemented by swidden agriculture.
1.90	Swidden agriculturalist with poor family establishment.
1.49	Swidden agriculturalist with land tenure problems.
1.49 1.49	Newly established families which raise cattle.
0 0	Newly married families practicing commercial agriculture.
0	Hunter/laborer.

Diversity among Fields

Diversity among fields (Table V), which combined the number and kinds of fields that a household tended, ranged from $H'_\beta = 0$ to 3.87, the latter having a total of thirteen fields with some of each type. Established swidden agricultural families had the greatest field diversity. Newly- and poorly-established families had the least diversity with only one or two fields planted. Alternative farming systems incorporating cattle, wage-labor, or hunting had intermediate field diversity below the maximum.

Area under Cultivation

The total area cultivated (Table VI) ranged approximately from one-fifth to 2¼ hectares per household and from one-fortieth to one-third of a hectare for each man, woman and child. The trends within these tremendous ranges were slightly different than with field diversity in that poorly established families subsist on a small amount of good land claimed in the past, while new families that have requested land only recently and have been relegated to acid uplands which demand extensive farming. These newly established families had the most area under cultivation even though the field diversity was low. Well established swidden farmers had an intermediate area of land cultivated with a combination of upland and lowland fields, well diversified.

CAUSES OF VARIATION

Throughout the different analyses, the same fields repeatedly emerged as atypical. The owners of these fields were questioned specifically about the causes of these anomalies. Recurring reasons were given by household informants as to why their fields were larger/smaller, more/less diverse, or atypically cropped or rotated. This allowed us to generalize about sources of variation, although we did not abandon detailed explanations or individual examples. Explanations combined and overlapped, so our categories were not exclusive.

Time of Planting and Harvesting

The stage of a field at the time of sampling affected many analytical results. Although we sampled fields at discrete times so that field comparisons would be meaningful, households planted, rotated and harvested their fields at different times. Time of planting was generally established by the seasons, but variation among households was common due to difficulties in weather forecasting and non-agricultural commitments. In very young fields percent cover was low and densities were low if farmers had not finished planting or high if they had just finished planting, with mortality yet to take its toll. Later in the rotational cycle, we saw variation in transitional stages depending on harvest and planting dates; crop rotations were relayed so there were gradients between categories with variables of intermediate value. When plantains were first established, fields had densities and diversities intermediate between plantain and cassava rotations. In old fields we saw a gradual domination of the bush fallow as plantains were weeded less frequently and finally not at all; concomitantly, species and height diversities for planted crops fell well below the mean for plantain rotations. Such variability in timing seems to be inherent in Amuesha slash-and-burn agriculture, with flexibility being one of its real strengths.

FIG. 7. Maize must be planted at least once a year: chickens need to be fed and seed for the next crop produced. Maize, as the first crop in a lowland rotation demanded opening much fallowed land and influenced the course of the cropping cycle.

Cropping Cycle

It was common to find variation in the Amuesha cropping cycles especially toward the final stages. Plantain fields were kept in production for anywhere from one to eight years, while some cassava fields were never rotated into plantains. Ironically, a recurrent agricultural rationale for the late-rotational variability was the need for maize, the first stage of the maize-cassava-plantain cropping cycle.

Maize must be planted at least once a year (Fig. 7), as chickens need to be fed and seed for the next crop produced. Maize, as the first crop in rotation, demanded opening much fallowed land. An extreme in maize production was a family which tended dry- and wet-season maize fields with different varieties specific for the two periods. After maize, the land continued in rotation and needed to be weeded over the 3–5 years during which lowland fields continued to produce cassava and plantains. When the time required for cultivating the ever-increasing amount of land became too great for the household to manage, fields were left to fallow before the full rotational sequence was completed. Thus, the result of the need for maize was seen in a field which was never rotated from maize to cassava, and a cassava field which was left to fallow without ever planting plantains. Again, variability in cropping cycle was an indication of flexibility which was available in the later stages of the cropping cycle.

Intercropping

All cropping systems of Amuesha agriculture were intercropped, with a total of 96 crops recorded (Salick, 1989). However, intercropping was practiced to different degrees and with different spacing from clumped to uniform. A regular association between crop species, such as beans and maize dibbled simultaneously into the same holes, was less common than the haphazardly mixed field (Fig. 1). A slightly less regular intercrop was cacao planted under plantains and later underplanted by *Inga* spp. which would eventually shade the cacao when the plantains failed. Lowland cassava fields had very haphazard mixes of many minor crops. Sporadic intercropping was found in rice fields and unassociated maize fields. Old fruit trees were sometimes left in newly cleared fields which had gone through a complete cropping cycle previously, where fruit trees were left to grow in a bush fallow. The established fruit trees, which were too valuable to warrant cutting, caused very aberrant statistics for maize and rice fields. Thus, types and patterns of intercropping were diverse, which in turn added much diversity to within and among field patterns.

Microsite Utilization

The flexibility in Amuesha agriculture to respond to microsite variation was important. Especially demanding crops such as papaya would only grow in pockets of rich, loose soil or where ash had accumulated after the burn. Plantains were grown in the uplands at the base of hills where topsoil had accumulated, in stream drainages, or on unusually fertile soil outcrops. Agriculturalists who were better at using microsites raised their crop and dietary diversity. It was illuminating to walk over a newly burned field with an Amuesha who was looking for the right place to plant her papaya or watermelon seeds.

Seed Storage

With 6000 mm of rain annually in the Palcazu Valley, both fungal growth and stored-grain insects abound. In such a climate the advantages of vegetatively propagated crops were obvious. Additionally, planning was not always perfect, as with the man who ate all his peanuts, or the family who raised too many chickens and fed all their maize before the next harvest. With a shortage of seed, seeds were borrowed or fields were not planted as usual, causing variations in the cropping patterns. The Amuesha sometimes practiced an alternative to seed storage by planting and raising an unproductive crop in the rainy season, knowing that the harvest would be poor but sufficient for replanting in the favorable, productive dry season; the objective in the rainy season was thus to replace only the quantity of seed planted and avoid losses during storage. Both problems with seed storage and measures to counteract them caused agricultural variation.

Experimentation

The Amuesha enjoyed experimenting within their traditional agricultural systems: trying new crops, varieties, and new ways of planting. They enthusiastically discussed who gave them a new variety or where they stole it; plants were memorabilia to the Amuesha as well as productive. Experimentation usually took place around the house in yard gardens or occasionally, in upland fields for which the Amuesha were searching continually for cultivars which would survive the acidity and aluminum toxicity. One farmer tried unsuccessfully to plant plantains and beans on an upland site just to make sure that what he had been taught was in fact true. Some of the variation in cropping systems was attributed to this experimentation. Until recently, experimentation had been initiated on a small scale, with a few seeds brought in from another region or taken from a German colonist. The recent change in scale of this activity will be discussed later.

Farming Systems

Farm-level decisions or major subsistence strategies caused variation in Amuesha cropping systems. Some households put more effort into obtaining meat: production of chickens, ducks and cattle or hunting wild game and fishing (Fig. 8). Other households were nearly vegetarian. The crop and field diversity of the latter were considerably above average, while the household of one superior hunter/fisher tended only upland fields and apparently thrived on a cassava/meat diet supplemented with gathered wild fruits. Flexibility in the choice of farming system has been lost with the apparently recent depletion of fish and game; the hunting, gathering, and starch production mode of subsistence was consequently fairly rare.

Land Tenure

When Puerto Laguna was originally founded in 1970 all families had sufficient land to farm, for pasture, and to leave in fallow rotation, and a "loose" land tenure system functioned. The population grew, until young families were forced to exclusively farm uplands. The most severe land tenure problem had befallen a family, the members of which belong to Loma Linda (the sister community across the Palcazu River), but whose land was part of Puerto Laguna. They were considered to be "borrowing" the land from Puerto Laguna, and only had managed to keep their historical claim to it by continuously cultivating it. They were not, however, allowed to expand their holdings or let them go fallow. In the long run, this was obviously an untenable arrangement and they were aware that they needed either to join the community or move. Their fields were not undergoing typical succession: one field had been in rice continuously for several years and another was in perpetual "cassava transition." They said that if they rotated to plantains they would not have had enough field space for cassava or the minor crops produced with cassava. They reported declining yields and weed invasions. Their field diversity was low and their yard gardens depauperate because they feared that a Puerto Laguna family would take their homestead, in which case home improvements would be wasted.

Family Establishment

Newly married families often lived for a period with the wife's family until the couple began having their own children, when they would home-

stead. This appeared to be a very precarious period in household subsistence. New fields were opened without established fields to produce a guaranteed income; yard gardens were not yet planted. Difficulty in establishment was probably part of the reason for living with the extended family and adding labor to that pool for a time; the new family then had rights to harvest from the extended family's fields until their own new fields began producing dependably. Within- and among-field diversity was low for newly established families both because of the unavailability of lowlands which supported a diversity of crops (see Land Tenure), and because of insufficient time to accumulate a full component of variable aged fields. Recently, new families often chose wage labor over the precarious period of field establishment, however, this only delayed the establishment of fields and gained them no long term economic security.

Family Stability

Once a household was established, how smoothly it functioned became critical. Death was a dominant cause for agricultural anomaly; the widow's household and another family which had experienced several deaths both had low field diversity though for different reasons. Subsistence for single women (Salick, 1985) is very difficult and very different from subsistence in fully functioning extended families. In general, single women cultivated smaller fields more intensively, and for longer periods of time before fallowing. Heavy work such as slash-and-burning, upland cultivation and plantain planting was minimized and delayed until an opportunistic moment when male labor was available. Weeding, intercropping, yard cultivation, and opportunism were practiced meticulously. Cash subsidies were used when available.

The family with several deaths had moved frequently; the custom of the Amuesha upon a death within the household is to move from the evil spirits (Smith, 1977). This had totally disrupted their land tenure, field succession, intercropping and yard gardens as well as a general will to carry on.

Labor and Time Allocation

When time and labor were lacking, crops sometimes did not get planted. This affected mi-

nor intercrops and whole rotations, depending on the time constraints. Cassava was the only crop which seldom seemed neglected (Fig. 9). Activities which compete with crop cultivation in this way include: cash labor, working with community administration, cooperative forestry, and cattle raising. "We didn't have time," was an extremely common explanation for agronomic anomalies.

Commerce

Commercial agriculture was already being tried by several Amuesha households on a small scale. Crops intentionally planted for commerce included cacao, pineapple, achiote (colorant), cassava, and maize to raise chickens for sale. Market acumen was variable; in the worst case, a bull was traded for a transistor radio. The effects of such commerce on cropping systems were also extremely variable, with diversity, density, cover, and field rotations either increasing or decreasing depending on whether the additional crops were planted into a diverse subsistence field or if a special field was dedicated to a near monocrop for commercial production.

Opportunism

A long standing and often advantageous response to agricultural decision making was to take advantage of an unforeseen opportunity. The range of examples was extensive. If there were five days of dry weather, a definite anomaly in the Palcazu, the opportunity to burn was taken, whether or not there had been plans to do something else. An unknown crop or variety or just some extra seed received was planted. Sporadic resources and labor were used by the widow so effectively, for example, that she raised her children without family support. Opportunities to work on a contract basis were taken universally by Amuesha men when extra time was available or when unforeseen needs arose. Spatial opportunism took the form of utilizing microsites. Examples were extensive and opportunism was central to Amuesha response and flexibility.

External Introduction

Introduction of agricultural species is an historical process continuing today. Precolonially, maize apparently was introduced originally from Mexico. The early Spaniards brought rice, plantains and chickens, originally from Asia. More recently in the Palcazu, the European colonists who had broader contacts were often responsible for introducing species and varieties from outside the immediate environment of the Amuesha farmer. Most recently, the development project offered the Amuesha hundreds of pijuayo (for palmheart and fruit production) and cacao plants, as well as sheep units. The immediate effect was to increase species and height diversity. A longer range effect in the case of cacao was to change the cropping succession by prolonging field cultivation. Cacao needs the more fertile alluvial soils and so was planted under plantains which provided the shade cacao needed. With this system plantain fields were not left to fallow on the usual schedule with two deleterious results. The Amuesha household had to weed an increased area under cultivation and the amount of older fallow available for new cultivation was reduced a few years later. Where lowlands were limited the amount of land in fallow was critical.

CHANGE IN AMUESHA AGRICULTURE

Agricultural Lands

Although change featured prominently in much of the variation in Amuesha agriculture, the only direct measure of change available was comparing land-use determined from airphotos taken in 1974 and 1980 and from our surveys in 1985 (Table VII). Land under cultivation along the Palcazu River stayed fairly constant and balanced with fallow lands. Change was limited to land given the Palcazu Project for an extension office, nursery, and experimental fields. With the school, church, and medical post in the original settlement area, most families still kept a house and some fields near the Palcazu, although many spent less time there due to land disputes over

←

FIG. 8. Construction of a fish trap. Subsistence agriculture was often supplemented by fishing, hunting and gathering.
FIG. 9. As labor and time were allocated, cassava was seldom neglected. Here, the skilled health post practitioner was preparing cassava to plant in his family's subsistence plot.

Table VII

Estimated hectares cultivated and under pasture in various sections of the Native Community of Puerto Laguna during three different years. Aerial photos were used for estimates from 1974 (1:25,000 negatives enlarged to 1 m × 1 m) and 1981 (1:30,000 negatives enlarged to 1 m × 1 m) and our own ground measurements used in 1985. ± indicates difficulty in precise measurement

Section	1974	1981	1985
Carachama River	8 ha	8	8
Inland	0	0.86	16.9
Raya River	18.5	53	±35
Palcazu River	±3	3.0	2.9
Road	0	0	7.6
Pasture around village (communal enterprise since 1978)	20	45	74

the limited rich alluvial soils. With a significantly increasing population, Amuesha were moving away from the community center on the Palcazu: first, up the River Raya to new areas of alluvial lowlands, and later, onto uplands along the new road and further interior.

Decrease in Raya cultivation in 1985 reflected a reduction in private pastures for several reasons. People joined the communal cattle enterprise (Fig. 10) and so did not need private pastures. There was also a switch from extensive private cattle management as a commercial enterprise to more restricted management as capital savings for unseen emergencies or for celebrations where a cow was substituted for a more traditional feast of wild game (i.e., female puberty rites, Fig. 11). This cutback of pasture land came about with low profits for individually owned cattle which had to be marketed through the European colonists who controlled the commerce and transportation of cattle. Subsequently,

FIG. 10. Communal cattle enterprise pastures surrounded the village of Puerto Laguna. While these communal pastures were expanding in area, private pastures were disappearing because of rabies and because individually marketed cattle were seldom profitable.

in 1986, the reversion of pasture lands became even more pronounced after rabies infected many privately owned cattle which farmers could not afford to vaccinate regularly. Private pastures were fallowed with no animals alive to graze. In contrast, the communal enterprise kept their cattle vaccinations current, and communal pasture lands surrounding the village increased with the growth of the enterprise.

There were different reasons given for field or family movement including opposite strategies of retreat from the road and the population center, and of active advance toward the road and market access. Opinion in Puerto Laguna was divided on the benefits and liabilities of the road and the approach of the greater Peruvian economy.

Land Tenure

A change in land tenure system came along with population growth and community titling of limited land. Originally, in 1970, land was not owned by individuals and the community as a whole supervised their lands. All the produce from any given field belonged to the household which cultivated it, but never the land itself, so that during fallow, land rights reverted to the community (usufruct rights). Since 1970, the community has grown substantially, with children born in that year marrying in 1985, and with land tenure becoming a hotly contested issue at community meetings. In 1985 the community decided to allot lands to all households, so that even if the land officially was owned communally, thereafter each household felt that it owned its parcel of land. Still, land could not be equally divided, and several young families ended up with upland parcels of poor soil quality, which influenced their subsistence strategies. These young families had the largest areas of cultivated land per member and yet were supported by extended families. This was the clearest indication that land pressure already was experienced in the community, even though it was only titled one year before this study.

Commercialization

At the time of land parcelling, several families requested land along the new road, specifically for commercial agriculture; other families incor-

FIG. 11. Young Amuesha woman dressed and painted for puberty rites represents the mixture of indigenous and modern influences on Amuesha life. For the feast of the puberty rites festival a cow was killed rather than the more traditional wild game. Not long after, the young woman left Puerto Laguna to begin her secondary education in Lima.

porated commercial products into their subsistence fields. Itinerant buyers sometimes forced all but the most stalwart farmers to sell their surplus produce, which in the past was kept for times of dearth, traded, lent or given to needy families, or if nowhere needed, let succumb to fallow. Profits were uniformly meager but welcomed by a basically self-sufficient society which wanted a few amenities and faced occasional crises. Imbalances were an occasional outcome of growing commercialism when stores were reduced, when animal breeding stocks or planting stocks were jeopardized, or when labor and time for the additional cultivation of cash crops became a significant burden.

Labor and Time Allocation

Community business, availability of off-farm labor, and expanded commercialism caused the greatest changes in allocation of Amuesha labor and time. Community administration and labor took increasingly significant amounts of time, especially with expanded cooperative ventures including a school, health post, store, and cattle, sheep, and forestry enterprises. Non-paid posts like community leader and coordinator for the cattle enterprise became difficult to fill and occasionally corrupted.

Off-farm labor became more convenient with road transportation to nearby coffee regions, to the development project, and to large farms. The road made commercial agriculture feasible. Both off-farm labor and commercial agriculture competed for time and labor with subsistence production which was reflected in variations in agricultural practices.

Family Structure and Stability

Changes in families are very difficult to document. Certainly, pairing, separating, birth and death have always been common occurrences (Smith, 1977). How has convenient transportation and off-farm labor of both men and women affected past trends? How much additional stress has been put on the extended family by these trends, by land pressure, and by commercial ventures? We cannot answer these questions directly but only document several cases where families were severely stressed or broken under the strain of these new patterns of labor and commerce. We do see that family stability has been a powerful positive force in agricultural production and in getting through hard times, but the effects of destabilization remain speculative and subjects for discussion.

External Influences

Again, external influences are not new to the Amuesha but the road and the development project have changed their nature and scope. "Improved varieties," new crops, and animals not previously tended have appeared at a rapid rate and in large numbers. Off-farm labor which previously was limited to the neighboring colonists' ranches or to the neighboring coffee growing region has extended to Lima with the draw toward Peru's centralized economy. The same was true of education which was limited previously to the community schools, but has been sought more recently in Lima (Fig. 11). Generally, external influences were reaching Laguna more immediately through the recent convenience of travel, and through the media of radio, magazine and newspaper.

Changes of Scale

Opportunities in hunting and gathering activities were presented by a fruit tree, flocking birds, or a beehive with honey. In swidden agriculture there was favorable weather, nutrient rich microsites, or new varieties of cultivars. In contrast, opportunities in capital/labor activities brought a change of scale in the opportunities offered and increased risks. If a cassava merchant was supplied from needed family reserves or if the family invested in a commercial cassava field which was significantly larger than most swiddens, these opportunities carried hazards of a different scale than those of an opportunistic hunter. The same was true of experimentation, where small scale experimental plantings of crops or varieties were recently superceded by large scale plantings of cacao and pijuayo seedlings given by the development project. These large scale plantings became a serious drain on time and fallow lands. The Amuesha were probably always experimenters and opportunists, but they have never faced the opportunities now offered, nor the long-term consequences of economic marginalization that could come if failure accompanied this change of scale.

Discussion

With recent attempts to understand and develop sustainable agriculture, Amuesha agriculture takes on some special significance. Amuesha swidden agriculture may have been sustained for some 4000 years (Allen, 1968; Jimenez, 1986). Under very difficult environmental conditions—with rainfall over 6000 mm/year, with acid, infertile, aluminum-toxic soils, and with tropical pest and pathogen pressures—it is no small feat to sustain agriculture. There may be lessons to be learned in the practice of Amuesha agriculture and in the cultural and biological evolution of the agricultural systems. Variation in Amuesha agriculture was great and due to numerous agronomic and social factors. From this variability and from these factors, perhaps two generalizations can be distilled: Amuesha sustainable agriculture is flexible and it responds to change. It is likely that Amuesha agriculture has been sustainable in the face of change due to its flexibility. Nonetheless, we worry whether it can and will continue to be so, because, recently, the Amuesha were experiencing not only qualitative change, but also changes in scale. The final question to be raised is whether traditional responses to change can deal with these changes in scale.

VARIATION AND CAUSES

Variation in Amuesha agriculture was found at all levels. Crops differed, spacing and density could be altered, field size and cover were modified, and diversity varied including species, height, field and cropping system diversity. Yard gardens differed, location and area under cultivation varied, and both cropping and farming systems were more and less diversified.

Reasons given for this impressive variability included complex family histories, great expectations, and simple differences of opinion. Field variability arose from variable agricultural practices from time of planting, intercropping techniques, and microsite utilization to whole farming system decisions. Social causes of variation ranged from an individual's time allocation to family stability and community land tenure policy. Furthermore, these variable causes of variability were themselves undergoing change. Thus, interpreting the limitations on and flexibility in Amuesha agriculture needs appreciation of agricultural detail and, simultaneously, of larger social, political and economic trends.

LIMITATIONS

Agricultural options of the Amuesha were limited by environmental variables common to many farmers, such as climate and soil fertility, but these limitations were probably more severe for the Amuesha because of the environmental extremes of the Palcazu Valley. Furthermore, the Amuesha lacked appropriate technology to ameliorate limitations from seed storage to weather forecasting. For example, time of planting was limited by people's ability to forecast the weather. Optimally, a field was cut during the end of the rainy season and the slash dried over an early dry spell at the end of which the field

was burned. Variability in the time of planting a new field most often reflected a limitation in people's ability to accurately predict the earliest prolonged dry spell. The Amuesha experienced other limitations less important to non-subsistence farmers for whom fish and game or extended family support is less necessary. Some limitations are of near universal concern to less developed farmers, like land tenure, time and labor allocations, costs of transportation and profit margins.

FLEXIBILITY AND BUFFERS

Limitations in Amuesha agriculture were overcome with flexibility and buffering. Cropping system flexibility was found in time of planting and harvest, cropping cycle, intercropping practices, and microsite utilization, as well as an overall flexibility in the farming system. Additional flexibility was actively sought by Amuesha farmers through opportunism and experimentation. It was obvious that the Amuesha depended heavily on the flexibility of their agriculture to get through difficult times.

Buffers to risk were used by the Amuesha ranging from intercropping, diversity and overproduction to reliance on the extended family. Diversity and overproduction assured a harvest of something or some amount even when disaster would strike. A well-established family has a developed support network to help them out in times of need. Buffering is a component of flexibility for Amuesha sustainable agriculture.

CHANGE

Change included historical processes such as opportunism and agricultural introductions; it included limitations such as the increasingly insufficient land base causing land tenure strife; and it included the entrance into commercial agriculture. When these changes took place slowly adaptation was possible; in the past communities divided and groups established new centers where more land was available, as with the founding of Puerto Laguna in 1970. Now that land is titled external colonists cannot expropriate Amuesha lands, but neither can growing Amuesha populations settle new lands, even within their traditional homelands (Barclay & Santos, 1980; Santos & Barclay, 1985; Smith, 1983, 1985).

As a distinct example of change, the scale of change itself was seen to change. External introduction of new varieties and crops has always taken place on a limited scale and has been essential to cropping system evolution. Recently, the development project offered opportunities on a much larger scale with the introduction of hundreds of pijuayo for palmheart and fruit production, with hundreds of cacao for chocolate manufacturing, and with units of African hair sheep when sheep were little known to the Amuesha. Past experimentation with a new subsistence crop on a limited scale was very different than recent experimentation with commercial crops which need a minimum quantity to be saleable. Even cassava grown for sale was planted as a whole new field which needed to be tended in addition to the subsistence components of Amuesha agriculture and represented substantially more time allocated to cultivation. In addition to this change in cultivation, there was the need for gaining market knowledge, difficult because of the unpredictable and volatile nature of the national Peruvian market.

Previously, the Amuesha met change through experimentation, opportunism and flexibility. Can these methods of cropping system adaptation be applied successfully to cacao orchards and sheep units? Individuals have attempted to do so, but the cost in time, labor and land allocation is high. A cacao plantation was established under the plantain rotation fitting this commercial production nicely into the subsistence rotations. However, the household soon realized that it then had an extra field to tend which would ordinarily have passed to fallow, and that it thus had less of its land in fallow for reopening later. A household which planted achiote for commercialization did not tend the planting when the price for the colorant dropped, but would undoubtedly try to rehabilitate it as the price rises. Chasing illusive market fluctuations proved to be a risky business.

With buffering too, the question arises whether traditional responses will be adequate with changes of scale. An extended family may buffer against a crop failure, but can it buffer against commercial marginalization within the national economy (e.g., Chevalier, 1982; Shoemaker, 1981)? With other indigenous groups, breakdown in family structure (e.g., Dieter-Heinen, 1972, 1975; Henley & Muller, 1978) and frag-

mentation of society (Dieter-Heinen, 1972, 1975; Murphy & Steward, 1956) have been attributed to the additional strain put on the community and the extended family during periods of change. Fortunately, in Laguna, we had yet to see symptoms of such disintegration.

At the time of this study, changes in Amuesha agriculture subsequent to the introduction of the Peruvian Amazon Perimeter Road and the large development project were in their initial stages. We were viewing the first tentative changes and a great diversity in initial attempts at adaptation to the changing environment. Our analysis differed from previous studies, which viewed change as a "fait accompli" (e.g., Davis, 1977), or which concentrated on great changes (e.g., MacDonald, 1984; Moran, 1981). We, instead, viewed the initial stages of change, the options and problems people perceived, and the choices they were making. This was possible partly because we were present at the moment the road arrived and partly because, with land titling and limited immigration, change was retarded. Additionally, the Amuesha have never been encouraged to give up subsistence agriculture. Swidden agriculture is not a system aimed at high yield, but at food security and maximizing production in relation to labor inputs (Ellen, 1982). Missionaries and colonists alike have found it advantageous to count on a stable, self-sufficient population which has provided a sporadic work force in exchange for minimal material goods. Although popular opinion in Peru holds that subsistence economies should be commercialized (e.g., the newspaper *El Comercio*), many agriculturalists in the Palcazu Valley appreciate the advantages of low input farming and subsistence security in Peru's present economic situation. With the unpredictable social, economic and ecological environment, a strong subsistence-agricultural base could give the Amuesha the option of independence from the consumer society, albeit attractive to the Amuesha.

CHOICES

Finally, to weave the components of variation and change in Amuesha agriculture into the real fabric of Amuesha life in Puerto Laguna, we can reconstruct the choices being made. The inhabitants of Laguna were very aware of many choices facing them, if not all the consequences. The components of their mixed strategies were easily identified: subsistence farming, fishing, hunting, gathering, borrowing, and cash enterprises including skilled and unskilled labor, commercial animal and crop agriculture, and, rarely, retail marketing. The selection among these were made by extended social processes and were neither absolute nor irreversible.

A major consideration for the Amuesha household was whether or not they would pursue capital and material goods. Options were simplified by the choice to continue on a largely subsistence economy, the attraction of which had brought many of the inhabitants to the Palcazu Valley and Puerto Laguna who were originally from upper valleys more connected with the Peruvian economy. Subsistence households stayed on the original Laguna settlement site along the Palcazu or they moved. The rich alluvial lands under cultivation along the Palcazu had stayed fairly constantly balanced between cultivated and fallow lands. That balance had been maintained through communal land tenure discussions. To avoid dispute many families moved, either up the Raya River or, less frequently, upland on poorer clay soils. The Raya River sites supported a diverse subsistence agriculture whereas the upland farms had limited production and were supplemented with fishing, hunting, and wage labor. The families which chose to retreat from land disputes much as did their ancestors (Smith, 1977) will not have this option in generations to come. This tactic of non-confrontational retreat is discussed by Lathrap (1970) and Golob (1982) as a native Amazonian response, but the potential for continued retreat will be limited by fixed boundaries of the native communities. The artificial nature of bounded native communities and the negative consequences are discussed by Barclay and Santos (1980), Smith (1983, 1985), and Santos and Barclay (1985).

Households which pursued capital did so through agriculture/forestry, labor, or retail sales. Commercial agriculture was usually pursued through cattle or chicken production, by opening a commercial field on upland soils along the road, or by incorporating commercial crops into subsistence fields. These commercial agriculturalists continued for a few years to evaluate their profits; if profitable and not overly demanding in time, labor or land, they continued commercialization and possibly expanded. If their venture did not succeed, the agriculturalist who had maintained a subsistence base reverted to low capitalization, while the agriculturalist with mainly commercial fields or cattle moved to the labor force or began again with subsistence farming. During this period of change, dependence on the extended family often increased, especially if the subsistence base was minor in comparison with their commercial venture.

Those reverting from commercial agriculture most often became laborers, skilled or unskilled. The opportunity for gaining a skill and practicing it in the community of Laguna had been limited so far to an Amuesha who occasionally worked as a commercial artist and to one health worker (Fig. 12), a profession which allowed the man to capitalize enough to open the only store in the community. Unskilled labor was sold or bartered at one time or another by nearly all males and several female members of the community. Some members worked for wages on the road construction or the development project, which drastically reduced the amount of effort they could put into subsistence agriculture. In the past, most people had worked part-time on debt-peonage agreements (discussed by Cope, 1972; Corry, 1976) for German colonists (Barclay & Santos, 1985) or on coffee plantations, much as in the rubber-boom era (Camino, 1984), repaying for basics lent them: machetes, cloth, medicines, pots, shotguns, needles, scissors, fishing gear, alcohol, etc. Recently, as workers sought out purchased goods they became more dependent on the national Peruvian market and the risk of economic marginalization increased (e.g., Chevalier, 1982; Shoemaker, 1981). When they no longer produced their basic necessities agriculturally, they were forced to turn to the extended family or available jobs, often the most menial. True marginalization had yet to take place in Laguna, but there were several families who had reduced their agricultural production and depended on jobs that were temporarily available with the small economic boom brought by the development project and road construction.

A major factor that influenced family strategy or buffered the consequences in choosing between subsistence and capital, and between agriculture and other labor was the degree of family establishment and the reliability of the extended family. When a family was well established with continuous land claims they often continued in

FIG. 12. The health worker extracting a tooth was one of the few professional workers in Puerto Laguna, where there were limited opportunities for gaining or practicing skills.

agriculture and might limit capitalization to a few cattle, small animals, a limited amount of commercial crops or a single member of the extended family working for wages. Poorly established families—the new families, the widow, the family that had experienced several deaths and hence had moved several times—had difficulties in even developing or maintaining a basic agricultural system. Young families just starting were easily lured into wage labor, because it took time to establish a full complement of subsistence fields. Once they took a job, less time was devoted to establishing subsistence fields, so the family continued to delay establishment of their subsistence agricultural base. Extended families were important in starting out and in getting through rough times.

In Laguna, the options discussed above were not absolute. For example, a move upland or upriver was sometimes temporary or relative since most families spent some time in the Palcazu River area because the school, church and healthpost are located there. All families vacillated to some extent in their choices: a laborer suddenly left his job for a week to take care of his family fields, and subsistence farmers sold several chickens or went to work on the coffee plantations. The healthpost worker planted two hundred pijuayo seedlings toward agricultural commercialization. Aiming for buffered flexibility, diversification was the preferred option along

with maintenance of the extended family, by providing capital and food to relatives in need.

Ultimately, the Amuesha were groping for their own means of development by exploring options, evaluating aberrations, taking advantage of opportunities, and experimenting. Responses to change which were previously favorable, such as opportunism or retreat, had unpredictable consequences under new conditions where opportunities were greater, but riskier and the land base was limited. The Amuesha were no strangers to change or adaptation, but their accustomed modes of change were being put to stringent tests. The Amuesha were directly limited by lack of experience and time to explore agricultural, social and economic alternatives which could affect their future as an indigenous group, family members, and individuals.

Acknowledgments

The primary author gratefully acknowledges generous funding provided by the Mellon Foundation and the United States Agency for International Development/Peru; and the secondary author, the Wahlberg Memorial Fund, the Vega Fund, Aixon Johnsson's Fund, and the Swedish Agency for Research Cooperation with Developing Countries. This is publication number 138 in the series of the Institute of Economic Botany of the New York Botanical Garden.

Literature Cited

Allen, W. L. 1968. A ceramic sequence from the Alto Pachitea, Peru: Some implications for the development of tropical forest culture in South America. Ph.D. Thesis. University of Illinois, Urbana.

Barclay, F. & F. Santos G. 1980. La conformación de las comunidades Amuesha (La legalización de un despojo territorial). Amazonia Peruana 3: 43–74.

——— & ———. 1985. El secreto de don Guillermo. Amazonia Indigena 5: 4–5.

Camino, A. 1984. Pasado y presente de las estrategias de subsistencia indigenas en la Amazonia Peruana: Problemas y posibilidades. Amazonia Peruana 5: 79–89.

Chevalier, J. 1982. Civilization and the stolen gift: Capital, kin, and the cult in eastern Peru. University of Toronto Press.

Cope, P. S. 1972. A contribution to the ethnography of the Colombian Macu. Ph.D. Thesis. University of Cambridge, Cambridge.

Corry, S. 1976. Towards Indian self-determination in Colombia. Survival International Document 2, London.

Davis, S. 1977. Victims of the miracle: Development and the Indians of Brazil. Cambridge University Press, Cambridge.

Dieter-Heinen, H. 1972. Adaptive changes in a tribal economy: A case study of the Winikina Warao. University Microfilms, Ann Arbor.

———. 1975. The Warao Indians of the Orinoco delta: An outline of their traditional economic organization and interrelation with the national economy. Antropologica 40: 24–55.

Ellen, R. F. 1982. Environment, subsistence and system: The ecology of small-scale social formations. Cambridge University Press, Cambridge.

Golob, A. 1982. The upper Amazon in historical perspective. Ph.D. Thesis. Cornell University, Ithaca.

Henley, P. & M.-C. Muller. 1978. Panare basketry: Means of commercial exchange and artistic expression. Antropologica 49: 29–130.

Izaguirre, P. Fr. B. 1923–1929. Historia de las misiones Franciscanas y narración de los progresos de la geografia en el oriente del Perú: Relatos originales y producciones en lenguas indigenas de various misioneros. Talleres Tipográficos de la Penitenciaria, Lima.

Jimenez, J. 1986. Preliminary archeological survey of the Palcazu Valley. USAID, Lima.

JRB Assoc. 1981. Central Selva Resources Management Project. 2 vols. USAID, Lima.

Larrabure y Correa, C. 1905–1909. Colección de documentos oficiales referente al departamento de Loreto. 18 vols. Lima.

Lathrap, D. 1970. The upper Amazon. Praeger, New York.

MacDonald, T. 1984. De cazadores a ganaderos: Cambios en la cultura y economia de los Quijos Quichua. Abya-Yala, Quito.

Moran, E. 1981. Developing the Amazon. Indiana University Press, Bloomington.

Mueller-Dombois, D. & H. Ellenberg. 1974. Aims and methods of vegetation ecology. Wiley, New York.

Murphy, R. & J. Steward. 1956. Tappers and trappers: Parallel processes in acculturation. Economic Development and Cultural Change 4: 335–353.

Ortiz, P. D. 1967. Oxapampa. Vols. I and II. Imp. San Antonio, Lima.

PEPP (Proyecto Especial Pichis Palcazu). 1981. Proyecciones y avances del Proyecto Especial Pichis-Palcazu. CENCIRA, Lima.

Peru. 1974. D.L. 20653. Ley de comunidades nativas y de promoción agropecuaria de las regiones de selva y ceja de selva. Lima.

———. 1978. D.L. 22175. Ley de comunidades nativas y de desarrollo agrario de las regiones de selva y ceja de selva. Lima.

Reiser, H. 1943. Indios. Braunschweig.

Salick, J. 1985. Subsistence and the single woman. Shupihui 10: 323–333.

———. 1989. Ecological basis of Amuesha agriculture, Peruvian upper Amazon. Adv. Econ. Bot. 7: 189–212.

Santos, F. 1980a. Vientos de un pueblo. Thesis for Licenciatura. Pontifica Universidad de Catolica, Lima.

———. 1980b. Belaúnde y la colonización de la Amazonia: De la fantasia a la realidad. Amazonia Indigena 1: 7–18.

——— & F. Barclay. 1985. Las comunidades nativas: Un etnocidio ideologico. Amazonia Indigena 5: 3–4.

Shoemaker, R. 1981. The peasants of El Dorado: Conflict and contradiction in a Peruvian frontier settlement. Cornell University Press, Ithaca.

Smith, R. 1977. Deliverance from chaos for a song: A social and religious interpretation of the ritual performance of Amuesha music. Ph.D. Thesis. Cornell University, Ithaca.

———. 1983. Las comunidades nativas y el mito del gran vacio Amazonico. AIDESEP, Lima.

———. 1985. Politica oficial y realidad indigena: Conceptos Amuesha de integración social y territorialidad. Amazonia Indigena 5: 11–16.

Stoll, D. 1985. Pescadores de hombres o fundadores de imperio? Desco, Lima.

Survival International USA. 1981. A road to destruction. SUISA News 1.

Swenson, S. & J. Narby. 1986. The Pichis-Palcazu Special Project in Peru—A consortium of international lenders. Cultural Survival Quart. 10: 19–24.

USAID/Peru. 1981. Central Selva Natural Resources Management Project Identification Document. USAID, Washington, D.C.

Varese, S. 1973. La sal de los cerros. Retablo de Papel Ed., Lima.

Ethnobotany of Contemporary Northeastern "Woodland" Indians: Its Sharing with the Public Through Photography

Judith G. Schmidt

Table of Contents

Abstract

SCHMIDT, J. (Institute of Economic Botany, The New York Botanical Garden, Bronx, New York 10458-5126, U.S.A.). Ethnobotany of contemporary Northeastern "Woodland" Indians: Its sharing with the public through photography. Advances in Economic Botany 8: 224–240. 1990. The present article is a discussion of the material developed for and presented in a four-projector slide presentation titled "Respect for Life," which demonstrates that information can still be obtained from persons in "Woodland Indian" communities of the northeastern United States and southeastern Canada on the uses of wild plants for food, ceremonial purposes, and technology (such as basketry). While the region is "species-poor" compared to the rich diversity of, for example, the Amazon, the contribution of knowledge of the usefulness of plants to the Native Americans of the northeast has been significant.

To forward the goal of sharing ethnobotanical knowledge with the general public, photography was used to record the research into several technologies still employed to make use of ash wood and birch bark in particular, and "weed" species in general. These have proved to be particularly effective for conveying a message on the waste involved in not using plant species around us. Alerted in this way to the potential of wild plants close to home, an audience also becomes more receptive to the urgency of protecting plants and animals both locally and internationally.

Key words: Ash wood; *Fraxinus*; birch bark; *Betula*; basketry; porcupine quill work; wild food plants.

Introduction

Is the Amazon region the only part of the Americas which is of value for or in need of ethnobotanical research today? No, not at all! Despite the diversity and richness of species, and the rapid and often devastating changes occurring to both the flora and fauna and the indigenous cultures of the tropical rain forest, which all urgently needs documenting, there are still valid reasons for research in North America— even in the generally by-passed and "species poor" Northeast. In seeking a region where ethnobotanical research could be carried out, it was quickly discovered that the "Woodland Indians" (both words are European designates) of the Northeast (Map: Fig. 1) offer the validity for some of us staying close to home base for research, even in the 1980's.

How did the original Americans of what is now Connecticut affect their natural environment? In 1974 this question prompted the choice for independent undergraduate research—in Connecticut, but through a mid-western college. From the first tentative studies it became evident that this is not a subject only of the past—there *are* indeed still many remnants of the "Woodland Indians" in the Northeast today. That initial study was cultural, superficial, imbued with a sense of discovery and wonder, and, personally, on totally new ground; yet it did plant the seed for future research, in ethnobotany.

Later, in deciding on the focus for graduate studies, it was tempting to plan to travel afar. But remembered was the light of interest and wonder in the eyes of both children and adults when the daily discoveries made about those "local Indians" were enthusiastically shared. It was realized that if a deeper knowledge was gained, its sharing could be educational and useful to society. Ethnobotany was chosen for this focus. The seed was sprouting.

GOAL OF THE RESEARCH

The goal of communication with the general public by bridging the gap commonly existing between it and the scientific community has given this study an unusual purpose from its inception. A camera was chosen as the primary tool for this communication. From the photographs, taken over eight years of fieldwork, a four-projector slide show titled "Respect for Life" has been developed.

Due to the unusual goal of the research, this paper is not written in the traditional scientific format. Instead it is presented as the story of how one scholar has attempted to bridge the gap between scientific studies and the general public. The informal style of this paper, although less personal, enthusiastic, varied, and anecdotal, approximates the style used for the public slide show. Built into the content are many of the messages conveyed to the lay audience, as well as indications of the method being used to cross this bridge. Only two utilizations of plant species are described in detail, and these are technological.

It was personally, and forcefully, stated at the start of this study that no attention would be paid to any political aspect of the Indian question. One is quickly disabused on that score. There can be no such separation of the American Indian from politics! Consequently, this type of forced personal learning connected with such a study has enhanced the awareness of bias, and how an average audience, when they arrive to see the slide presentation, could be viewing the Native Americans.

"Are there still Indians in the Northeast?" Such

FIG. 1. Map. Geographic area of Northeastern Woodland Indians.

wonder and surprise shown by the lay audience when they receive some current information about the Woodland Indians of today has frequently been equalled by the experienced scholars of Amazonian ethnobotany who live in New England. These scholars, familiar with the cultures of remote tribal peoples living in the tropical rain forests, may know very little or nothing of the indigenous population in their home state. Some of them have even marvelled that *any* ethnobotanical information can be gleaned from live Woodland Indians in the 1980's.

FOCUS OF THE STUDY

The ethnobotanical focus of this study has been on the uses of wild plants by the Woodland Indians of northeastern North America for food, ceremony and technology. The intent has been to garner whatever information one can today from anywhere within this geographic region (see Fig. 1)—in contrast to the "Plains Indians" for example—and therefore tribal distinctions have not been stressed.

Ethnobotany provides a channel through which desired messages about the culture of some of the Indian nations, and about the environment, can be delivered as a natural part of a slide show. For example, important fundamental and continuously useful aspects of the indigenous cul-

tures are illustrated, including the people's conceptual allegiances to the land. Inherent in this view of the land is an awareness of its value and of the need for its nurture, and a respect for all forms of life, along with the maintenance of a workable, sustainable balance. It is often stated by the members of the Indian nations that each individual has the responsibility of thinking and planning for the seven generations yet unborn—a guidance that inevitably changes one's treatment of the land!

Not surprisingly, many indigenous societal customs of the Northeast have now been abandoned. The plant uses most commonly retained are in the ever important field of healing, although for many years the traditional practice of medicine by the Indians was discouraged by the dominant governments. However, neither medicines nor hallucinogens have been a part of this particular study. Those two uses of plants are already the more thoroughly studied aspect of the ethnobotany of the region, and they are less suited to a slide show for people of varied ages and backgrounds. However, a few medicinally important uses are mentioned and the species pictured, to indicate their vital place in the cultures.

All aspects of these indigenous cultures are closely interrelated, including those of food, ceremony and technology. But, to the extent one can presume to categorize, it appears from the research that the utilization of wild plant species for food was the first of these traditional customs to be dropped. In some instances wild plant species are still in use for traditional ceremonies; this occurs both where these customs have actually continued since European settlement, and also in a number of instances where they have been revived today. Some technological uses of plants have been passed on from one generation to another, both for the purpose of making items for sale, and often just for the pride of workmanship itself. The technology was an aspect of the culture that was more free from the ridicule or attempted control by the dominant sector of the society.

Members of an audience will frequently be ignorant not only of the Native population of their state today, but also of its recent history. They often do not know that soon after European contact the indigenous population was decimated by disease, and then by warfare, nor that in the mid

1800's the government forced most of the tribal remnants to move to Indian Territory, which is now the state of Oklahoma.

In embarking on this doctoral study in 1978 it was soon evident that the indigenous cultures of the region have undergone erosive changes in the last 350 years. Today, in outward form, the life style of the indigenous peoples still living in the East approximates that of the dominant society. Nevertheless, fundamental and mainly internal differences do remain. Respect for some of these differences is conveyed through the slide show. Just how deeply rooted and basic some of these differences are is not fully realized until one lives with the people over extended periods of time. It was encouraging to discover the strength of the surviving pockets of indigenous cultures in the Northeast, even after three and a half centuries of oppression, and continuous attempts at cultural genocide in efforts to exorcise the so-called Indian problem.

Examples of Wild Plants Used by Northeastern Indians Today

FOODS

In endeavoring to learn directly from the people themselves, rather than to depend on already written sources, it has indeed been found possible to amass knowledge enough of wild plant foods to ensure dietary variety, and even a person's survival, during the growing season. Methods of preservation would need to be understood more fully in order to use traditional means to store sufficient foods for winter use, and this specific knowledge is difficult to obtain orally today. However, in existing literature there is such information, notably on the storage of cultigens (Waugh, 1973: 41–44).

Particularly significant to the purpose of the study has been learning the usefulness of many of the common "weeds," which most of us have made concerted efforts to eradicate. A message strikes a receptive chord when the audience can relate to it, so weed species make a particularly effective bridge for sharing scientific information with the public in the slide show.

In order to increase my own understanding of the processes, and to be able to share the resultant knowledge with an audience, a number of plant products have been successfully dried or frozen, or cooked for storage as preserves. Mushrooms were found to store well when dried, and they can also be frozen although some species are tough after being thawed. Bark, roots, leaves and flowers have been dried for beverage teas.

A partial list of plants experimentally preserved during this study would include:

Plant Dried for Beverage:

Cichorium intybus (Chicory): root, roasted & ground.
Ledum groenlandicum (Labrador Tea): leaves.
Matricaria matricarioides (Pineapple Weed): flowers, leaves.
Melilotus alba (White Sweet Clover): flowers, leaves.
Mentha spp. (Mint): leaves.
Rhus typhina (Staghorn Sumac): fruits.
Rubus allegheniensis (Blackberry): leaves.
Sassafras albidum (Sassafras): bark.
Taraxacum officinale (Dandelion): root, roasted & ground.
Trifolium pratense (Red Clover): flowers.

Plants Dried for Seasonings:

Asclepias syriaca (Common Milkweed): flowers.
Mentha spp. (Mint): leaves.
Myrica pensylvanica (Northern Bayberry): leaves.

Fruits Dried Most Successfully:

Vaccinium angustifolium (Blueberries).

Vegetables Frozen Most Successfully:

Phytolacca americana (Pokeweed): leaves.
Portulaca oleracea (Portulaca): leaves, stems.
Taraxacum officinale (Dandelion): leaves, blossoms.

Fruits Cooked for Preserves:

Cornus canadensis (Canada Bunchberry).
Prunus pensylvanica (Pin or Fire Cherry).
Sambucus pubens (Red Elderberry).

A Few of the Edible Plants Described in More Detail:

Typha angustifolia. T. latifolia (Cattail) (Fig. 2). These plants are often referred to as "the su-

FIG. 2. *Typha latifolia* (Cattail).

permarket of the swamps" because they provide a variety of serviceable and nourishing foods—which was demonstrated for this study by both Menominee in northern Wisconsin and Ojibwe in Michigan.

Edible material can be obtained from the starchy roots, which are available year round. The lower portion of the young stem is crisp and flavorful, both raw and cooked. For an early summer vegetable the male flowers can be lightly steamed and buttered while still on the cut stem. Moreover, these male flowers quickly develop an abundance of pollen which is gathered as a useful, light and nutritious flour. An Ojibwe woman spoke enthusiastically of using a fifty-fifty combination of Cattail pollen and store bought flour to make light and delicious pancakes. Additionally, the down from the well known brown seed spikes is one of the best of the wild insulators, traditionally stuffed into pillows, comforters, jackets and boots, as well as providing a soft baby's bed. One example of the down still being used to insulate boots in winter was obtained from an Ojibwe family on a lake in western Ontario. Yet, in spite of the still potential usefulness of the Cattail plants thousands

of acres of marshland have been drained and filled, destroying their habitat.

Sagittaria spp. (Arrowhead). This is another staple food of wet places, as among Ojibwe families near the shore of Lake Michigan. There is a valuable starchy tuber which, when loosened from its bed in the bottom mud, will float to the surface of the water for convenient harvesting.

Tragopogon porrifolius (Salsify). A plant of meadows and grass verges, Salsify has a thick root, which is garnered today for a cooked vegetable, as by some Ojibwe people in the midwestern states.

Taraxacum officinale (Common Dandelion). This is an introduced variety that the Native people quickly learned to utilize. There is also an indigenous species, *T. ceratophorum,* which is smaller. All parts of these "weeds" are good for food. The roots can be roasted and ground to make a coffee-style hot beverage. The leaves and blossoms are eaten raw and cooked. Gourmet quality dandelion blossom omelets can be a welcome addition to springtime menus. This plant is one of the few sources of body-soluble copper. It also provides yellow dye, and has important medicinal properties. Yet what have we done about it, other than to attempt to eradicate it? It is a prime example of the wasted usefulness of the plant species around us.

The uses of each of these species for food are still commonly known by members of the indigenous groups contacted across the Northeast.

In Michigan, Wisconsin and eastern Canada Native persons were found using some of the less well known fruits, including *Sambucus canadensis* (Black Elderberry), and *S. pubens* (Red Elderberry), the fruits of which must be fully ripened to be safe for eating. The fruits of both species can be made into a tasty preserve. The same necessity for full ripeness is true of the fruits of *Podophyllum peltatum* (Mayapple), which are then free of toxicity to man and can be eaten raw, or they also make a delicious preserve or sauce.

In the northeastern United States it is more generally the familiar varieties of wild fruits that are used today, and in abundance. These include fruits such as: *Vaccinium* spp. (Blueberries and Cranberries); *Rubus* spp. (Blackberries and Raspberries); *Fragaria* spp. (Strawberry).

Quercus spp. (Oak). The acorns, fruits of all species of oak, have been a prolific staple for flour all across this continent, wherever oaks grow. However, due to the lengthy process of leaching with water, necessary to remove the high content of tannin found in most oak species acorns, their use declined once milled flour became readily available. There was more then one traditional method of leaching. It has not been possible to get information on the personal practice of this process from indigenous people of the Northeast today. One of the typical methods is described in the literature (Weiner, 1980: 160, 161), in which the acorns are first ground to meal, the meal then placed in a shallow depression in porous material, and water percolated through it until the tannin is removed, which requires two or more hours.

Ceremony

It was with some of the Ojibwe people that I experienced the most ceremonies during this research. Ceremonial use has continued for several wild plant species, notably: *Juniperus virginiana* (Cedar); *Hierochloe odorata* (Sweetgrass); *Arctostaphylos uva-ursi* (Bearberry); and *Artemisia* spp. (Sage). One common method of use is to dry the leaves of these species for the burning mixtures known as Kinnikinnik (Algonquian), or "much mix." These may be burned in pipes, and also in the natural depressions of offering stones. *Verbascum* spp. (Mullein) makes a good igniter in these burning mixtures. Each of these herbs continue to be important for ceremonies. They are frequently employed in sweat lodges, morning and evening prayers, traditional weddings, vision quests and naming ceremonies, as well as in connection with the practice of healing.

Healing

Before discussing the technological uses of wild plants, a brief mention will be made of the healing art itself, which indeed cannot be entirely separated from the study of any other part of the Native American culture, and about which unsolicited information is frequently received.

First, it should be stressed that spiritual insight is an essential ingredient. Many medicine persons use plants; all who use plants combine this use with prayer; and some use prayer alone in

their life-long and successful practice of healing. Second, it is not a matter of an identical mixture of herbs employed in an identical manner every time a particular kind of ailment needs healing. Each case is treated as unique, and may require a different mixture, or other approach, for healing. The thought of the patient is generally considered, as much as are the physical symptoms. Third, from the choosing, gathering and preparation of herbal medicines, to their administration, there is no separation from spiritual vision. Fourth, respect for the land, and for all living things, as well as for their Creator, is integral to the healing practice. This leaves no room for decimation of a species, for rape of the soil, or for any other act which upsets the balance of life.

This attitude towards the plants, and its practice *has* lessened among the Native Americans of the Northeast, but it has not vanished. As a Montagnais in Quebec said recently: "You do not have to change because you become modern, have a house, a car, a T.V. We hunt with modern weapons, but we still have respect for the animals." His is a culture still closely associated with hunting and trapping (though in comparative moderation today), as well as with wild plants such as *Betula papyrifera* (White Birch) for technology.

Among the indigenous peoples there are many degrees and views of acculturation versus traditionalism, and of ways to deal with the dichotomy of this challenge. These searching questions apply not only to the speaking of their language and the practice of the spiritual aspects of their interrelated cultures—including ceremony and medicine, each of which was so long barred, or actively discouraged, by the government. But these questions even apply to the technologies such as basket making and the uses of birch bark.

Technology

DYES

Traditionally this was one of the important plant-use technologies. Since the introduction of aniline dyes most of the Indians have chosen to use them instead. However, it is still possible to find a few people using natural products for dye.

Dandelion has already been cited as a source

of yellow dye. *Alnus* spp. (Alder) provides bark from which is obtained a reddish-brown color. *Sanguinaria canadensis* (Bloodroot) stems and leaves give an orange-red dye. However, this plant is becoming scarce in some regions of its range, and for this reason its use has been discontinued by some of the Native artisans, including one Chocktaw basket maker who had still been accustomed to gathering the plant for his work. A yellow dye is derived from the roots of *Coptis trifolia* (Goldthread). *Phytolacca americana* (Pokeweed) has black shiny berries which produce a blue-black color.

There are other plants known to provide dye, some from the bark, others from nuts. There are a variety of lichens, too, which have been used for this purpose. Some of the young Indians today, for example a young Passamaquoddy woman in Maine, are anxious to re-learn the traditional uses of wild plants for dyes, so there has been a slight increase in this practice in the last decade.

FIBERS

Fibers have been crucial in the life of the indigenous people of the Northeast. Four of the most prized species of plants from which it has been obtained are: (a) *Tilia* spp. (Basswood), the inner bark of which can be used. A Menominee method of getting this fiber is to soak the bark in water for about two weeks, by which time the other parts of the bark have begun to decompose and it is easy to separate the wanted fiber; (b) *Asclepias incarnata* (Swamp Milkweed), which has fibrous stems; (c) *Urtica* spp. (Stinging Nettle), of which it is again the stem that yields the fiber (and a particularly fine one) from which a linen-like fabric used to be woven, as well as some of the more delicate nets and snares; and (d) *Apocynum canabinum* (Indian Hemp), which has a strong covering on its stem that can itself be used to tie things. By pounding the stem an abundance of fiber can also be obtained from the inside. During this study these fiber uses were demonstrated by individual Native Americans in Wisconsin, Michigan and southern New York.

As an ethnobotanist it had been hoped that some of the work would be in the woods and fields with knowledgeable people. But most of the persons who retain this traditional knowledge are now too elderly to go out to botanize. It is sometimes necessary to take fresh specimens directly to an informant to elicit accurate information on the plants. Indeed, it has often been easier to gain information on general history and social customs than to obtain information on the plant uses, as these have faded in importance for any of the Native people of the Northeast. This demonstrates the timeliness of my study. This is probably the last decade in which some of the information would have been available orally from persons who have practiced many of these traditional plant uses.

OTHER CRAFTS

Two examples of the utilitarian function of wild plants will now be explored in some depth, as is done through photography in the slide presentation. Technological uses of wild plants has not been necessary for survival in the Northeast for much of the period since the conquest by the settlers from Europe. In early colonial years there was a ready market for the items made from the wild, although the payment was pitiful. Family labor was easily divided for the gathering, preparation and crafting of wild plants into these useful and saleable items. It was not necessary to incur any direct monetary exchange in the process. Today, with families dispersed in search of jobs, and availability of necessary species declining, which the Indians say is due to damage from acid rain as well as land "development," the gathering is more difficult and there is an increasing monetary expense involved.

For example, when I accompanied a Passamaquoddy elder in Maine on an expedition to obtain a log of *Fraxinus nigra* (Black Ash), for basket making, we drove a 60 mile round trip. This man used to harvest trees within walking distance of his home and be able to utilize the whole tree. Today he gets only the six foot log, due to the logistics of transportation. Although he is a relative of the basket maker, he now expects to be paid for his time and gas.

Basket Making with Ash

Upon delivery of the log the next stage could begin. The basket maker paid another man to remove the bark and to pound the log up and down its length with the blunt back side of an axe (Fig. 3). "Hardwood is ring-porous, the cells

FIG. 3. A Maliseet pounding a log of *Fraxinus nigra* (Black Ash) with the back side of an axe to separate the annual growth rings. This gradually makes it possible to pull apart the log in strips, called splints. Tobique, New Brunswick.

FIG. 4. Pat Meuse, Maliseet, dividing the splint of Black Ash. When the annual growth ring is wide it can make two thinner splints for constructing "fancy" (more delicate) baskets.

laid down in the spring of the year, a time of rapid growth, are more coarse and less dense than those laid down during the summer. These cells collapse when pounded, allowing an easy separation of the grains or splints" (Bardwell, 1986). This pounding gradually separated the annual growth rings (ash being ring-porous), which could then be pulled off in long, narrow strips, called splints. These are used to make the sturdy ash splint baskets.

There has been some question about the indigenous nature of this craft. Ted Brasser (1975) contends the knowledge was imported from Sweden by early colonists, and adopted by the Indians to appeal to the settler's market. In a more recent study Kathryn Bardwell (1986) compellingly suggests that such a craft did exist here in precontact days. It seems logical that this unique feature of the ash—the easy separation of the annual growth rings—would have been utilized by the indigenous people. And they certainly did make and use baskets.

An English village tradition is the Morris dance, in which sticks are clashed together. Recently, a team of dance enthusiasts from Maine needed new sticks for their performances. They happened to cut some Black Ash. After only a few dances, they were surprised to discover that their new sticks began to separate into strips. An easy discovery indeed! Bang an ash stick a few times, and see how the annual growth rings separate, providing natural strips. There are so many uses for such sticks, it taxes the reason to suppose that Indians would not have made that discovery long before the Europeans arrived and used it for making some of the necessary storage containers.

Ash splint basket-making continues throughout the Woodland region today (Figs. 3–8). Particularly common in Maine, New Brunswick, Nova Scotia and New York, this craft is being

FIG. 7. Ash splint baskets, made with unshaven splints. Left: Small potato basket style with traditionally splint-bound rim (today the large potato baskets, made in hundreds, are usually nailed to the wooden rim). Louis Paul, Maliseet, New Brunswick. Center: Pack basket, shaped over a wooden mold. Right: Clothes basket. Bill Altvater, Passamaquoddy, Maine.

practiced by the Passamaquoddy (Figs. 7, 8), Penobscot (Fig. 8), Maliseet (Fig. 7), Micmac and Mohawk.

The rugged potato baskets used by the farmers of Maine and New Brunswick are constructed with thick, rough ash splints (Fig. 7). Pack baskets, long valued items of equipage for any trip into the woods, are also very sturdy, and the strong baskets for laundry and other utilitarian purposes last for years. The Native people also craft what they refer to as "fancy" baskets. For generations one of these most popular types— sewing baskets with fitted lids—have been turned out for the tourist trade (Fig. 8). The style, design, and useful function of the fancy baskets is as varied as the number of basket makers.

For this fancy work the splints are made smooth on each side. This is accomplished by drawing the splint under a knife held against a padding on the knee of the craftsperson, thus shaving off the rough surface (Fig. 5).

If the year's growth was good the splint is thick enough to be divided again for making these more delicate baskets. The careful dividing of the splint immediately doubles the yield as the inner side is already naturally smooth. This dividing can be done by hand, starting with a sharp knife, and continuing with the fingers, maintaining an equal balance of thickness on both sides, a slow process. Some artisans have a clever tool for the same purpose. They feed a splint up the center of the wooden tool, again use a knife to start an

←

FIG 5. Smoothing the surface of a splint of Black Ash by drawing it under a knife held against a padded knee, Maliseet.

FIG. 6. Forming the bottom of an ash splint basket with a double layer of splints. The "spokes" will then be bent up to form the standards for the basket sides, Maliseet.

equal division, and then pressing their knees firmly on either side of the wooden stand, and holding the two sides of the splint in either hand, in seconds they can pull it up through, and it is evenly divided along its entire length (Fig. 4).

The splints are trimmed into even widths. These widths vary from an inch or more to ones so narrow they can be used for finish sewing. A common method used for this cutting is to draw the splint across a gauge. The gauge is a wooden tool in which are embedded sharp metal blades equi-distant from each other. These blades may be at least an inch apart, or so close together they cut the very narrow splints used for sewing. The basket makers have a collection of various sizes of gauges. Another common tool is a wooden mold for forming the basket shape. These can be of many sizes and shapes. Some baskets are also made free hand. The choice of the method is individual.

The heavier ash splint baskets are generally made of ash alone. Even the top rim and the curved handles are bent from green ash wood, often White Ash (*Fraxinus americana*), which is shaved to the desired thickness on a shaving horse, using a draw knife. The fancy baskets are woven in more varied and decorative designs. The splints are often dyed (though seldom with natural dyes today). Sweetgrass (*Hierochloe odorata*), is frequently added after the basket is woven, for fragrance and decoration. It is used either in narrow straight bundles, or in thin braids added onto the splint basket. The sweet fragrance is activated by the drying, and is everlasting. Some basket makers can gather the Sweetgrass they need within reasonable distance of their home. Other artisans find it necessary to buy Sweetgrass from Indians who still know a wild source in their area; again adding to the basket makers' costs for materials.

Although ash splint baskets are strong, utilitarian and attractive, the market for them is not sufficient to offer much encouragement to the craft person. The price charged by the Indian may cover out of pocket expenses for materials, but it will pay little toward the hours of skilled labor. A poor self image has discouraged the Native craft person from placing a true monetary value on their work. Often a patronizing, denigrating bias on the part of a purchaser perpetuates the negative self image of the Indians, and blinds the buyer to the intrinsic value of the baskets. Very few young people are learning the craft; there is little incentive to do so. Yet it is worthy of continuance, and is one of the few remaining traditional and indigenous art forms of the Northeast.

Use of White Birch Bark

Another craft still practiced is the use of the bark of the *Betula papyrifera* (White Birch). Most of the artisans found continuing this work live in Canada. The Montagnais, Manawan and Cree of Quebec produce particularly sturdy containers, in a variety of shapes and sizes. There are even a few makers of birch bark canoes, notably in Quebec and Ontario. And occasionally a wigwam of birch bark is constructed, though it would generally be in response to a request for exhibition purposes.

If the outer whitish-grey bark of the birch is taken carefully, and at the right time of year, it can be removed without damaging the inner bark or cambium layer, which is the life blood of the tree. Such care in gathering not only enables the tree to survive but eventually the outer bark will grow back, ensuring another harvest for one's children.

The "right" time of year is determined by the seasonal flow of life-providing sap. Warming temperatures instigate the flow and thus govern their exact timing. When there is a plentiful supply of sap between the two layers of bark the

→

FIG. 8. Ash splint-sweetgrass "fancy" baskets. These splints are shaved smooth. The ash splint (*Fraxinus nigra*) curls are added to the finished basket on right; and bundles of Sweetgrass (*Hierochloe odorata*) are bound onto the edge of the lid. Narrow braids and a handle of Sweetgrass are added to the larger basket. Passamaquoddy, Maine.

FIG. 9. *Betula papyrifera* (Paper or Canoe Birch). Birch bark items. Left: Carrying basket with design made by scratching away all dried and darkened sap except where it is wanted for picturing leaves and flowers, and Center: Container, clearly showing root from White Spruce (*Picea glauca*) used for sewing. Both were made by the Siméon family, Montagnais, Quebec. Right: Moose call, through which moose-like bellowing calls are made, attracting the hunted moose in close. Tête de Boules, Quebec.

FIG. 10. Arthur Smith, Tête de Boules, using root from White Spruce (*Picea glauca*) to sew together sections of a birch bark wigwam for a museum in Ottawa.

FIG. 11. Gérard Siméon, Montagnais, Quebec, pouring water into a birch bark ladle he made in a few minutes from freshly gathered, pliable bark of *Betula alba*. The carefully executed design of folding allows the container to hold water, for carrying or drinking.

outer bark lifts off with surprising ease. The sap adhering to the inside of the outer bark looks colorless, but it soon dries to a chocolate brown. This provides a ready means of decoration. It is this inner side of the bark that is turned outward to form the outside of a container or canoe. A design can be made on it by scratching away small areas of the dried and darkened sap to make light colored forms surrounded by a dark background. A preferred method is to remove all the dry sap *except* the area forming the shape of the moose or other animal, or bird, flower, leaf, or geometric pattern. This forms a dark picture on a light background (Fig. 9).

The main sap flow occurs in conjunction with an early spell of hot weather. For the Montagnais this is often in June. With abundant sap flowing up through the entire cambium layer the supply of nutrients for the tree is not noticeably interrupted when the outer bark is removed. The ac-

tion of the sap itself will aid in the quick formation of a protective seal.

When only water is flowing it is still easy to remove the outer bark. But the water cannot form a protective seal. The water so essential to the tree is then continually evaporated from the exposed inner bark. The cambium will soon dry, and the tree will die.

Should the whole tree be wanted at any time during the growing season then there is no harm in taking this useful outer bark before felling it. The wood itself is used by the Native Americans for a variety of purposes, including the frames of snowshoes. When the tree is cut down then the cambium can also be removed, and used for tea or a yellow dye.

Birch bark has lines in it which run parallel to the ground. In the finest bark these lines are very short and infrequent. If there are many of the lines and they are several inches long the bark

FIG. 12.   An Ojibwe woman, Jean Pegahmagadow, in Ontario inserting a porcupine (*Erethizon dorsatum*) quill to decorate a birch bark box.

will be less strong, and more likely to split when under stress.

During the traditional gathering of bark by the Montagnais a number of pieces are placed on the ground in a growing pile. The pile is then rolled and tied, and can be carried by a simply formed tump line placed against the forehead; a method that leaves the hands free for axe carrying and pushing aside low branches as the gatherer walks out of the woods.

Once home the bark is laid flat for storage, with weights on it. If possible it is kept in a damp storage area. When it is fresh the bark is pliable and easily molded to a desired shape (Fig. 11). After lengthy storage, which causes the bark to become dry, it can still be worked, but it must first be soaked in water for several days. The time suggested for soaking varies with the artisan, from several days to weeks. Some Indians say that if heat is applied with care to bark that was stored and has become stiff, it will regain its pliability.

Exact methods of gathering, storing and using birch bark do vary among the distinctly defined groups of aboriginal people of the Northeast. Most of the information on birch bark cited in this paper is from the Montagnais of Quebec. There are minor variations in the procedures followed by the Cree, Manawan and others.

The root of White Spruce (*Picea glauca*) is the most widely used sewing material for birch bark, from the smallest containers (Fig. 12), to the largest canoe, or a wigwam (Fig. 10). If seams need to be watertight they are sealed. A common sealant is formed from a mixture of bear's grease, charcoal and resin.

Thomas Siméon, a Montagnais woodsman, said it is better to get on one's knees to cut a tree. It is gesture of deference to the tree, it is less work to do the cutting, and it is safer because one is in the best position to really see what the tree is going to do.

The bark containers, and sometimes even the canoes, are decorated with both naturalistic and stylized designs, scraped on the surface (Fig. 9). A separate art is the application of detailed designs made with porcupine (*Erethizon dorsatum*) quills generally on small birch bark boxes (Figs. 12, 13). This is an art that was almost lost. Fortunately, there is some revival of it today, particularly among the Ojibwe in northern Michigan

FIG. 13.   Quill box. Porcupine (*Erethizon dorsatum*) quill decoration on birch bark box. The quills are in naturally round shape and black and white shades. Ojibwe, Michigan.

and in Ontario. Many of the quills are left in their natural white and black shades, but some artisans will dye quills to add colors to portions of their designs. Again, aniline dyes are used today. The quills are inserted to form animal, plant and geometric patterns.

Once more one sees an example of lessening availability of a useful species as forests in the areas where the Native Americans live are clear cut, and the birch are gone. Also, the Indians say there has been a weakening of the White Birch by acid rain. The art of using birch bark is ingenious. The items made are useful today. But will the next generation continue to make them?

Conclusion

Whether or not an individual Native American chooses to continue the practice of traditional ceremonies (healing, religion), languages or crafts, the knowledge that he now has the legal right to make his own choice is an important link in his forward progress. It enhances his ability to survive as an individual of dignity. An Indian can keep with him these proven strengths, the true values and spiritual realities, and build on these traditional perceptions which do not change. These are still alive for each individual Native American, wherever they go in today's society, wherever and at whatever they work. It is essential that their own innate self-worth and dignity be restored so they can grapple with the challenges of today from a position of strength, both individually and collectively.

The slide show "Respect for Life" awakens an awareness of another culture in our midst. It provides another societal view, and also insight into a respect for and a care of the land which can help to halt the destruction of our natural environment. The audience sees the lessons we can learn through an understanding and appreciation of the Native Americans. As this awareness about a culture and plants close around them occurs among the general public, it inevitably makes the same people more easily and intelligently aware of, and responsive to, the implications, of such vital issues as the destruction of the tropical rainforests on a global scale. Indeed, it can make them more sensitive to any abuses of the delicate balance of life on our planet.

The indigenous knowledge of the uses of wild plants in the northeastern United States, and across the border into Canada, obtainable from written records and orally from the people themselves, now does not equal the fund of working data available, for example, from the Yanomami tribal people of Venezuela and Brazil. But even the knowledge collected orally during this study, since 1979, has provided far more information than can be used in one slide presentation. And this show has proved an effective vehicle for communicating with the general public the values of the wisdom of the Native Americans, and the vital necessity of re-gaining a respect for plants and recognizing their importance to us all.

As ethnobotanists we each accumulate important data. It seems essential that many of the insights which are afforded to us during this research—into fundamental life needs, and the means of preserving them—be effectively communicated to a larger audience. This paper has touched on information used in one method of accomplishing this goal.

This study has both documented some of the remaining ethnobotany of Native Americans and its presentation to the public. It is as important for the Institute of Economic Botany to be involved with the cultural remnants in our own society as to work with the more pristine Amazonian tribes. I plan to follow this work with the preparation of a visual presentation of the Institute of Economic Botany's research with South American Indians, because of the importance of the interpretation to the public of this type of information.

This is publication number 89 in the series of the Institute of Economic Botany of The New York Botanical Garden.

Annotated Bibliography

This bibliography contains published writings closely related to the content of this paper. Four of the entries are by Native Americans.

Black, Meredith Jean. 1980. Algonquin ethnobotany. An interpretation of aboriginal adaptation in southwestern Quebec. National Museum of Canada, Ottawa.
    Significant records of current fieldwork, with comparative quotations from prior written sources adding a certain perspective of value. One could wish the two sources were separable at a glance.
Bardwell, Kathryn. 1986. The case for an aboriginal origin of Northeast Indian woodsplint basketry. Man in the Northeast No. 31. State University of New York, Albany.

A persuasive reply to Ted Brasser and other doubters.

Brasser, Ted J. 1975. A basketful of Indian culture change. No. 22. Mercury Series. National Museum of Man, Canadian Ethnology Service, Ottawa.

The development and function of basketry among the Eastern Woodland Indians as an adaptive response to the Colonial market, and the argument that the splint basket making knowledge was not indigenous.

Gidmark, David. 1980. The Indian crafts of William and Mary Commanda. McGraw-Hill Ryerson Limited, Toronto.

The making of snowshoes, birch bark canoes and other traditional items as still practiced and described by the Commandas. These skills provide the livelihood of this Algonquin couple of Maniwaki, Quebec.

Kavasch, Barrie. 1979. Native harvests. Recipes and botanicals of the American Indian. Vintage Books, New York.

A variety of wild foods have been selected, with recipes for their use according to Native customs. There is a brief section on medicinals and cosmetics.

Keewaydinoquay. 1983. Direction we know: Walk in honor. Miniss Kitigan Drum, Box 176, Leland, Michigan.

Anishinaabeg (Ojibwe) ceremonial traditions, mainly related to passing over.

———. [no date]. Min: Anishinaabeg Ogimaawiminan. Blueberry: First fruit of the people. Miniss Kitigan Drum, Box 176, Leland, Michigan.

Traditions of use, and respect for a fruit vital to the Ojibwe people.

Nicholas, Joseph A. Ca. 1980. Baskets of the Dawnland people. Maine Indian Education, Box 412, River Road, Calais, Maine.

A brief and carefully illustrated story of ash splint baskets by a Passamaquoddy educator.

Simeon, Thomas & Louise. 1979. 1. Pusslagan le Casseau d'Encorce. Hull, Quebec (Lac St. Jean, Pointe Bleue, Quebec).

A thorough step by step description from the gathering of the bark to the finished decoration of a birch bark container as made by the parents of the Montagnais author.

Waugh, F. W. 1916 (1973). Iroquois foods and food preparation, fascimile ed. Government Printing Bureau, Ottawa.

A concentration on agricultural foods, including their preparation and preservation.

Webber, Alika Podolinsky. 1973. Birch bark baskets. Imperial Oil Review No. 4, Vol. 57, Toronto.

A brief, clearly illustrated article on the craft as practiced by some Têtes de Boule of Quebec today.

Weiner, Michael A. N. d. Earth medicine-earth food. Plant remedies, drugs, and natural foods of the North American Indians.

Principally devoted to medicines, with a small but detailed section on wild plant foods.

Whitehead, Ruth Holmes. 1982. Micmac quillwork. The Nova Scotia Museum, Halifax.

An exhaustive and honest study of porcupine quill decoration abundantly and excellently illustrated with color and black and white photographs.

Appendix

Some of the Indian nations from which information in this article was obtained:

Ojibwe	(Michigan, Ontario)
Menominee	(Wisconsin)
Passamaquoddy	(Maine)
Penobscot	(Maine)
Maliseet	(New Brunswick)
Mohawk	(New York)
Wampanoag	(Massachusetts)
Montagnais	(Quebec)
Cree	(Quebec)

Plant species mentioned in the text include (besides those in actual lists under the section on food):

FOOD

Podophyllum peltatum (Mayapple)
Quercus spp. (Oak)
Sambucus sp. (Elderberry)
Taraxacum officinale (Dandelion)
Typha angustifolia (Cattail)
T. latifolia (Cattail)

CEREMONY

Arctostaphylos uva-ursi (Bearberry)
Artemisia spp. (Sage)
Hierochloe odorata (Sweetgrass)
Juniperus virginiana (Red Cedar)
Verbascum thapsus (Mullein)

TECHNOLOGY

Dyes:

Alnus spp. (Alder)
Coptis trifolia (Goldthread)
Phytolacca americana (Pokeweed)
Sanguinaria canadensis (Bloodroot)

Fibers:

Apocynum cannabinum (Indian Hemp)
Asclepias incarnata (Swamp Milkweed)
Tilia spp. (Basswood)
Urtica spp. (Nettle)

Basketry:

Betula papyrifera (Canoe Birch, Paper Birch)
Fraxinus americana (White Ash)
F. nigra (Black Ash)
Picea glauca (White Spruce)

Commercial Management for Palm Heart from *Euterpe oleracea* Mart. (Palmae) in the Amazon Estuary and Tropical Forest Conservation

Jeremy Strudwick

Table of Contents

Abstract

STRUDWICK, J. (CUNY Graduate School, Dept. of Biology, Lehman College, Bronx, New York 10468, U.S.A.). Commercial management for palm heart from *Euterpe oleracea* Mart. (Palmae) in the Amazon estuary and tropical forest conservation. Advances in Economic Botany **8**: 241–248. 1990. *Euterpe oleracea* is a common multi-stemmed palm in the seasonally flooded, river-margin forests of the Amazon estuary. It is now the major world source of palm heart. Many palm heart factories are dotted along the rivers of the region, obtaining palm heart from either private individuals or from personnel supervised by the factory. Where the forest is carefully managed, a steady, renewable source of palm heart can result while at the same time helping to conserve a framework of native forests and soils. In this paper one such commercial management situation, currently in operation by a palm heart firm, is described. Economic forces are probably the major factor in the destruction of tropical forests. Conservationists should not be closed-minded to these forces. Perhaps if they would integrate their ideals with commercial ventures they could be more successful.

Key words: Agroforestry; Amazon estuary; conservation; economic botany; *Euterpe oleracea*; palm heart; palms.

Introduction

In a previous paper the ethnobotany, economic botany, habit and habitats of *Euterpe oleracea* Mart. (Fig. 1) in the Amazon estuary were discussed in detail (Strudwick & Sobel, 1988). To-gether with the use of the fruits, palm heart is the most important use in the region in terms of geographic extent and economics. 1982 figures (I.B.G.E., 1984) showed a total of almost 100,000 tons of palm heart production for Brazil. Of this, almost 93,000 tons came from the eastern part of the state of Pará where the Amazon estuary is located. Within this region lies the large island of Marajó, and it is upon here that this paper focuses. The Marajó area accounts for 85,000 tons or 85% of Brazil's total production of palm heart.

In this paper some possible benefits of planned commercial management systems for the palm heart industry and for soil and forest conservation in general are discussed and a working example of a commercial management scheme is described. It is not suggested that this or similar schemes are the answer to the complex subject of tropical forest conservation. For instance, the example presented does not allow for conservation of all plants (and therefore probably other organisms linked to those plants) within the area, since selective cutting is practiced. However, the scheme described does involve a fairly large commercial enterprise manipulating and using a native forest in situ without the introduction of exotic species. Clearly, this is a form of land use preferable, from the conservationist viewpoint, to the felling of forests and in their place cultivating the land and introducing exotic monoculture crops. The author suggests that this 'symbiosis' of commercial ventures using native forests is a realistic and valuable model towards meeting the interests both of developing countries and of conservationists and environmentalists.

Since the 'heart' of a palm includes the apical meristem of the palm shoot, and most palms are unable to branch, harvesting a palm heart kills the stem from which it is taken. Thus, in single-stemmed species, harvesting the palm heart results in the death of the entire palm. The genus *Euterpe* has long been the major world source of palm heart. The wild populations of certain single-stemmed species, such as *E. edulis,* have suffered because of this. However, since *E. oleracea* is multi-stemmed (Fig. 2), removal of palm hearts from this species does not kill the entire plant. The palm heart industry in the Amazon estuary is largely supplied by the cutting of this species from the native, seasonally flooded, river-margin forests and this is now the major world source of palm heart. Many palm heart factories are dotted along the numerous rivers in the estuary region in areas surrounded by these forests.

Palm hearts may be brought to the factories by both private individuals and by individuals employed or supervised by the factory management. Without a planned scheme of management, in certain cases all the palm heart in a given area may be cut, resulting in the complete removal of forest cover. In order to maintain a steady supply of palm heart, and thus ensure the continued operation and success of a factory, it is clearly preferable to manage an area of forest so that it will yield a continued supply. Studies have been made in both Brazil and Venezuela (Calzavara, 1972; Urdaneta, 1981) of methods of management that might best supply the demands of the industry. These have involved either plantations or wild stands of *Euterpe oleracea.* Such studies in Brazil may, in fact, have led to the development of management schemes such as the one described in this paper. Legislation has been introduced in Brazil in the past (Carneiro, 1970) to regulate the cutting of the palm and to attempt to enforce replanting. Other factors, besides its commercial importance for palm heart, place demands upon the industry to ensure the palm's continued abundance in the Amazon estuary. Since the açai liquid prepared from the fruits is such an important staple of the region (Strudwick & Sobel, 1988), the palm heart industry has been blamed locally for leading to a demise in *E. oleracea* populations in the past and consequent increases in the price of the liquid.

A primary conservation issue is the preservation of tropical rain forests. Clearly, one of the driving forces in the continuing destruction of these forests is economic growth in developing countries. Ideal conservation strategies for such areas might include satisfactory economic return combined with conservation of native biota and soils. Laws governing what individuals may do or not do are difficult, if not impossible, to police and enforce in tropical wildernesses. Were a commercial enterprise to police and conserve parts of an ecosystem for economic gain, a good 'symbiosis' could result, which would satisfy conservationists, private enterprise and politicians.

In October 1984 the author witnessed a situation, such as that described above, in operation on the island of Marajó. This enterprise is described in this paper as an example of an apparently successful commercial/conservation synthesis. Certainly, from a short visit the author cannot make conclusions about the long-term

FIG. 1. *Euterpe oleracea,* showing typical habit and habitat in the Amazon estuary.
FIG. 2. *Euterpe oleracea,* showing its characteristic multiple stems.

effects of this management system, nor of such details as what forest species are eliminated, how reproduction of species is affected, how wildlife presence is affected, etc. However, personal observation showed the area to have retained the overall appearance of a natural forest and healthy soil. It is not suggested here that this management system be used like a 'monoculture' throughout the region. It is suggested, however, that such enterprises deserve serious consideration in a realistic approach to conservation. The author is unable to say what percentage of the palm heart industry uses this scheme at present but enquiries with various management personnel of the factory involved revealed that it was being used by them in other areas also. To my

knowledge, this scheme has not been hitherto reported in the English literature. Describing it here may be useful for agroforesters and conservationists interested in suitable schemes for this region. Furthermore, by showing a scheme in practice, rather than just a theoretical situation, it demonstrates greater viability and credibility.

Often, business interests are seen as being in conflict with conservation interests; a common attitude seems to be to view commercialism and private enterprise as 'the enemy.' Such an attitude, it is argued here, is unproductive, unhealthy and impractical, leading to a polarization of interests. By presenting the example described here, it is suggested that both interests might be served in certain situations by private enterpris-

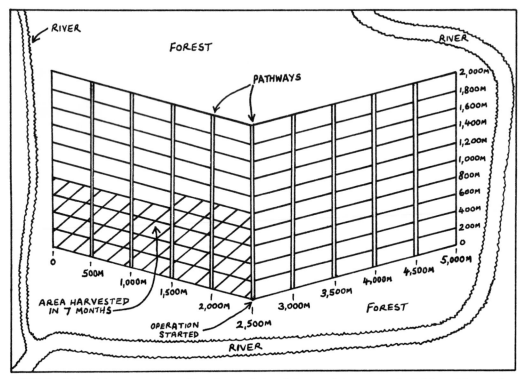

FIG. 3.   Map of the managed area. Note the proximity of rivers and the systematic division of the area into rectangular units.

es, given thoughtful management strategies. Most importantly, ways should be sought to accommodate both, especially in developing countries where economic improvement is often, and with good reason, the key driving force.

Methods

Description of the management scheme is based upon personal observations, conversations with different levels of management and workers of the palm heart factory and a map of the managed area. The work was carried out in October 1984.

Management Scheme

The palm heart factory, which serves as the center for the operation described here, is located in a town on the large island of Marajó in the Amazon estuary, Brazil. The town is small and isolated (as are all towns of this region) and is situated well into the interior of the island. Palm

hearts are delivered by boat here from the surrounding regions and processed as described previously (Strudwick & Sobel, 1988). Some of these are brought by private individuals who cut them from areas not under the control of the factory. Other land, however, is controlled by the factory and managed specifically for the production of palm heart. The area described here is actively managed by the palm heart company and is located about six hours by small motorized passenger boat upstream from the factory, well into the interior of Marajó.

The area under management (Fig. 3), as is common with *E. oleracea* habitats in the region, is subject to seasonal flooding. In fact, palm heart production from the factory is greatest from January to May since waters are higher and boats used to transport palm heart are able to get closer to the areas of production, enabling more palm heart to be brought in. During this period the managed land is flooded to an average depth of ca. 75 cm by the water from the adjoining rivers (Fig. 3). At other seasons, palm hearts must be bundled and carried by hand (Fig. 4) or secured in 'backpack'-like baskets (Fig. 5) to get them out of the forest to the riverside where they can be loaded into a boat for transport to the factory. This takes considerably more time and effort.

The land is managed by the factory under contract from other owners, including a timber extraction company. Not long ago (probably several decades) the area was virgin forest. Prior to management for palm heart only *Virola* and other selected timber trees were removed from the area. Besides providing good wood, this also promotes better development of *Euterpe oleracea*, probably due in part to better light penetration.

In order to manage the palm heart extraction area it is divided up into a grid so that workers can be moved through it in a systematic fashion without retracing their steps or missing an area. When they return they will move through it in the same chronological sequence, allowing equal time for the redevelopment of harvestable palm heart in all parts. The area is a slightly hinged rectangle measuring 5000 m in one direction, 2000 m in the other and is divided into smaller rectangles of 200 m by 500 m (Fig. 3).

The factory pays workers to clean the area of undesirable vegetation and debris at the same time as harvesting palm heart. The purpose of this is to get it into an optimum state for the future production of palm heart. A forest framework remains which is not dense, allowing sufficient light penetration for good palm heart production (Fig. 6). Other workers are paid on a daily basis solely to cut paths and keep them clear so that harvested palm heart can be carried out of the forest during the dry season (Fig. 5). In the rainy season access is by canoe. The workers who harvest the palm hearts are called palmiteiros (Fig. 7) and are able to cut and process about 100 stems per day. On average there are approximately 700 harvestable stems per 100 × 100 m area. The number of palmiteiros who come to the site varies between three and 11 per day but the average is about six.

When observed the project had been in operation for about seven months. It was estimated that it would take a total time of just over two years to completely harvest the entire area. Upon completion the project would go to another area, to return here when the palm heart grew to a sufficient size to harvest again (probably only a year or two later). It would begin again where it

FIG 4. Bundle of *Euterpe oleracea* palm hearts in protective surrounding leaf sheaths ready for carrying. A twisted leaf of the palm has been used to tie it.

first started. This point and the area covered in the first seven months are indicated on the map (Fig. 3).

To encourage the palmiteiros to follow this management scheme they were paid 70 cruzeiros per palm heart delivered to the factory at the time of observaton. Palm heart brought from non-factory supervised schemes fetched only 30 cruzeiros (the exchange rate during this period was approximately 2800 cruzeiros per U.S. dollar). Obviously this method of payment is a strong incentive for palmiteiros to work for the factory and follow a rational management scheme. As mentioned above, palmiteiros operating on the factory-managed land must prepare the forest for optimal future productivity. All stems of sufficient size to yield a good palm heart are cut and processed (Fig. 8) ready for transport to the factory (described in detail, Strudwick & Sobel, 1988). All the leaves from the crown of the palm are removed during this process and remain in the forest as an organic mulch (Fig. 9). At the

FIG. 5.   *Euterpe oleracea* palm hearts being carried from the forest in 'backpack'-like basket.
FIG. 6.   Managed area several months after palm heart harvest. Note its clean, open nature. The narrow trunks are *Euterpe oleracea*. The base of the large tree shows that a forest framework is maintained.

same time as harvesting the palm hearts, the palmiteiros must remove lower layers of brush, small shrubs and saplings, to allow young plants and seedlings of *Euterpe oleracea* to survive and grow. In the tropical heat and humidity the cut vegetation (stems, palm leaves and debris left after removal of the outer leaf sheaths from the palm hearts) takes only three months to rot down, leaving the forest floor clear (Fig. 6). Only a small portion of the stem containing the palm heart leaves the site, therefore the amount of organic matter on site will remain similar. Once the area is cleaned and the organic debris has rotted down, greater germination of *E. oleracea* appears to occur (Fig. 10), increasing the frequency of the palm in the forest. Cut, fallen and rotting palm leaves are cleared from the base of each palm to allow free growth of young shoots. Other tree species of over 2 cm width are left to become the future framework of the forest and for possible use as timber.

To ensure that the management scheme is ad-

→

FIG. 7.   One of the 'palmiteiros' who harvests the palm heart and cleans the forest.
FIG. 8.   Palmiteiros cutting outermost sheaths from the palm heart. These sheaths add to the organic refuse in the forest.
FIG. 9.   Leaves from the crown of *Euterpe oleracea* left to rot down after harvesting palm heart.
FIG. 10.   Profusion of *Euterpe oleracea* seedlings where forest was cleared by palmiteiros.

hered to the factory has a supervisor at the site full time. The palmiteiros spend two weeks at the site, then return down river to the factory and town, where they get paid and have their homes and families. Some of their wives are employed inside the factory. They have four days break before returning upstream.

Conclusion

The palm heart factory described here has been in operation for only a short time but was the only visible sign of large scale commercial employment for local people. As such it probably is an important source of income for some families in the town. Because the company's management system is carefully planned and supervised it will probably provide a steady source of palm heart so that the industry can be sustained in the area, as well as the jobs it supplies. This is significant because several factories in the Amazon estuary were seen to have closed, most likely due to fluctuating supplies of palm heart cut without management from the wild. This managed system also appears to help conserve native forest cover. Even if the conserved species are selected while others are removed, the manipulated forest can be viewed as a framework which, it can be surmised, aids in soil conservation and the preservation of a percentage of other native flora and fauna. The system also allows selective harvesting of timber. Other palm heart operations of the same company in different areas, using similar management methods, have been in practice for over ten years, demonstrating the apparent sustainability of the system. Since the system regulates cutting of palm heart, confining it to a designated area, as well as discouraging palm heart cutting from the wild by paying much less for material cut at random from the wild, it presumably frees up a great deal more plants of *E. oleracea* for fruit harvesting for the important açai liquid, both for subsistence use and for the commercial cottage industry (as described in Strudwick & Sobel, 1988). The company operating this system is one of the major palm heart companies in Brazil. This fact might be a reflection of the success of its palm heart forest man-

agement methods. It is not suggested here that this is, or should be, the only management system used for palm heart production from *Euterpe oleracea*. It does however provide a good role model for combining apparent economic growth and stability with conservation efforts in the region. Conservation of tropical forests and soils as well as economic growth of developing countries are real and important issues today. Commercial management systems which satisfy the aims of both these causes should be carefully considered by conservationists. Ways should be sought to work with commercial enterprises wherever possible and to work within the realism of economic development forces.

Acknowledgments

The author wishes to thank Gail L. Sobel for her invaluable support, advice, encouragement and enthusiasm. I also wish to thank the personnel of the palm heart company for their friendly help and hospitality. Funding was provided by Projeto Flora Amazonica (NSF grant BSR 8106632), which Dr. G. T. Prance made available to me. During the course of the field-work the author was supported by an Andrew Mellon doctoral fellowship with the Institute of Economic Botany. This is publication number 16 in the series of the Institute of Economic Botany of The New York Botanical Garden.

Literature Cited

Calzavara, B. B. G. 1972. As possibilidades do açai-zeiro no estuario amazônico. Bol. Fac. Agrar. Pará, Belém 5: 1–103.

Carneiro, N. 1970. Legislação. Palmito já tem manejo e incentivo para reflorestamento. Brasil Florestal 1(1): 55–60.

I.B.G.E. 1984. Produção Extrativa Vegetal 10. Fundação Instituto Brasileiro de Geografia e Estatística.

Strudwick, J. & G. L. Sobel. 1988. Uses of *Euterpe oleracea* Mart. in the Amazon estuary, Brazil. Adv. Econ. Bot. 6: 225–253.

Urdaneta, H. Finol. 1981. Planificacion silvicultural de los bosques ricos en palma manaca (*Euterpe oleracea*), en el delta del rio Orinoco. Fac. de Cienc. For., Univ. de los Andes, Mérida, Venezuela.

A Review of Sources for the Study of Náhuatl Plant Classification

David E. Williams

Table of Contents

Abstract

WILLIAMS, D. E. (Institute of Economic Botany, The New York Botanical Garden, Bronx, New York 10458-5126, U.S.A.). A review of sources for the study of Náhuatl plant classification. Advances in Economic Botany 8: 249–270. 1990. The broad knowledge of plants attained by the ancient Mexicans is well known, yet this branch of indigenous science has not received the attention it deserves. Few systematic treatments of Náhuatl botany have been undertaken. Náhuatl language and culture were dominant in Mesoamerica for centuries, and books using a pictorial writing system were widespread at the time of the Conquest. A small number of the surviving pictorial manuscripts provide important yet incompletely studied information regarding Aztec plant knowledge. The ethnohistoric sources of actual or potential value to the study of Náhuatl botany are reviewed, as are the modern interpretations of the subject. There is evidence suggesting a prehispanic system of scientific plant classification based on pictorial representations of the plants and plant groups. The reconstruction of Náhuatl plant classification through ethnohistoric and ethnographic research would facilitate identification of many plants whose uses are now forgotten.

Key words: Aztec botany; codex; ethnohistory; Náhuatl; plant classification.

Advances in Economic Botany 8: 249–270, 1990
© 1990 The New York Botanical Garden

Resumen

Aunque esta bien conocido el amplio conocimiento botánico que alcanzaron los antiguos mexicanos, esta parte importante de la ciencia indígena no ha recibido la atención que merece. La cultura y el idioma Náhuatl eran dominantes en Mesoamérica durante siglos, y el uso de libros y un sistema de escritura pictográfica eran ampliamente distribuidos en el tiempo de la conquista. Un numero reducido de estos manuscritos pictóricos han sobrevivido, los cuales nos proporcionan mucha información importante, aunque incompletamente estudiada, sobre el ordenamiento azteca del mundo vegetal. Se revisan las fuentes etnohistóricas de importancia comprobada o potencial en el estudio de botánica Náhuatl, asi como las interpretaciones modernas de las mismas. Hay evidencias que sugieren la existencia prehispánica de un sistema científico de clasificación vegetal basado en las representaciones pictóricos de las plantas. La recontrucción de los sistemas Náhuatl de clasificación, mediante investigaciones etnohistóricas y etnográficas, facilitaría la identificación de muchas plantas cuyos usos han sido olvidados.

Introduction

The purpose of this paper is to acquaint botanists with the rich body of Náhuatl botanical literature and to provide some background information that puts the study of Mesoamerican plants in its proper scientific, cultural, and historical context.

The Mesoamerican culture area, as defined by Kirchoff (1952), was one of the world's cradles of civilization, a distinction in the New World shared only with the Andean culture area. Mesoamerica distinguished itself from other culture areas by numerous unique traits, among which are several involving plants. Other distinctive culture traits include hieroglyphic writing systems, a written vigesimal numerical system, books folded screen-style, historical chronicles and maps, and a complex and extremely accurate calendric system. These traits formed part of a cohesive shared culture which encompassed numerous ethnic groups (Peterson, 1962). Mesoamerica, at the time of the Spanish invasion, extended from north-central Mexico to present-day Costa Rica (see Fig. 1).

Not only an independent cradle of civilization, Mesoamerica was also a world center of plant domestication (Vavilov, 1931). Archeobotanical excavations have revealed that man had begun actively manipulating and modifying the region's flora as early as 6500 B.C., and by about 2300 B.C. an agriculture based on maize, beans, squashes, chile peppers, and amaranth was practiced throughout Mesoamerica (MacNeish, 1967). By the time of the Spanish Conquest, peoples of the region had developed both agriculture and horticulture to a high degree, cultivating over 70 species of plants originating in Mesoamerica, as well as a number of species introduced from other areas (Dressler, 1953).

Náhuatl Culture and Language

The Aztec civilization was the last to dominate the central highlands of Mexico before the cataclysmic arrival of the Spanish conquerors. The Náhuatl culture of the Aztecs was shared by several peoples of the central highlands, and had evolved in the area long before the arrival of the Aztecs.

In Central Mexico an apogee of cultural sophistication was achieved in the Classic Period (300 A.D.–900 A.D.). Teotihuacan, in the Valley of Mexico, was the New World's first great city, and its influence was felt as far away as Guatemala. Important advances in the arts and sciences were made there before the eventual decline and fall of the city around 850 A.D. It is not known what language was spoken by the Teotihuacanos, but it may have been Náhuatl (Peterson, 1962).

Out of the ashes of Teotihuacan, and perhaps even incorporating some of its descendants, the first indisputably Náhuatl empire was established with the founding of Tula, in the present-day state of Hidalgo. The founders of Tula, known as the Toltecs, were Náhuatl speakers who came to Central Mexico some time during the ninth century. The name Toltec became synonymous with all things civilized, and the Toltecs were remembered by their Aztec successors as being skilled in architecture, agriculture, weaving, sculpture, pottery, astronomy, and, not least of all, warfare. The cosmopolitan Toltec culture was in contact with all of civilized Mesoamerica (Vaillant, 1962). A combination of drought, crop failure, and invading barbarians caused the abandonment of Tula around the year 1200 A.D. (Peterson, 1962). The Toltecs scattered and formed small city-states, while maintaining their high cultural standards and militaristic ways. Some of the invading barbarians chose to adopt the civilized customs of the Toltecs; among these invaders were the Mexica, better known as the Aztecs.

After their arrival in the Valley of Mexico around the year 1250 A.D. and the founding of their capital, Tenochtitlan, around 1325 A.D., the Aztecs made a spectacular rise to power through alliances and conquests, so that by the time of the Spanish arrival they commanded the most powerful and extensive empire in the long history of Mesoamerica. The Aztecs freely and enthusiastically incorporated the trappings of Toltec civilization in their own militaristic and bloodthirsty style and expanded the sphere of influence of Náhuatl culture (Vaillant, 1962). Although they had varying political allegiances, other groups such as the Texcocans, Chalcans, Tlaxcaltecs, Huexotzincans, and Cholultecs were also members of the Náhuatl culture of the central highlands.

Because it was the Aztecs who came into direct contact with the Spanish conquerors, it is their form of Náhuatl culture that is best known and their form of the language, now known as Classical Náhuatl, which became the standard against which all other dialects were compared. In this paper, however, the term Náhuatl is used in the broad sense, to include all dialects.

Náhuatl has been spoken in Central Mexico, and in various parts of Central America, from at least Toltec times (ca. ninth century) to the present. At the time of the Conquest it was the lingua franca of Mesoamerica. Bilingualism among Indians was so widely reported in the late sixteenth century that there were probably few localities throughout Mesoamerica where Ná-

huatl would not serve (see Fig. 1). In contrast to other Indian languages, Náhuatl seems to have been highly respected and was considered a refined tongue. It has been estimated from ethnohistorical sources that on the eve of the Spanish Conquest nearly 19 million Náhuatl speakers lived in Central Mexico alone (Harvey, 1972).

Although the Aztecs did not possess a true alphabet, a literary tradition existed in ancient Mexico. One of the first tasks of the Christian missionaries in Mexico was to reduce the Náhuatl language to the Spanish alphabet. Thousands of post-Conquest documents were produced in Náhuatl (transcribed in the Spanish alphabet) (León-Portilla, 1963), leaving us with a considerable body of Náhuatl literature—a unique situation for an Amerindian language.

In the 450 years following the Conquest, various socio-political factors brought about the displacement of the Classical Náhuatl dialect by the language of the conquerors (Garibay-K., 1970). Modern Náhuatl speakers pertain to numerous disjunct groups, each speaking different dialects incorporating Spanish loan words in greater or lesser degree.

Náhuatl speaking peoples today constitute Mexico's largest indigenous group (Madsen, 1969), estimated in 1946 to number over one million (Whorf, 1946), and today numbering probably more than 750,000 (León-Portilla, 1971).

The Role of Plants in Mesoamerican Life and Culture

The importance of plants in Mesoamerican life and culture cannot be overestimated. The particular geography of the region encompasses nearly every ecological zone on earth, all within relatively short distances of one another. As a consequence, the floristic diversity available to the Mesoamericans ranks among the richest in the world (Rzedowski, 1981). Mesoamerica was an important world center of plant domestication and its inhabitants had carried the art of agriculture to high levels, employing complex systems of polyculture, terracing, irrigation, raised fields, and the so-called floating gardens or *chinampas*. In a practice that dates from pre-agricultural times, the collection of wild plants persisted, as it has even to present times (Wilken,

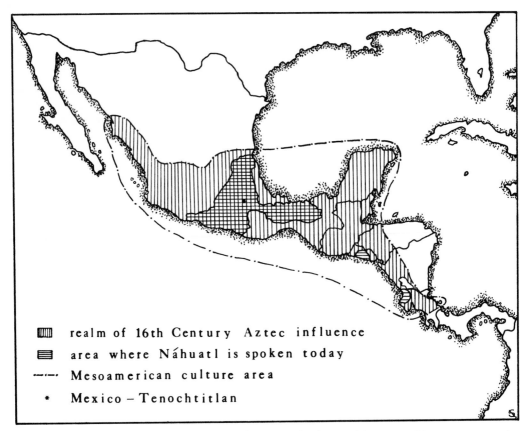

FIG. 1. The Aztec empire dominated much of Mesoamerica in the 16th century. Those areas not under direct Aztec control were influenced by the Aztecs through trade relations and cultural exchange. Náhuatl remains the first language of several hundred thousand people in Central Mexico.

1970). An extensive pharmacopoeia of medicinal plants was used in ancient Mexico, a portion of which remains intact, especially in rural areas.

As Standley (1920) pointed out, the Aztecs not only had knowledge of the economic qualities of plants but also a well-developed aesthetic appreciation of plants. Flowers were an essential element of daily life and were equally important in religious ceremony. Cut flowers were abundantly sold in the markets. Moreover, the metaphoric significance of flowers even came to represent a genre of Náhuatl literature. In the *xochicuicatl* or poetry (literally "flowery songs"), there are many references to flowers and their attributes (León-Portilla, 1985). In fact, one of the traits of Náhuatl poetry was the preoccupation with flowers (Wasson, 1973). The deep-seated love of flowers by the Mexican people has persisted to the present day.

Much has been written about the botanical gardens of ancient Mexico. Paso y Troncoso (1885) maintains that botanical gardens were not a recent development in Central Mexico but rather were the perpetuation of a tradition dating back at least as far as the Toltecs (ca. ninth century A.D.).

Probably the most complete account of the gardens of ancient Mexico is that of the archeologist Zelia Nuttall (1925). Located at numerous sites around the central highlands, the gardens were sometimes large and elaborate. They predated and were superior to contemporary sixteenth century botanical gardens of Europe (Dressler, 1953; Standley, 1920). Moctezuma's gardens at Ixtapalapa were noted with enthusiasm by both Cortéz and Bernal Díaz (1955). Nuttall reports that Moctezuma kept sumptuous gardens in which he ordered his physicians to conduct experiments with medicinal herbs. The famous garden at Huaxtepec was supplied with lowland species imported from distant provinces and planted in artificially maintained conditions of warmth and humidity. Native gardeners from the respective regions were brought with these exotic plants to ensure their well-being. The early chroniclers were filled with admiration for this garden and all state that it was unsurpassed even in Spain.

Nezahualcoyotl, the poet-king of Texcoco, was a patron of the arts and sciences who possessed numerous gardens, some of which he inherited from his royal antecedents. He constructed a noteworthy botanical garden not far from Texcoco, at Tezcutcingo. Tezcutcingo was an elaborately terraced hill, irrigated by a massive aqueduct. It incorporated religious shrines and sculptures, and had baths and stairways cut into the living rock.

Under Nezahualcoyotl, Texcoco flourished as the seat of Náhuatl culture. It was the site of libraries as well as a *calmecac,* a school of higher learning. In the *calmecac* the fine arts of codex painting and reading were taught, along with other scholarly and religious pursuits. It appears that some of the *calmecac* were specialized in the training of professionals in different fields (Calnek, 1988) and there is evidence to suggest that the Texcoco *calmecac* was a center of botanical study. The Spanish naturalist, Francisco Hernández, writing between 1570 and 1577, recorded that Nezahualcoyotl had devoted himself to the study of plants and animals and, unable to have living specimens of many of the tropical species, had pictures of them painted from nature and copied upon the walls of his palace. These depictions of exotic plants were so excellent that Hernández was able to make use of them (Nuttall, 1925).

Standley (1920: 9) affirms that

"the true history of botanical activity in Mexico begins at a much earlier date [than 1519], for the native inhabitants, who had already reached a high degree of civilization, may be said to have begun scientific researches. No other primitive people, probably, ever took so great an interest in botanical matters, and at the time of the Conquest none of the nations of Europe were much superior to the Mexicans in botanical knowledge."

The picture books (codices) that survived the Conquest and those produced shortly thereafter contain ample evidence of the degree to which plants permeated the daily and spiritual lives of the ancient Mexicans. Important agricultural plants such as maize (*Zea mays* L., Gramineae) and maguey (*Agave* spp., Agavaceae) were venerated through their respective gods, *Cinteotl* and *Mayahuel.* Grain amaranths (*Amaranthus hypochondriacus* L. and *A. cruentus* L., Amaranthaceae) were intimately involved in certain religious rituals, a fact which has been cited as a motive for a Spanish campaign to eradicate these cultigens (Sauer, 1950). Flower glyphs accompanying speech glyphs in the codices indicated the utterance of song and poetry rather than rhetoric. Illnesses were cured and gods were appeased with herbs. The codices and other records left by the sixteenth century chroniclers are our richest sources for the reconstruction of the botanical knowledge of the Náhuatl peoples. Ethnographic investigation also has its place, as vestiges of this ancient knowledge are still utilized by many Mexicans today. The ethnohistoric and ethnographic sources each yield data important to the problem that the other cannot.

Ethnohistoric and Ethnographic Sources

THE ETHNOHISTORIC SOURCES

Ethnohistory is the field of study concerned with the reconstruction of native cultures from written sources. This is an especially fruitful field of study in the case of Mesoamerican cultures, which were unique among Amerindian groups because they already possessed an indigenous body of written history and literature at the time of European contact.

Indirect evidence such as painted polychrome ceramics, bark paper beaters, murals, and hieroglyphic writings on stone and wood suggests that paper making and manuscript painting were practiced as early as the Classic Period (300–900 A.D.). Painted books were certainly in extensive use by the time of the Toltecs. Books were known at the time of the Conquest from Central Mexico to as far south as Nicaragua, and few early accounts fail to mention their existence (Glass, 1975). It is said that Moctezuma learned of Cortez's arrival on the shores of Mexico, and of the outcome of subsequent battles, via written reports brought to him by runners (Peterson, 1962).

Prehispanic Painted Books

Prehispanic painted books are called codices, a term which has been extended to include ancient historical sources, regardless of date of origin, that are based on Indian paintings and reflect the native pictorial tradition. They were usually produced on paper (*amatl*) made from the bark of the *amacuahuitl* (=paper tree, *Ficus* spp., Moraceae). Other codices were produced on animal hide, cloth, and later, on European paper. There were various formats including single sheets, long strips, rolls, and most commonly, screenfold books painted on both sides (Nicholson, 1975).

The pre-Conquest codices were produced by scribes known as *tlacuilohqueh* (plural of *tlacuiloh*) at the schools of higher learning or *calmecac*, where they were trained and employed. These scribes may have held hereditary positions and they enjoyed a high social status. The famous chronicler, Bernal Díaz del Castillo (1955), who accompanied Cortéz, reported that the conquerors frequently found libraries, called *amoxcalli* (=book house), complete with librarians and scribes.

Bark paper was an important tribute item. It is known from tribute lists that a total of 480,000 sheets of paper were paid annually to Tenochtitlan by various towns. This is a good indicator that a large quantity of literature was being generated and Gates (1939) feels certain that some of this was botanical description.

Few prehispanic codices survived the Conquest. Cortéz burned many libraries in the towns he conquered, not always the express purpose of doing so—he simply burned the whole town. A much more systematic program of destruction was implemented, however, by the offices of the Holy Inquisition. Bishop Landa in Yucatán remarked, "We found a large number of books of these characters, and as they contained nothing but superstition and lies of the devil, we burned them all, which the Indians regretted to an amazing degree and which caused them great anguish" (Peterson, 1962: 240). The first Archbishop of Mexico, Juan de Zumarraga, burned a large number of codices in the Valley of Mexico, probably including the great libraries of Netzahualcoyotl in Texcoco. Zumarraga stated of the books, "All of them contained figures and characters that represented rational and irrational animals, herbs,

trees, stone, mountains, waters, and other things of like tone, being understood that it was a demonstration of superstitious idolatry" (Peterson, 1962: 240).

The reported acts of colonial destruction probably represent only a fraction of the true total (Gibson, 1975). The Indians themselves very likely destroyed some books lest they be tried as idolators and burned at the stake or strangled by the Inquisitors (Peterson, 1962). Most of the codices that escaped detection by the Inquisition undoubtedly fell victim to the ravages of time, neglect, and political upheaval (Glass, 1975).

But, as Peterson (1962) points out, the missionary zeal was a "two-edged weapon," destroying yet investigating and documenting Indian customs. Historians such as Sahagún and Hernández not only utilized existing codices but also commissioned pictorial manuscripts for their own writings. Indians as well as Spaniards found that maps, tribute registers, and census documents made in the native tradition satisfied a common need. But by the early years of the seventeenth century the pictorial manuscript as a traditional form of communication was virtually extinct (Glass, 1975).

Of the thousands of codices which were known to exist, only sixteen putatively pre-Conquest survivals have come down to us. Three of the pre-Conquest books are Mayan and not applicable to the present study. Of the remaining thirteen, none is indisputably pre-Conquest and none deals specifically with plants; all are ritual-calendric in their subject matter (Glass, 1975). Peterson (1962) believes from indications and hints of the chroniclers that the ancient Mexicans also had books of botany, zoology, and medicine. Clavigero (1945) shares this opinion, but no such prehispanic books have survived.

Perhaps as many as 500 colonial pictorial manuscripts in the native tradition have survived (Nicholson, 1975). Glass (1975: 4), in a major contribution to the study of Mesoamerican ethnohistory, compiled a census of pictorial manuscripts which provides "a comprehensive and descriptive inventory as well as a guide to the bibliography of the surviving corpus of Indian manuscript drawings and paintings that are in the native artistic and historical traditions." An invaluable tool for students of ancient Mexico, the census includes most known pictorial manuscripts or codices, numbering 434, nearly all of which were produced during the sixteenth century.

A review of Glass's (1975) census and of those codices (copies or facsimiles) available to the present author has revealed only nine manuscripts which deal to some extent with plants and which are or may be useful for the study of Náhuatl plant classification.

The nine manuscripts fall into two categories of origin, as determined by Glass (1975). Manuscripts produced under Spanish patronage are in the first category. The Catholic Church, backed by viceregal authority, promoted the systematic investigation of native customs and religion in order to satisfy the curiosity of Renaissance Europe. These documents rank second in importance only to pre-Conquest survivals as they were the product of conscious efforts to document native affairs and knowledge. Four manuscripts produced under Spanish patronage deal specifically with plants: the *Florentine Codex,* the *Libellus de Medicinalibus Indorum Herbis,* the *Codex Mendoza,* and the *Matrícula de Tributos.*

The remaining pictorial manuscripts are grouped under the second category, those of mixed colonial origin. These manuscripts were destined for use as records of the economic and mundane affairs of the colonial world and concerned primarily with tribute and land titles. Five mixed colonial manuscripts have specific references to plants: the *Oztoticpac Lands Map,* the *Humboldt Fragment 14,* the *Codex Osuna,* the *Relación de la Grana Cochinilla,* and the *Historia Natural de Nueva España.* It is likely that other manuscripts not discussed here, particularly those dealing with tribute payment, may also yield valuable information regarding plant use and classification.

Documents Produced Under Spanish Patronage

Florentine Codex. Also known as the *Historia general de las cosas de Nueva España,* this encyclopedic work produced by the Franciscan friar, Bernardino de Sahagún (1499–1590), represents the single most important record of pre-Conquest life available to scholars. The original is housed in the Biblioteca Medicea Laurenziana in Florence, Italy and was produced in Tlatelolco, D.F. around 1575–1580. It comprises three large volumes containing twelve books. The end result of nearly a lifetime's work, the final product is essentially three interdependent parts: (1) a Náhuatl text, (2) a Spanish text, and (3) a pictorial account (D'Olwer & Cline, 1973). In Book X, entitled The People, chapters 18, 19, 21, 22, and 24–26 deal with foods and herbs sold in the markets. In Chapter 28, paragraphs 2–6 deal with medicinal plants. Book XI, entitled The Earthly Things, is a long summary of zoology, botany, and related topics of natural history (Cline, 1973). This remarkable book contains over one-half (965) of all the pictorial images (1864) in the entire codex (Glass, 1975). At least 450 of the illustrations in Book XI are of plants, many of them representing more than one species (Fig. 2). The botanical paintings, although sometimes incomplete or uncolored and obviously executed by more than one artist, are often rendered with great morphological detail and striking beauty.

No discussion of the *Florentine Codex* would be complete without some mention of the life of Sahagún himself. The Franciscan Fray Bernardino de Sahagún arrived in Mexico to begin his apostolate in 1529, only eight years after the fall of the Aztec Empire. Upon his arrival, Sahagún became interested in the Indians not only as potential converts but equally as bearers of high culture and a rich historical past. He quickly gained the reputation of master without peer in the Náhuatl language. Around the year 1559, Sahagún was ordered to "write in the Mexican language all that which may seem useful for the indoctrination, culture, and religious conversion to Christianity among the natives of New Spain, to aid the workers and missionaries toward their indoctrination" (D'Olwer & Cline, 1973). He took up the mandate with enthusiasm and spent the following 32 years carrying it out. He meticulously cross-checked his data three times, empirically employing a rigorous method of ethnographic research which produced an extremely objective and reliable work. After 60 years of productive missionary life, Sahagún died in Mexico City at the age of 90 (D'Olwer & Cline, 1973).

The *Florentine Codex* was practically unknown until after 1882 and no significant part of the manuscript was published until about 1889 (Glass, 1975). One of the most useful and accessible editions is that edited and translated (into Spanish) by Garibay-K. (1956). It has been eclipsed only by the eleven volume effort of Dib-

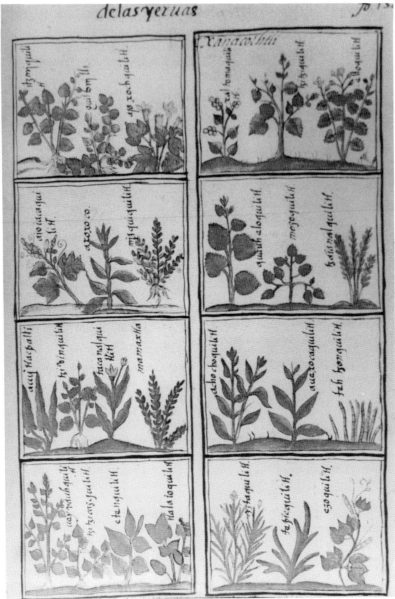

FIG. 2.   Book XI, Chapter 7, folio 134 recto of the *Florentine Codex* (Sahagún, 1979) illustrating edible plants belonging to the Náhuatl genus *quilitl,* some of which remain unidentified.

ble and Anderson (1950–1969) which provides the complete Náhuatl text with an English translation accompanied by black-and-white photographs of the color lithographs published by Paso y Troncoso (1905–1907). In recognition of its place in the nation's patrimony, the Mexican government commissioned a highly faithful fac-

simile edition of the *Florentine Codex* (Sahagún, 1979). The publication of this facsimile has, for the first time in 400 years, made Sahagún's *magnum opus* readily available in its original form to a wide variety of scholars.

Libellus de Medicinalibus Indorum Herbis. Also known as the *Badianus Manuscript,* the *Codex*

FIG. 3. Folio 39 recto of the *Libellus de Medicinalibus Indorum Herbis* (de la Cruz, 1964), showing use of native pictorial elements, including stylized soil glyphs.

Barberini, and by other names, this book was prepared in 1552 by two Indian students of the Colegio de Santa Cruz in Tlatelolco, D.F. The manuscript was discovered in 1929 in the Biblioteca Apostolica Vaticana in Rome, where it remains today. The *Libellus,* which has been described as an "Aztec herbal" (Gates, 1939), contains ca. 184 exquisitely rendered colored drawings of Mexican plants and herbs by the native physician Martin de la Cruz. Some of these drawings incorporate elements of native symbolism (Fig. 3). The drawings are accompanied by a Latin translation of de la Cruz's orignial Náhuatl text (now lost) made by Juan Badianus, an Indian from Xochimilco. The names of the illustrated plant species are given in Náhuatl only, as are the numerous other plants mentioned but not depicted. The text and glosses give medical and pharmacological remedies for the treatment of diseases. Dedicated to Francisco de Mendoza, son of the Viceroy Diego de Mendoza, the *Libellus* was apparently intended for presentation to Charles V (Glass, 1975).

The *Libellus* is America's earliest extant work in botany, and in it are included the first illustrations known of many New World plants.

Of the three editions of the *Libellus* published since its discovery, those of Gates (1939) and Emmart (1940) are based on an accurate copy of the original; the color reproductions in Emmart are more useful than the black-and-white version published by Gates. The Emmart facsimile is followed by a Latin transcription, an English translation, and meticulous notes and comments including a painstakingly cross-referenced treatment of the plants mentioned, their tentative identifications, and etymologies of the Náhuatl names.

A third and most recent edition of the *Libellus* (de la Cruz, 1964) is a photofacsimile of the original document. It is without a doubt the most useful for students of Náhuatl plant classification. This edition is accompanied by valuable commentaries from several prominent scholars. The commentaries include the most up-to-date botanical treatment of the codex, written by Faustino Miranda and Javier Valdés (1964), drawing upon the works of Emmart (1940), Hernández (1959), and Reko (1947), as well as on their own botanical expertise.

Codex Mendoza. This manuscript, produced around 1541 or 1542 in Mexico City, contains 71 beautifully painted folio pages and is divided into three parts. Part I is a historical account, probably copied from an earlier chronicle now lost. Part II is a pictorial record of the tribute payments made by the different provinces of the Aztec Empire with a Spanish interpretation. It closely resembles the *Matrícula de Tributos* (q.v.) of which it may have been a copy, but it preserves five pages lost from the *Matrícula*. Both may have been derived from the same prototype. This section of the codex includes tribute items, many of which are plant products, along with their places of origin and the quantities delivered, which permits the evaluation of the relative importance and distribution of each. Part II also contains over 600 toponymic glyphs, approxi-

mately 130 of which have botanical components (Ortiz de Montellano, 1976), and each is accompanied by a Náhuatl gloss. Part III is a unique ethnographic portrayal, comparable in style and importance only to portions of the later *Florentine Codex* (Glass, 1975; Ross, 1984).

Matrícula de Tributos. Produced on sixteen leaves of *amatl* paper in Mexico City sometime in the sixteenth century, the *Matrícula de Tributos* is closely related to Part II of the *Codex Mendoza*. It is an itemized pictorial list of tribute paid to the ruling cities in the Valley of Mexico by the different provinces of the Aztec Empire. The document is a major source for the study of tribute, hieroglyphic place names, and political economy and geography of the Aztecs (Glass, 1975). It was not examined by the present author.

Documents of Mixed Colonial Origin

Oztoticpac Lands Map. This map was produced in Texcoco, Mexico, around 1540 and is comprised mostly of plans of fields with native measurements and hieroglyphic place names. The drawings are accompanied by Náhuatl and Spanish glosses and three long Náhuatl texts. At the lower left are unique depictions of fruit-tree grafts, showing branches grafted to tree trunks, and 20 fruit trees identified by glosses (Glass, 1975). This document was not examined by the present author.

Humboldt Fragment 14. A record of economic tribute from Central Mexico produced around 1562, this document was one of 16 purchased by Baron Alexander von Humboldt in Mexico in 1803. It contains drawings of cacao beans, turkeys, and other goods and foodstuffs with numerical signs (Glass, 1975). This document was not examined by the present author.

Codex Osuna. This codex consists of seven discrete documents on 40 leaves of European paper produced in and around Mexico City in 1565 as part of a government inquiry. Documents 2–4 concern unpaid accounts for services and fodder supplied by the Indians to the Viceroy. Document 5 presents unpaid accounts for services and food. The documents include Náhuatl and Spanish texts (Glass, 1975). A sample page reproduced in Wolf (1959) depicts Indians cultivating a variety of plants on a Spanish estate.

Relación de la Grana Cochinilla. Around 1599,

Gonzalo Gómez de Cervantes wrote this book in Central Mexico. It concerns the production of the cochineal insect which was important in colonial times as a dye. Seven pages of drawings illustrate the care and cultivation of the cochineal cactus (*Nopalea cochinellifera* Salm-Dyck, Cactaceae) on which the cochineal insect (*Coccus cacti*) is grown. The paintings also depict the birds and other insects which prey on the cochineal insect. According to Glass (1975), the style of the drawings reflects an earlier period of native style than the 1599 date on the manuscript suggests. This document was not examined by the present author.

Historia Natural de Nueva España. This scientific milestone was produced by Francisco Hernández between the years 1571 and 1577. Although the pictorial portion of this manuscript has been lost, the immense historical and scientific significance of Hernández's work deserves mention.

Francisco Hernández, the personal physician of King Philip II of Spain, was ordered to prepare an account of the natural history of New Spain. As indicated by Maldonado-Koerdell (1946), Hernández's trip to Mexico was the first scientific expedition ever organized for a systematic investigation of the flora and fauna of any country. Hernández holds the distinction of being the first European botanist to study the plants of the New World (Emmart, 1940). After seven arduous years, Hernández produced 16 folio volumes: six volumes of Latin text describing the animals, plants and minerals, and ten volumes of drawings. After being duly handed over to Philip II, the manuscripts were deposited in the Royal Monastery of the Escorial where, except for a few fragments, they were destroyed by fire in 1671 (Standley, 1920).

The original manuscript was illustrated with drawings made by Indian artists. Nieremberg (1635), familiar with the original work, described

→

FIG. 4. Examples of illustrations published by Nieremberg (1635) which were copied from the original work of Francisco Hernández before it was destroyed in the Escorial fire of 1671. The use of water (A), stone (B), and soil (C) glyphs reveals the native iconographic character of the original illustrations. The plants illustrated are (A) *atatapalacatl*, (B) *nopalli,* and (C) *teoamatl.*

the character of the drawings and published some copies (Fig. 4) which indicate their native origin (Glass, 1975). The text is known through various publications based on extracts, a summary, and a contemporary copy lacking the drawings. The edition of 1651 (Hernández, 1651) reproduced hundreds of the drawings, mostly of plants, through Italian copies in which all elements of native style and iconography have been removed (Glass, 1975).

Despite the tragedy that befell the *Historia Natural,* it remains today a work of foremost importance in the study of Aztec botany and medicine. Containing some 3000 Náhuatl plant names (Nuttall, 1924) which correspond to at least 1200 species of medicinal plants (Emmart, 1940), as well as a multitude of botanically-derived place names, the book represents a wealth of botanical information which has remained, to a large extent, untapped.

The Pictorial Writing System

Books and hieroglyphic writing are distinctive traits of Mesoamerican cultures and were unique to them in the New World. Unfortunately, the few surviving prehispanic codices permit only an approximate concept of their writing system.

Before the Mesoamerican writing system evolved, communications and the transmission of ancestral knowledge were limited to oral traditions. These eventually became supplemented, but never entirely replaced, by a writing system based on glyphic symbols. León-Portilla (1961) divides the glyphs into five principle types: numeric, calendric, pictographic, ideographic, and phonetic. As in what appears to be the genesis of all writing systems, the most primitive written symbols were pictographic—pictures of actual objects or actions. The hieroglyphic writing of Central Mexico relied considerably on pictographs which, taken alone, can be considered true writing only in the broadest sense. To some degree, the Mesoamerican writing system can be termed ideographic, in the sense that one is concerned with the qualities, attributes or ideas associated with the pictured object. Náhuatl is much given to metaphor, and the interpretation of symbols is often culturally determined. The two forms, picto- and ideographic were written intermingled on a single written page and are often difficult to distinguish. Some researchers have combined the two into one context: picto-ideographic (Dibble, 1971).

It is apparent that there was a certain standardization and schematization in the pictographic symbols that simplified their execution and assisted in their rapid identification (Ortiz de Montellano, 1976). The ideographic glyphs, also standardized to a certain extent, represented ideas symbolically and were used extensively in plant descriptions, primarily to indicate the habitat (e.g., stony, watery) where a plant is found. The ideographic symbols for stone and water are commonly associated with plant illustrations in the *Libellus* and in the *Florentine Codex,* for example.

The last and most advanced class of glyphs were exclusively phonetic (León-Portilla, 1961). These phonetic symbols were used in what has often been referred to as "rebus writing," where objects are depicted for their sound and are aggregated, syllabically, to form words (Thompson, 1959). Some words were written using as many as five elements, and the elements tend to be arranged in the same order as the syllables. Dibble (1971) estimates that the advancement from the picto-ideographic system to the beginnings of phonetic writing and syllabic writing took place in the last 50 years of the Aztec Empire.

Colors, rather than purely illustrative, were also symbolically significant and were an extremely important element in the hieroglyphic writing (Paso y Troncoso, 1885; Peterson, 1962). Often representing abstract concepts such as emotion, association with deities, or other religious connotations, colors could also indicate cardinal direction, sex and social status. The abstract significance of the colors used in the pictorial writings is only imperfectly understood today.

The combination of colorful numeric, calendric, pictographic, ideographic, and phonetic glyphs permitted the permanent recording of numbers, dates, objects, abstract ideas, and sounds. Nevertheless, this system of writing had obvious limitations for the transmission of complex concepts. To overcome this limitation, a dual system of written symbols and oral traditions evolved. The oral traditions from pre-literary times, instead of being supplanted by the writing system, were refined and reinforced as they assumed a complementary role. Just as the oral traditions served to explain the heiroglyphics, so did the pictorial images serve as mnemonic devices to aid in the recall of lengthy oral texts. The two interwoven systems became an eloquent hybrid of literature and oratory by which history, government, religion, education, and art were conserved and perpetuated.

Dibble (1971: 323) states, "That these codices were also accompanied by oral traditions to elucidate and amplify the hieroglyphic record is . . . clear. . . ." There is direct testimony of the conquistadors and missionary chroniclers that in every *calmecac* the priests and teachers explained the paintings of the codices in their libraries, making the students learn their commentaries by rote (León-Portilla, 1971). In this way, the students "sang their paintings":

> *"I sing the paintings in the book,*
> *I am unfolding the book,*
> *I am as a flowery macaw,*
> *I make the codices speak,*
> *Inside the house of paintings."*

(From *Cantares en idioma mexicano,* facsimile reproduction by A. Peñafiel, 1904; cited in León-Portilla, 1961: 64. Translation by D. E. Williams.)

Thus, the memorization of texts using the pictorial books was the principal means of instruction at the *calmecac.* This greatly facilitated the compilation of much information by Sahagún and other chroniclers because their learned informants were capable of reciting, verbatim, extensive written texts that had been carefully memorized.

Prose Sources

The early missionaries transcribed large amounts of oral text into the Latin alphabet while maintaining the native literary style. Moreover, many of the pictorial documents produced after the Conquest were annotated with glosses in Náhuatl or Spanish, or both. These glosses are often instrumental in deciphering the hieroglyphic inscriptions. Many of the Náhuatl texts are of literary value per se and the vast majority contain elements of genuine prehispanic tradition (Glass, 1975).

León-Portilla (1985) recognizes two genres of Náhuatl literature. The first, known as *cuicatl,* is made up of songs, hymns, and poetry. Particularly in the *xochicuicatl,* or "flowery songs," we find many references to flowers and their attributes as well as to birds, butterflies, symbolic colors, tobacco, chocolate, previous objects, and musical instruments. The second genre, *tlahtolli,* is roughly equivalent to our prose and includes, among a vast array of other subjects, fields of knowledge such as natural history. León-Portilla cites sections of Books IX, X, and XI of the *Florentine Codex* as examples of the latter. While prose sources dealing specifically with plants are few, they represent, nonetheless, a vital area for study. Perhaps more importantly, especially in the context of the *xochicuicatl,* the spiritual and aesthetic importance of plants and flowers in the Náhuatl world view becomes evident. An adequate appreciation of this aspect of Náhuatl culture is fundamental for the reconstruction of its plant taxonomy.

The Ethnographic Sources

In the more than 450 years that have elapsed since the fall of the Aztec Empire, Náhuatl culture has undergone sweeping changes and much has been lost. However, abundant examples of Náhuatl culture have also persisted, in many cases modifying the European culture rather than succumbing to it. Plant-related knowledge and terminology is an example of the persistence of ancestral knowledge in many areas of Mesoamerica. This extant body of knowledge is a fertile field of investigation for students of Náhuatl botany.

Today, an estimated three quarters of a million people speak some dialect of Náhuatl as their first language. These people, collectively referred to as Nahuas, presently live in the Mexican states of San Luís Potosí, Hidalgo, Mexico, Puebla, Distrito Federal, Tlaxcala, Morelos, Guerrero, Veracruz, Jalisco, Michoacán, Nayarit, Oaxaca, and Tabasco. The largest monolingual group is located in the northern part of the state of Puebla (Madsen, 1969). A Náhuatl-speaking people known as the Pipil may also still exist in El Salvador and Costa Rica.

Ironically, although the Nahuas today constitute Mexico's largest autochthonous group, the ethnographic investigations of contemporary Náhuatl-speaking peoples have been meager as yet in comparison with research on other Mexican groups (Madsen, 1969). Some botanical work on modern Náhuatl has been done by Baytelman (1977), Dehouve (1974), Ramirez and Dakin (1979), and D. E. Williams (1985).

There are obvious advantages regarding the type of information which ethnographic investigation yields when compared to that offered by the ethnohistorical record. As in the study of codices, the ethnographic data represent a primary source. But in ethnographic studies, the informants may be interviewed, sometimes at great length.

The transcendence of Náhuatl plant taxonomy is overwhelmingly evident in modern Mexican Spanish (see Martínez, 1979; Robelo, 1948; and Santamaría, 1974). Modern Mesoamerican peoples continue to place an inordinately high degree of importance on the abundance of plants and plant products used in the various aspects of their lives. This is particularly true of Náhuatl speakers who, by and large, are the rural poor whose means of subsistence consists primarily of farming, often supplemented by plant gathering. Typically, plants are also a primary source of medicines and ritual materials. The intimate interaction between these people and their plant environment, in combination with their cultural conservatism, has resulted in the persistence and perpetuation of ancestral plant knowledge, which may be elicited by the judicious researcher.

Assuming that Aztec botany was at its zenith on the eve of the Spanish Conquest, the ethnographer's task is that of reconstructing a system after nearly half a millenium of change. Perhaps for this reason, all previous efforts in the elucidation of Náhuatl botany have been based on one or more of the handful of extant sixteenth century codices dealing with plants.

To take full advantage of the rich body of ethnographic information pertaining to Náhuatl plant classification, a multidisciplinary undertaking would be necessary, similar to that of the Berlin et al. (1974) study of Tzeltal folk taxonomy.

Interpretations

When seen in the light of the sheer volume of raw information available regarding Náhuatl botany, surprisingly few modern scholars have attempted to describe it in more than the most limited or general terms. Any comprehensive treatment of the subject will require proficiency in the Náhuatl language, great familiarity with ancient Mexican history, culture, religion, and writing, and formal training in both linguistics and botany—skills not often encountered in one person (Maldonado-Koerdell, 1946).

The famous ethnohistorian Francisco del Paso y Troncoso (1885) offered what is probably the first attempt to specifically define Náhuatl plant classification. Paso y Troncoso claimed that the synonymy so frequently encountered in Náhuatl plant names is the result of the confusion between the common and technical names of plants. He maintained that there existed two classification systems, one popular (common names), and one scientific (technical and therapeutic names). The popular names were the source of much synonymy due to the different regional names provided by many different informants for a single plant species, as were the technical names which may have differed depending on whether the names were being used for therapeutic or purely botanical purposes. At the present time, the problem of synonymy is further aggravated by the fact that many of the original names have become adulterated during the intervening centuries, making accurate etymologies difficult.

To further confuse matters, it is apparent that in the therapeutic classification of plants, a great deal of homonymy was created by the replication of one name for numerous different species based on their shared effectiveness in combating a particular ailment. For example, Hernández (1651) lists 37 different plants called *iztacpahtli* (white medicine) and 21 called *cihuapatli* (woman's medicine).

The Aztecs, in their far-ranging travels, brought exotic plants back to the central plateau where they were kept in special gardens. According to Paso y Troncoso (1885), these plants were named in two ways. The most common means was on the basis of similarities with plants already familiar to them, using the known plant name as the type and applying that same name with a qualifying affix to the exotic species. Numerous examples of qualifying terms are cited, many of which describe morphological characters of the plants which reveal a high degree of detailed observation. The second means of naming exotic plants was by translating into Náhuatl the name for the plant used in its place of origin. Paso y Troncoso points out that this latter system of naming was common practice when applying Náhuatl names to newly conquered towns.

Paso y Troncoso (1885) stresses the importance of the pictorial writing in the study of Aztec botany. He reminds us of the fact that iconography was the principal descriptive resource not only for the Aztecs but also for the pre-Renaissance European botanists such as Dioscorides, and even including the work of Francisco Hernández whose brief legends seem quite deficient now that their accompanying illustrations are lost. The Indians relied solely on pictographic images for plant description. All those properties that remained unexpressed by the name itself or its synonyms had to be conveyed through the figure-writing. The pictorial images in turn became a very efficient mnemonic and didactic device.

Paso y Troncoso (1885) distinguishes three kinds of hieroglyphic representations of plants—the figurative, the symbolic, and the syllabic, often appearing in complex combination. The figurative representations were most commonly used when dealing with plant parts. The symbolic representations were most commonly used for entire plants. The syllabic representations were the least common, and believed to be an incipient and recent development in the evolution of the Mesoamerican writing. The introduction of symbolic representations is regarded by Paso y Troncoso as a positive advance that permitted the expression of generalizations through conventional signs, grouping numerous objects that are linked together by common properties. He believed that the use of standardized generic symbols in the pictorial writing gave rise to the classification and nomenclatural system. Paso y Troncoso demonstrates through examples that the Aztecs had symbolic signs that applied to certain plant groups and, through various determinatives, also served to designate all of the species that formed that group. A single hieroglyph condensed various attributes that could not have been expressed through the language without employing a large number of synonyms. Yet Paso y Troncoso maintains that the primary function of the generic symbols was not to indicate the features that the group members had in common, but for the ordering and arrangement of the plants themselves. Thus, the pictorial aspect of a plant's name was a primary element in its systematic classification.

Paso y Troncoso affirms that the pictorial writing alone is sufficient to establish that the Indians had developed a systematic classification for their plants. This conclusion is also supported by analysis of the language. He points out that there is an intimate relationship between the classification system and the nomenclature, the two co-existing, with the classification dependent upon the nomenclature. Consequently, irregularities in the nomenclature will be reflected in the classification system. Paso y Troncoso believed that two classification systems existed: an artificial (folk) system and a natural (scientific) system, depending on the characters used as the basis for their classification. These have been confused as one system by most researchers, giving rise to many discrepancies in their subsequent interpretations of Náhuatl plant classification.

Unfortunately, Paso y Troncoso's discussion of his efforts to distinguish and segregate the two systems was interrupted and his treatment of this key element of Náhuatl botanical classification was never published.

Nuttall (1924) notes that in Náhuatl prefixes and suffixes are used in plant names to express their qualities, characteristics, and habits. She cites as an example the word *xochitl* (flower) which always appears in the names of plants valued for their flowers, accompanied by modifiers which characterize them. She goes on (p. 594) to affirm that, "This ingenious system of botanical nomenclature reveals great observation, knowledge and experience and which, in such an original way, refers to the relationship of the plants with the human race." In a later paper (Nuttall, 1925), she gives a series of Náhuatl terms for gardens from which she infers a prolonged familiarity with horticulture:

xochitla	= garden, flower place
xoxochitla	= place of many flowers
xochitepanyo	= walled garden
xochitepancalli	= flower palace
xochichinancalli	= house garden enclosed by reeds

Gates (1939) also attempts to specifically define Náhuatl plant classification. While an ethnohistorian of stature and a collector of ancient manuscripts, Gates was not a botanist. If one can overlook the abundant botanical errors (Miranda & Valdés, 1964) in this interpretation of the *Libellus,* Gates remains very clear on the importance of the pictorial writing system in Aztec plant classification.

Nevertheless, Gates's interpretation is instructive. He divides the plant kingdom into two great "natural" orders: *cuauh-* (woody) and *-xiuh* (herbaceous). Each of these orders is then divided into four great artificial classes according to use: *quilitl* (edible greens), *pahtli* (medicinal), *xochitl* (flowers), and an economic convert complex *sensu* Berlin et al. (1974). Although he is inconsistent in the "equivalent" taxonomic ranks he assigns, Gates lists the following families:

xocotl	=	acidulous fruits [e.g., Myrtaceae/*Psidium*]
tzapotl	=	sweet fruits [e.g., Sapotaceae/*Manilkara*]
etl	=	beans [e.g., Fabaceae/*Phaseolus*]
ayotl	=	gourds [e.g., Cucurbitaceae/*Cucurbita*]
tomatl	=	tomatoes [e.g., Solanaceae/*Physalis*]

Gates mentions the following Náhuatl genera:

huaxin	=	pod [e.g., *Leucaena*]
mizquitl	=	legume [e.g., *Prosopis*]
etl	=	legume seed [e.g., *Phaseolus*]
huacalli	=	striated thing [?]
coyolli	=	rattle [e.g., *Acrocomia*]
tecomatl	=	vase [e.g., *Lagenaria*]
tollin	=	reeds, rushes [e.g., *Typha*]

(Determinations listed in brackets provided by D. E. Williams.)

In explaining the Náhuatl taxonomic system, Gates drastically oversimplifies the rules of Náhuatl morphology as outlined by Garibay-K. (1970), by affirming that the adjective precedes the substantive, as in English, and ignoring other forms of syntax. In Gates's words (1939: xix):

> *"The last member of a compound Aztec term represents the genus, and is connotative, giving the dominant characteristic and thus the basis of the Classification. It thus also retains the final −tl of the Aztec noun, while all the preceding species-qualifiers, limiting as adjectives the final element of the term, lose the −tl and show only the simple stem, as* cuauh *instead of* cuahuitl; acatl *is a cane, but* aca-zacatl *is cane-grass. As seen in the pictorial nomenclature, we find the major groups shown by this final member as giving us the orders, genera and species, each developing the lower members of the class by these adjectival descriptives, exactly as we do, save that the Aztec word order of the compounds is as in English, not the Latin."*

To illustrate his concept of word position as the basis of the classification system, Gates gives a list of plants in his reed genus, *tollin: tepetollin* (mountain reed), *tliltollin* (black reed), *petlatollin* (mat reed), *atollin* (aquatic reed), *nanacetollin* (triangular reed), *aitztollin* (sharp-leaved reed),

tzontollin (hairy reed), *ixtollin* (eye reed), *caltollin* (house reed), *popotollin* (broom tule). In the adjectival position he cites *tolcimatl* (tule root) and *tolpahtli* (tule remedy).

Gates goes on to discuss synonymy, homonymy, and toponymic iconography, briefly explaining the composition of 33 place glyphs from the *Codex Mendoza* which incorporate plant symbols.

Emmart (1940: xx), in her efforts to identify the plants mentioned in the *Libellus,* found that the common Náhuatl names for many plants have changed over the years and that in some cases the modern names were of little help. She states that many plants must remain unidentified pending further exploration, "when the ethnologist works hand in hand with the botanist." She laments that of the many hundreds of plants known to the Aztecs and referred to by Hernández, Sahagún, and de la Cruz, comparatively few are known botanically, much less utilized, by modern society. Cline (1973) reports that Emmart, building upon the information compiled while studying the *Libellus,* was identifying the plants in Book XI of the *Florentine Codex,* but this work, if published, has not been seen by the present author.

Rivera-Morales (1941) published a lengthy attempt at identifying each plant mentioned in the Spanish text of Book X of the *Florentine Codex.* None of the many additional plants mentioned in the Náhuatl text of the same book are treated nor are any of the plants mentioned in Book XI. No discussion of indigenous taxonomy is given. The systematically ordered list provided at the end of Rivera-Morales's paper does, however, give the reader a feeling for the degree of coincidence between the two systems of nomenclature.

More recently, Estrada-Lugo (1987) extracted the botanical references from the Spanish text of the entire *Florentine Codex* and placed them in a computerized database, permitting an ordination of indices under the headings of Náhuatl name, scientific name, plant family, and use. This work greatly facilitates access to much of the plant information contained in this important codex.

After Paso y Troncoso and Gates, the next researcher to specifically treat the subject of Náhuatl plant classification was Maldonado-Koerdell (1941, 1946). Unconvinced by what he considered to be romantic exaggeration of indigenous achievements in the sciences, and apparently influenced by the writings of Hewett (1936), Maldonado-Koerdell presents a forceful argument against the existence of any systematic classification of plants in Mexico before the arrival of Francisco Hernández in 1571. He maintains that in the superstitious context of their world, "no really scientific interpretation could evolve" (Maldonado-Koerdell, 1946: 52). Doubting the existence of any science (e.g., astronomy, botany, medicine) in ancient Mexico, Maldonado-Koerdell (1946: 52) dismisses endeavors in these fields as "merely a cultural process, extending through countless generations, being gradually reduced to schematic statements." In an earlier paper (1941: 81) he states that "for plants and animals they had an adequate nomenclature which incorporated one or more of their properties; the knowledge which they had reached was essentially empirical and utilitarian, in accordance with the [non-scientific] mental traits mentioned earlier."

Pictorial images of plants in the codices have been conspicuously neglected as subjects of study, per se. A notable exception is a treatment of *Agave* representations by Gonçalves de Lima (1956). In another paper, Gonçalves de Lima (1955) identifies a number of the food plants mentioned by Sahagún and recognizes four genera: *metl* (*Agave*), which includes at least 15 species (*tlacometl, teometl, iztacmetl, tlemetl,* etc.); *tomatl* (includes *Solanum, Physalis, Lycopersicon,* and *Saracha,* Solanaceae), in which Sahagún lists ten species, some still unidentified; *chilli* (*Capsicum,* Solanaceae), with at least 25 species listed; and *quilitl,* with 20 phylogenetically unrelated species listed.

Several researchers working on various groups of plants have used information gleaned from the codices to good advantage. Sauer (1950) was able to establish the extent and importance of grain amaranth cultivation as well as get some idea of the genetic diversity of this crop in prehispanic Mexico. Perhaps the most fruitful research in Náhuatl botany has been that carried out by various researchers studying psychoactive plants (López-Austin, 1965; Schultes, 1939, 1941; Wassén, 1960; Wasson, 1966; and others). These researchers, drawing upon the ethnographic, ethnohistoric, and botanical data available, effectively discovered or clarified the identity of a number of plants which were sacred to the an-

cient Mexicans and are now important in modern psychopharmacology. Wasson (1973, 1980) proposes that the Náhuatl word for flower, *xochitl,* which is so frequently encountered in the ancient poetry, was actually a metaphor for inebriating mushrooms in that context; and he believes the *xochitemictl,* or flowery dream, is a term for the hallucinatory experience. In his scholarly and convincing argument, Wasson demonstrates an uncanny appreciation for the relationship between the ancient Mexicans and their plants. He emphasizes the preeminent importance of plants in Náhuatl culture and points out that Náhuatl was rich in botanical terminology.

The most scientific analysis of Náhuatl plant classification to date is that offered by Ortiz de Montellano (1976, 1984). His studies are based primarily on linguistic analysis, using a model the hierarchical system of folk taxonomy developed by Berlin and collaborators (1974). While recognizing that the methodology used in his study was developed for an ethnographic application, Oritz de Montellano extends it to written names taken from sixteenth century sources, primarily Sahagún. He provides us with a number of very instructive hierarchical diagrams of the Náhuatl plant kingdom, including the number of lexemes found for each taxon. Also included is an extremely useful table of classificatory Náhuatl lexemes denoting color, texture, form, etc. Despite the fact that he is confronted by difficulties deriving from the application of his chosen model to ethnohistoric sources, Oritz de Montellano is able to conclude from his analysis that the Aztecs possessed an extensive system of taxonomic classification that corresponds closely to a natural system.

In his earlier work, Ortiz de Montellano (1976) supplements and supports his linguistic analysis with an analysis of some of the approximately 130 toponymic glyphs containing botanical elements found in the *Codex Mendoza.*

Discussion and Conclusions

It is clear that the ancient Mexicans possessed a great deal of knowledge about plants and that this knowledge was taught in the *calmecac* where scribes were proficient in the pictorial depiction of plants. It is likely that the city of Texcoco under Netzahualcoyotl (1402–1472 A.D.) was a center for botanical research. It is known from the reports of the chroniclers that the city was the home of a great *calmecac*, great libraries, and numerous botanical gardens. Netzahualcoyotl himself was interested in the study of plants, and maintained what may have been an herbarium. The coincidence of an overriding interest in and abundant knowledge of plants, together with schools of higher learning and great libraries is a very strong argument for the existence of botanical science in ancient Mexico.

That the ancient Mexicans were capable of systematic classification is clearly demonstrated by B. J. Williams (1981) in her study of soil glyphs. Admirably correlating comparative studies of sixteenth century pictorial documents with information gleaned from modern Nahua farmers, Williams was able to decipher the glyphs and segregate them into generic and subordinate taxa, as well as detect non-significant stylistic variants. She concludes (p. 214), "The Texcocan-Aztec *tlacuilos* (sic) successfully conveyed the structure and complexity of their soil domain." Gibson (1964) also discusses the scientific precision with which the Indians identified and recorded soil types. That the ancient Mexicans would have an elaborate and useful system of soil classification, yet would neglect to classify their plants is highly unlikely.

Lozoya (1984: 11) affirms that "The Náhuatl nomenclature of Mexican plants demonstrated the previous existence of a sophisticated method of botanical classification in which the use, properties and physical characteristics of each plant found a congruent systematization."

An entire classification system cannot be detected, much less dismissed, simply on the basis of the existence or non-existence of generic and specific epithets. In fact, in many instances the scientific and common names of plants with salient features are remarkably similar—even in different languages and different systems:

LATIN:
Chiranthodendron pentadactylon Larr. (=five-fingered hand-flower-tree)

NÁHUATL:
Macpalxochiquahuitl (=palm-of-the-hand-flower-tree)

SPANISH:
Arbol de manitas, Flor de manita (=little-hand tree, hand flower)

ENGLISH:
Hand-flower tree (=tree with flowers-like-hands)

Maldonado-Koerdell emphasizes the great cultural gulf and, instead of trying to understand the Náhuatl system in the context of its cultural milieu, he concludes that science was beyond the Mexicans' grasp. Furthermore, he says that the lack of "useful" writing was a "tremendous impediment for the accumulation, conservation and transmission of a system of scientific ideas" (Maldonado-Koerdell, 1946: 53).

Paso y Troncoso was certainly on the right track when he suggested that the key to the Náhuatl nomenclatural system was in the pictorial writing. As Gates (1939: xxvi) pointed out, the iconographic aspect of the writing "gave them a further, wholly distinct set of determinative or descriptive media, which we lack in our botany completely." In conjunction with the formal oral complement to the texts, prompted mnemonically, the hieroglyphics were capable of transmitting as much information as an orthographic alphabet. Thus, instead of actually *spelling out* the taxonomic rank of plants, as in our system, their classification was dependent on the painted and oral components of their writing system. These are precisely the components of Náhuatl botany which have been lost. Almost immediately following the Conquest, the ancient system of pictorial writing began to be supplanted by the Spanish alphabet. According to Garibay-K. (1970), this adaptation reproduced all of the advantages and disadvantages of said alphabet. In one process, the twin systems of oral tradition and pictorial writing were replaced. Quiñones-Keber (1988: 199) states that "an understanding of the possible structure or systematic nature of Aztec pictorial images was obscured by the general inability of 16th century Europeans to recognize it and by the subsequent loss of images that formed its constituent parts." The new orthographic system, as taught by the Spanish friars, was insufficient to encompass the scope of botanical knowledge, or at least the scientific system of classification, of the Indians. The new system was certainly incompatible with the traditional system of recording botanical information. Thus, the formal native system of plant classification undoubtedly perished in the mid-sixteenth century—literally lost in the translation.

Those native botanists who were conversant in formal botany were, necessarily, graduates of a *calmecac* and therefore of noble rank. Theirs was the social class that came into immediate and prolonged contact with the Spanish, being the first to be baptized, the first to succumb to Old World diseases, etc. As in our modern society, formal botanical knowledge was restricted to specialists, and the jargon of plant taxonomy was unknown to most other people, even physicians. With the acculturation and death of the native botanists, the closing of the *calmecac,* and the loss or destruction of the botanical texts, the perpetuation of Náhuatl plant lore was relegated to the comparatively uneducated and illiterate rural people for the next 400 years.

Much of the formal Náhuatl system of plant classification is probably irretrievable. Yet a general reconstruction should be possible and would be worthwhile, not only from a purely academic standpoint, but also for the retrieval of the many economically useful plants that are mentioned in the codices but remain unidentified. Modern researchers, particularly pharmacologists, attempting to identify useful species in the codices inevitably run into difficulties, sometimes due to the fact that the same Náhuatl name may be given for as many as 37 different plants (Ortiz de Montellano, 1975). A clearer understanding of Náhuatl plant systematics would resolve the identity of many of these.

Numerous cultivated plants referred to by Sahagún remain unidentified, and the history of cultivation of certain species could answer important questions about the evolution of these plants under domestication. This is particularly important because Mexico is a world center of plant domestication, a process that to this day remains imperfectly understood. As an example, the careful examination of the *Florentine Codex* reveals that the Aztecs recognized two species of *xaltomatl* (the name has been incorporated into our Linnaean system as *Jaltomata,* Solanaceae) whose descriptions coincide precisely with two varieties of *J. procumbens* recently reported by Davis and Bye (1982) and D. E. Williams (1985) as being semi-cultivated by native farmers. One of these varieties of *Jaltomata* may represent a relict cultigen (D. E. Williams, 1985) and may merit modern taxonomic distinction (T. Davis, pers. comm.).

The corpus of Náhuatl plant nomenclature that has come down to us is an agglomeration of common and formal names pertaining to separate systems of folk and scientific classification that coexisted in Mexico at the time of the Conquest. Therefore, it is not surprising that modern attempts to define the prehispanic taxonomy based solely on the nomenclature find evidence to support the existence of a scientific system, yet the system appears to be rife with inconsistencies. The confusion of the two coexisting systems was noted by Paso y Troncoso (1885). As long as scholars continue to ignore the pictorial aspect of the ancient system, the confusion is inevitable. As pointed out by Glass (1975), the manuscript paintings and drawings represent not only a body of art, but also a corpus of documentary information; and few studies exist that have analyzed them from either point of view.

As Maldonado-Koerdell (1946) admits, few individuals combine the training and inclination that would be necessary to single-handedly reconstruct the Náhuatl systems of plant classification. A more realistic approach to the problem would involve a multidisciplinary team of researchers. The works of Sahagún (1979), de la Cruz (1964), and Hernández (1651) require thorough re-examination by a team of Náhuatl scholars, ethnohistorians, and botanists. The remaining pictorial and prose manuscripts also need to be scrutinized for plant information. Special attention needs to be directed at those examples of extant pictorial representations of plants whose importance in this regard has been underestimated. Information gleaned in the library will then need to be articulated with that collected by a team of ethnographers, linguists, and botanists working in the field with modern-day Nahua groups.

Acknowledgments

The author is indebted to the following people who made suggestions and corrections on earlier drafts of this paper: Rupert Barneby, David Hammond, David Johnson, Nancy Murray, Christine Padoch, Mick Richardson, and Sandy Williams. Gratitude is expressed to the two anonymous reviewers for their helpful suggestions and for bringing to the author's attention some errors and important references. All remaining errors are the responsibility of the author. Special thanks go to the reference librarians of The New York Botanical Garden Library for their help in securing obscure documents and facsimiles, to Sandy Williams for technical assistance, and to Susannah Laskaris who prepared the map. This is publication number 129 in the series of the Institute of Botany of The New York Botanical Garden.

Literature Cited

Baytelman, B. 1977. Etnobotánica en el estado de Morelos. Secretaría de Educación Publica, Instituto Nacional de Antropología e Historia, Mexico City.

Berlin, B., D. E. Breedlove & P. H. Raven. 1974. Principles of Tzeltal plant classification. Academic Press, New York.

Calnek, E. 1988. The calmecac and telpochcalli in pre-conquest Tenochtitlan. Pages 169–177 in J. J. Klor de Alva, H. B. Nicholson & E. Quiñones-Keber (eds.), The work of Bernardino de Sahagun. Institute of Mesoamerican Studies, Albany.

Clavigero, F. G. 1945. Historia antigua de México. Mexico City.

Cline, H. F. 1973. Sahagún materials and studies, 1948–1971. Pages 218–239 in H. F Cline (ed.), Guide to ethnohistorical sources. Part 2. Handbook of Middle American Indians. Vol. 13. University of Texas Press, Austin, Texas.

Davis, T. & R. A. Bye, Jr. 1982. Ethnobotany and progressive domestication of *Jaltomata* (Solanaceae) in Mexico and Central America. Econ. Bot. **36(2):** 225–241.

Dehouve, D. 1974. Corvée des saints & luttes de marchands. Klincksieck, Paris.

de la Cruz, M. 1964. Libellus de medicinalibus indorum herbis. E. C. del Pozo (ed.). Instituto Mexicano de Seguro Social, Mexico City.

Díaz del Castillo, B. 1955. Historia verdadera de la conquista de la Nueva España. 2 vols. Editorial Porrúa, Mexico City.

Dibble, C. E. 1971. Writings in Central Mexico. Pages 322–332 in G. F. Ekholm & I. Bernal (eds.), Archaeology of northern Mesoamerica. Part 1. The handbook of Middle American Indians. Vol. 10. University of Texas Press, Austin, Texas.

——— **& A. J. O. Anderson.** 1950–1969. Florentine Codex: General history of the things of New Spain (by) Fray Bernardino de Sahagún. Translated from the Aztec. . . . Monographs of the School of American Research, 14. Parts 2–13. School of American Research and the University of Utah, Santa Fe.

D'Olwer, L. N. & H. F Cline. 1973. Sahagún and his works. Pages 186–207 in H. F. Cline (ed.), Guide to ethnohistorical sources. Part 2. Handbook of Middle American Indians. Vol. 13. University of Texas Press, Austin, Texas.

Dressler, R. L. 1953. The pre-Columbian cultivated plants of Mexico. Bot. Mus. Leafl. **16(6):** 115–173.

Emmart, E. W. 1940. The Badianus manuscript. The John Hopkins Press, Baltimore.

Estrada-Lugo, E. I. J. 1987. El Códice Florentino: Su información etnobotánica. Master's Thesis. Colegio de Postgraduados, Chapingo, Mexico.

Garibay-K., A. M. 1956. Historia general de las cosas de Nueva España por Fr. Bernardino de Sahagún. 4 vols. Editorial Porrúa, Mexico City.

———. 1970. Llave del Náhuatl. Editorial Porrúa, Mexico City.

Gates, W. 1939. The de la Cruz-Badiano Aztec herbal of 1552. The Maya Society, Baltimore.

Gibson, C. 1964. Aztecs under Spanish rule. Stanford University Press, Stanford.

———. 1975. A survey of Middle American prose manuscripts in the native historical tradition. Pages 311–321 in H. F. Cline (ed.), Guide to ethnohistorical sources. Part 4. Handbook of Middle American Indians. Vol. 15. University of Texas Press, Austin, Texas.

Glass, J. B. 1975. A survey and census of native Middle American pictorial manuscripts. Pages 3–252 in H. F. Cline (ed.), Guide to ethnohistorical sources. Part 3. Handbook of Middle American Indians. Vol. 14. University of Texas Press, Austin, Texas.

Gonçalves de Lima, O. 1955. Alimentos e bebidas no México prehispânico, segundo os manuscritos de Sahagún. Inst. Joaquim Nabuco do Pesq. Soc. Publ. Avul. No. 2. Recife.

———. 1956. El maguey y el pulque en los códices mexicanos. Fondo de Cultura Económica, Mexico City.

Harvey, H. R. 1972. The relaciones geográficas, 1579–1586: Native languages. Pages 279–323 in H. F. Cline (ed.), Guide to ethnohistorical sources. Part 1. Handbook of Middle American Indians. Vol. 12. University of Texas Press, Austin, Texas.

Hernández, F. 1651. Rerum medicarum Novae Hispaniae Thesaurus. Rome.

———. 1959. Historia natural de Nueva España. UNAM, Mexico City.

Hewett, E. L. 1936. Ancient life in Mexico and Central America. The Bobbs-Merrill Co., Indianapolis.

Kirchoff, P. 1952. Mesoamerica: Its geographic limits, ethnic composition and cultural characteristics. In S. Tax (ed.), Heritage of conquest. The Free Press, Glencoe, Illinois.

León-Portilla, M. 1961. Los antiguos mexicanos a través de sus crónicas y cantares. Fondo de Cultura Económica, Mexico City.

———. 1963. Aztec thought and culture. University of Oklahoma Press, Norman, Oklahoma.

———. 1971. Pre-hispanic literature. Pages 452–458 in G. F. Ekhold & I. Bernal (eds.), Archaeology of northern Mesoamerica. Part 1. Handbook of Middle American Indians. Vol. 10. University of Texas Press, Austin, Texas.

———. 1985. Nahuatl literature. Pages 7–43 in M. S. Edmonson (ed.), Literatures. Supplement to the handbook of Middle American Indians. Vol. III. University of Texas Press, Austin, Texas.

López-Austin, A. 1965. Descripción de estupefacientes en el Códice Florentino. Univ. Nac. Aut. Méx. **19(5):** 17–18.

Lozoya, X. 1984. Plantas y luces en México—La real expedición científica a Nueva España (1787–1803). Ediciones del Serbal, Barcelona.

MacNeish, R. S. 1967. A summary of subsistence. Pages 290–309 *in* D. S. Byers (ed.), The prehistory of the Tehuacan Valley. Vol. 1. University of Texas Press, Austin, Texas.

Madsen, W. 1969. The Nahua. Pages 602–637 *in* E. Z. Vogt (ed.), Ethnology. Part 2. The handbook of Middle American Indians. Vol. 8. University of Texas Press, Austin, Texas.

Maldonado-Koerdell, M. 1941. Los jardines botánicos de los antiguos mexicanos. Revista Soc. Mex. Hist. Nat. 2(1): 79–84.

———. 1946. Aztec botany and zoology. Chicago Naturalist 9(3): 51–58.

Martínez, M. 1979. Catálogo de nombres vulgares y científicos de plantas mexicanas. Fondo de Cultura Económica, Mexico City.

Miranda, F. & J. Valdés. 1964. Comentarios botánicos. Pages 243–284 *in* M. de la Cruz, Libellus de medicinalibus indorum herbis. E. C. del Pozo (ed.). Instituto Mexicano de Seguro Social, Mexico City.

Nicholson, H. B. 1975. Middle American ethnohistory: An overview. Pages 487–505 *in* H. F. Cline (ed.), Guide to ethnohistorical sources. Part 4. Handbook of Middle American Indians. Vol. 15. University of Texas Press, Austin, Texas.

Nieremberg, I. E. 1635. Historia naturae, maxime perigrinae. Antwerp.

Nuttall, Z. 1924. Los aficionados a las flores y los jardines del México antiguo. Soc. Sci. "Antonio Alzate" - Mémoires 43: 593–608.

———. 1925. The gardens of ancient Mexico. Ann. Rep. Smithsonian Inst. (1923): 453–464.

Ortiz de Montellano, B. 1975. Impirical Aztec medicine. Science 188(4185): 215–220.

———. 1976. ¿Una clasificación botánica entre los nahoas? Pages 27–49 *in* X. Lozoya L. (ed.), Estado actual del conocimiento en plantas medicinales mexicanas. IMEPLAM, Mexico City.

———. 1984. El conocimiento de la naturaleza entre los mexicas. Taxonomía. Pages 115–132 *in* A. López-Austin (ed.), Historia general de la medicina en México. Vol. 1. UNAM, Mexico City.

Paso y Troncoso, F. del. 1885. Estudios sobre la historia de medicina en México. An. Mus. Nac. Méx. 3: 137–235.

———. 1905–1907. Fray Bernardino de Sahagún: Historia de las cosas de Nueva España. Hauser y Menet, Madrid.

Peñafiel, A. 1904. Cantares en el idioma mexicano. Secretaría de Fomento, Mexico City.

Peterson, F. 1962. Ancient Mexico. Capricorn, New York.

Quiñones-Keber, E. 1988. Reading images: The making and meaning of the Sahaguntine illustrations. Pages 199–210 *in* J. J. Klor de Alva, H. B. Nicholson & E. Quiñones-Keber (eds.), The work of Bernardino de Sahagún. Institute for Mesoamerican Studies, Albany.

Ramirez, C. & K. Dakin. 1979. Vocabulario náhuatl de Xaltitla, Guerrero. Cuaderno de la Casa Chata, Mexico City.

Reko, B. P. 1947. Nombres botánicos del Manuscrito Badiano. Bol. Soc. Bot. Mex. 5: 23–43.

Rivera-Morales, I. 1941. Ensayo de interpretación botánica del Libro X de la Historia de Sahagún. Anales Inst. Biol. Univ. Nac. México 12(1): 439–488.

Robelo, C. A. 1948. Diccionario de aztequismos. Cuernavaca.

Ross, K. 1984. Codex Mendoza: Aztec manuscript. Productions Liber SA, Fribourg.

Rzedowski, J. 1981. Vegetación de México. Editorial Limusa, Mexico City.

Sahagún, Fr. B. de. 1979. Códice Florentino. El manuscrito 218–20 de la Colección Palatina de la Biblioteca Medicea Laurenziana (edición facsimilar). Secretaría de Gobernación, Mexico City.

Santamaría, F. J. 1974. Diccionario de mejicanismos, 2a edición. Editorial Porrúa, Mexico City.

Sauer, J. D. 1950. The grain amaranths: A survey of their history and classification. Ann. Mo. Bot. Gard. 37: 561–619.

Schultes, R. E. 1939. The identification of teonanacatl, a narcotic basidiomycete of the Aztecs. Bot. Mus. Leafl. 7(3): 37–54.

———. 1941. A contribution to our knowledge of *Rivea corymbosa,* the narcotic ololiuqui of the Aztecs. Botanical Museum, Harvard University, Cambridge, Massachusetts.

Standley, P. C. 1920. Trees and shrubs of Mexico. Contr. U.S. Natl. Herb. Vol. 23, Part 1.

Thompson, J. E. S. 1959. Systems of hieroglyphic writing in Middle America and methods of deciphering them. Amer. Antiq. 24(4): 349–364.

Vaillant, G. C. 1962. Aztecs of Mexico, 2nd ed. Doubleday.

Vavilov, N. I. 1931. Mexico and Central America as the principal center of origin of cultivated plants of the New World. Trudy Prikl. Bot. 26(3): 179–199.

Wassén, S. H. 1960. Glimtar au aztekisk medicin. Medicinhistoriska Museet. Medicinhistorisk Arsbok pp. 1–16, Stockholm.

Wasson, R. G. 1966. Ololiuhqui and other hallucinogens of Mexico. Pages 329–348 *in* A. Pompa y Pompa (ed.), Suma antropológica: En homenaje a Roberto J. Weitlaner. Mexico City.

———. 1973. The rôle of "flowers" in Náhuatl culture: A suggested interpretation. Bot. Mus. Leafl. 23(8): 305–324.

———. 1980. The wondrous mushroom: Mycolatry in Mesoamerica. McGraw-Hill, New York.

Whorf, B. L. 1946. The Milpa Alta dialect of Aztec, with notes on the classical and Tepostlan dialects. *In* H. Hoijer et al. (eds.), Linguistic structures of Native America. Viking.

Wilken, G. C. 1970. The ecology of gathering in a Mexican farming region. Econ. Bot. 24(3): 286–295.

Williams, B. J. 1981. Aztec soil glyphs and contemporary Nahua soil classification. Pages 206–222 *in* International Colloquium: The Indians of Mexico in Pre-Columbian and Modern Times. Leiden, June 9–12, 1981.

Williams, D. E. 1985. Tres arvenses solanáceas comestibles y su proceso de domesticación en el estado de Tlaxcala, México. Master's Thesis. Colegio de Postgraduados, Chapingo, Mexico.

Wolf, E. 1959. Sons of the shaking earth. University of Chicago Press, Chicago.

Index to Common Names

A page number in **boldface** indicates a primary reference. A page number followed by an asterisk (*) denotes an illustration or map.

Index to Scientific Names

Page numbers in **boldface** indicate a primary reference. Page numbers followed by asterisks denote maps or illustrations.